T0203065

Lecture Notes in Computer Science　　9819

Commenced Publication in 1973
Founding and Former Series Editors:
Gerhard Goos, Juris Hartmanis, and Jan van Leeuwen

More information about this series at http://www.springer.com/series/7409

Norbert Fuhr · László Kovács
Thomas Risse · Wolfgang Nejdl (Eds.)

Research and Advanced Technology for Digital Libraries

20th International Conference on Theory and Practice
of Digital Libraries, TPDL 2016
Hannover, Germany, September 5–9, 2016
Proceedings

 Springer

Editors

Norbert Fuhr
Universität Duisburg-Essen
Duisburg
Germany

László Kovács
Hungarian Academy of Sciences
Budapest
Hungary

Thomas Risse
Leibniz Universität Hannover
Hannover
Germany

Wolfgang Nejdl
Leibniz Universität Hannover
Hannover
Germany

ISSN 0302-9743 ISSN 1611-3349 (electronic)
Lecture Notes in Computer Science
ISBN 978-3-319-43996-9 ISBN 978-3-319-43997-6 (eBook)
DOI 10.1007/978-3-319-43997-6

Library of Congress Control Number: 2016947200

LNCS Sublibrary: SL3 – Information Systems and Applications, incl. Internet/Web, and HCI

Printed on acid-free paper

This Springer imprint is published by Springer Nature
The registered company is Springer International Publishing AG Switzerland

Preface

These proceedings contain the reviewed papers presented at the 20th International Conference on Theory and Practice of Digital Libraries (TPDL), which was held in Hannover, Germany, during September 5–9, 2016. The L3S Research Center and the German National Library of Science and Technology (TIB) organized the conference.

The TPDL conference constitutes a leading scientific forum on digital libraries that brings together researchers, developers, content providers, and users in the field of digital libraries.

The 20th conference in the series of ECDL/TPDL conferences is a milestone in the professional activity of the digital library community in Europe. Digital library research and technology are becoming mainstream. Information creation, acquisition, access, processing, distribution, evaluation, and preservation are no longer possible without the methods and technologies developed in our field.

Our community in Europe is rapidly changing. Great "pioneers"of the domain are now retiring, and newcomers are creating productive research atmosphere in which our community members meet with representatives of other communities. For the first time, the conference included two additional tracks on "Digital Humanities" and "e-Infrastructures," opening our door to these neighboring communities.

This year's lead topic at the conference was "Overcoming the Limits of Digital Archives" with a clear intention to absorb the problems of other domains in which techniques and technologies of digital archiving could be truly challenged.

From the 93 papers submitted to the conference, 46 long papers and 13 short ones belonged to the main conference track; also, nine papers were submitted for the Digital Humanities track, and five papers to the e-Infrastructures track. The Program Committee accepted 28 full papers and eight short papers. Furthermore, five posters were accepted.

In addition to these papers and posters, we also invited three keynote speakers: Jan Rybicki (Jagiellonian University, Kraków, Poland), Tony Veale (University College Dublin, Ireland), and David Bainbridge (University of Waikato, and Director of the New Zealand Digital Library).

The conference program also included a Doctoral Consortium, and three tutorials on "Introduction to Fedora 4," "Building Digital Library Collections with Greenstone 3," and "Text Mining Workflows for Indexing Archives with Automatically Extracted Semantic Metadata." Finally, three workshops were organized in conjunction with the main conference: "Videos in Digital Libraries: What's in it for Libraries, Publishers, and Scientists?" "NKOS – 15th European Networked Knowledge Organization Systems," and the "First International Workshop on Reproducible Open Science (RepScience2016)."

The success of TPDL 2016 is a result of the collegial teamwork of many individuals, who worked tirelessly to make the conference a top research forum. We acknowledge and thank the Program Committee members and the Program Committee Chairs for their extraordinary efforts in assembling this outstanding program: Gerhard Lauer and Milena Dobreva (Digital Humanities Track Chairs), Tobias Blanke and Laurent Romary (eInfrastructure Track Chairs), Stefan Rüger and Hannu Toivonen (Creativity and Multimedia Chairs), Heiko Schuldt and Peter Löwe (Workshops Chairs), Wolf-Tilo Balke (Panels Chair), Fabio Crestani and Adam Jatowt (Poster and Demonstration Chairs), Ingo Frommholz (Tutorial Chair), and Kjetil Nørvåg (Doctoral Consortium Chair). We would especially like to thank Uwe Rosemann, who served as the honorary chair.

We also express our deep appreciation of the outstanding work put in over many months by Nattiya Kanhabua (Publicity Chair), Gerhard Gossen (Publication Chair), and the local organization team comprising Marion Wicht, Nicole Petri, Katrin Hanebutt, and Miroslav Sheltev. Without their tireless efforts, this conference would not have been a success. We are also thankful to the many student volunteers from Leibniz University Hannover and L3S Research Center.

In addition, there are many other individuals whose contributions we warmly acknowledge. We benefited greatly from the sage advice provided by José Borbinha (TPDL Steering Committee Chair) and the fruitful discussions with the TPDL Steering Committee members.

We warmly acknowledge the financial support of our corporate sponsors: Elsevier B.V. (at the gold level) and Ex Libris (at the bronze level).

Finally, we thank all the authors, presenters, and participants of the conference. We hope that you enjoy the conference proceedings.

September 2016

Norbert Fuhr
László Kovács
Thomas Risse
Wolfgang Nejdl

Organization

Honorary Chair

Uwe Rosemann   TIB Hannover, Germany

General Co-chairs

Wolfgang Nejdl   Leibniz Universität Hannover and L3S Research
                             Center, Germany
Thomas Risse   Leibniz Universität Hannover and L3S Research
                             Center, Germany

Program Co-chairs

Norbert Fuhr   University of Duisburg-Essen, Germany
László Kovács   MTA SZTAKI, Hungary

TPDL-DH Track Co-chairs

Gerhard Lauer   University of Göttingen, Germany
Milena Dobreva   University of Malta, Malta

TPDL-eInfra Track Co-chairs

Tobias Blanke   King's College London, UK
Laurent Romary   Inria, France

TPDL-Creativity and Multimedia Track Co-chairs

Stefan Rüger   The Open University, UK
Hannu Toivonen   University of Helsinki, Finland

Workshops Chairs

Heiko Schuldt   University of Basel, Switzerland
Peter Löwe   TIB Hannover, Germany

Panels Chair

Wolf-Tilo Balke   TU Braunschweig, Germany

Poster and Demonstration Chairs

Fabio Crestani   University of Lugano, Switzerland
Adam Jatowt   Kyoto University, Japan

Tutorial Chair

Ingo Frommholz   University of Bedfordshire, UK

Doctoral Consortium Chair

Kjetil Nørvåg   Norwegian University of Science and Technology,
                               Norway

Publicity Chair

Nattiya Kanhabua   Aalborg University, Denmark

Publication Chair

Gerhard Gossen   Leibniz Universität Hannover and L3S Research
                               Center, Germany

Local Organizing Committee

Marion Wicht   L3S Research Center, Germany
Katrin Hanebutt   TIB Hannover, Germany
Nicole Petri   TIB Hannover, Germany
Miroslav Shaltev   L3S Research Center, Germany
Dimitar Mitev   L3S Research Center, Germany

Senior Program Committee

Maristella Agosti   University of Padua, Italy
Wolf-Tilo Balke   TU Braunschweig, Germany
Tobias Blanke   King's College London, UK
Jose Borbinha   INESC-ID, Instituto Superior Técnico, Universidade
                               de Lisboa, Portugal
George Buchanan   City University London, UK
Donatella Castelli   CNR-ISTI, Italy
Stavros Christodoulakis   Technical University of Crete, Greece
Panos Constantopoulos   Athens University of Economics and Business, Greece
Fabio Crestani   Università della Svizzera Italiana (USI), Switzerland
Sally Jo Cunningham   Waikato University, New Zealand
Nicola Ferro   University of Padua, Italy
Richard Furuta   Texas A&M University, USA

C. Lee Giles	Pennsylvania State University, USA
Marcos Goncalves	Federal University of Minas Gerais, Brazil
Matthias Hemmje	FU Hagen, Germany
Andreas Henrich	University of Bamberg, Germany
Adam Jatowt	Kyoto University, Japan
Carl Lagoze	University of Michigan, USA
Ray Larson	University of California, USA
Peter Löwe	TIB Hannover, Germany
Clifford Lynch	CNI, USA
Carlo Meghini	CNR-ISTI, Italy
Wolfgang Nejdl	Leibniz Universität Hannover and L3S Research Center, Germany
Erich Neuhold	University of Vienna, Austria
Christos Papatheodorou	Ionian University, Greece
Andreas Rauber	Vienna University of Technology, Austria
Thomas Risse	Leibniz Universität Hannover, Germany
Laurent Romary	Inria, France
Seamus Ross	University of Toronto, Canada
Pertti Vakkari	University of Tampere, Finland
Herbert van de Sompel	Los Alamos National Laboratory, USA

Program Committee

Trond Aalberg	Norwegian University of Science and Technology, Norway
Robert B. Allen	Yonsei University, South Korea
Werner Bailer	Joanneum Research, Austria
David Bainbridge	University of Waikato, New Zealand
Thomas Baker	DCMI, Germany
Christoph Becker	University of Toronto, Canada
Jöran Beel	University of Konstanz, Germany
Maria Bielikova	Slovak University of Technology in Bratislava, Slovakia
Pável Calado	INESC-ID, Instituto Superior Técnico, Universidade de Lisboa, Portugal
Jose H. Canos	Universitat Politecnica de Valencia, Spain
Vittore Casarosa	ISTI-CNR, Italy
Lillian Cassel	Villanova University, USA
Theodore Dalamagas	Athena Research Center, Greece
Lois Delcambre	Portland State University, USA
Giorgio Maria Di Nunzio	University of Padua, Italy
Boris Dobrov	Research Computing Center of Moscow State University, Russia
Shyamala Doraisamy	University Putra Malaysia, Malaysia
Fabien Duchateau	Universite Claude Bernard Lyon 1 – LIRIS, France
Pinar Duygulu	Hacetepe University, Turkey

Pierluigi Feliciati   Università degli studi di Macerata, Italy
Niels Ole Finnemann   University of Copenhagen, Denmark
Schubert Foo   Nanyang Technological University, Singapore
Edward Fox   Virginia Polytechnic Institute and State University,
                                  USA
Nuno Freire   The European Library, Portugal
Ingo Frommholz   University of Bedfordshire, UK
Manolis Gergatsoulis   Ionian University, Greece
Pablo Gervas   Universidad Complutense de Madrid, Spain
Jane Greenberg   College of Computing and Informatics, USA
Cathal Gurrin   Dublin City University, Ireland
Martin Halvey   University of Strathclyde, UK
Allan Hanbury   TU Wien, Austria
Preben Hansen   Stockholm University, Sweden
Mark Hedges   King's College London, UK
Annika Hinze   University of Waikato, New Zealand
Frank Hopfgartner   University of Glasgow, UK
Jieh Hsiang   National Taiwan University, Taiwan, R.O.C
Jane Hunter   University of Queensland, Australia
Antoine Isaac   Europeana and VU University Amsterdam,
                                  The Netherlands
Jaap Kamps   University of Amsterdam, The Netherlands
Sarantos Kapidakis   Ionian University, Greece
Michael Khoo   Drexel University, USA
Claus-Peter Klas   GESIS – Leibniz Institute for Social Sciences,
                                  Germany
Martin Klein   University of California Los Angeles, USA
Stefanos Kollias   NTUA, Greece
Alexandros Koulouris   TEI of Athens, Greece
Mounia Lalmas   Yahoo Labs, UK
Ronald Larsen   University of Pittsburgh, USA
Séamus Lawless   Trinity College Dublin, Ireland
Hyowon Lee   Singapore University of Technology and Design,
                                  Singapore
Suzanne Little   Dublin City University, Ireland
Haiming Liu   University of Bedfordshire, UK
Penousal Machado   University of Coimbra, Portugal
Joao Magalhaes   Universidade Nova de Lisboa, Portugal
Richard Marciano   University of Maryland, USA
Bruno Martins   INESC-ID, Instituto Superior Técnico, Universidade
                                  de Lisboa, Portugal
Philipp Mayr   GESIS – Leibniz Institute for Social Sciences,
                                  Germany
Cezary Mazurek   IChB PAN – PCSS, Poland
Dana Mckay   Swinburne University of Technology Library, Australia
Andras Micsik   MTA SZTAKI, Hungary

Keynotes

Pretty Things Done with (Electronic) Texts: Why We Need Full-Text Access

Jan Rybicki

Jagiellonian University, Kraków, Poland

Abstract. Stylometry, aka computational stylistics, is a field that has produced compelling visualizations of patterns of similarity and difference between texts, based on various quantitative, i.e. countable features. Even if these countable features are often very basic elements that have not been traditionally associated with "style" or "meaning" or "message" – more often than not, stylometrists work with frequencies of function words or of part-of-speech n-grams – the various statistical measures applied to them often yield results that "make sense" in terms of authorial attribution, chronology, genre, or gender – or simply from the point of view of traditional literary studies. It has now become quite simple to produce a "map," or in fact a network analysis, of 1000 novels in English that shows a very clear progression from early (green) to modern (purple) writing:

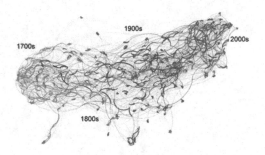

Whether or not this is a simple effect of linguistic change (there are reasons to think it is NOT), the feasibility of such approaches – apart from a plethora of methodological problems – relies on stylometrists' access to full texts. This is still a particularly unpleasant stumbling block: even when dealing with public-domain material, textual collections are dispersed or incompatible or unreliable or fragmentary (pick one or any of these), and stylometrists continue to struggle, often steering on the margins of reliability and of (copyright) laws. Perhaps the main reason is that they do not complain enough to the right people; digital librarians might be a good group to start with.

Jan Rybicki is Assistant Professor at the Institute of English Studies, Jagiellonian University, Kraków, Poland; he also taught at Rice University, Houston, TX and Kraków's Pedagogical University. His interests include translation, comparative literature and humanities computing (especially stylometry and authorship attribution).

He has worked extensively (both traditionally and digitally) on Henryk Sienkiewicz and the reception of the Polish novelist's works into English, and on the reception of English literature in Poland. Rybicki is also an active literary translator, with more than twenty translated novels by authors such as Coupland, Fitzgerald, Golding, Gordimer, le Carré or Winterson.

Metaphors All the Way Down: The Many Practical Uses of Figurative Language Understanding

Tony Veale

University College Dublin, Dublin, Ireland

Abstract. More and more of the content we consume on social networking platforms such as Twitter is computer-generated. If this content were just another form of spam then these networks would truly become the 21st century's version of Borges' Library of Babel, a world in which the content that is actually worth consuming is ultimately lost in a sea of random, meaningless noise. Yet there are encouraging signs that it need not be so. On Twitter, savvy users follow machine generators of content – called Twitterbots – knowing them to be machines and valuing their outputs all the more for their oddity and artificiality. These bots are not designed to fool humans, but to engage them in language games that explore the relation of form to meaning and provocatively flit along the boundary between sense and nonsense. As there is a long history of human artists doing precisely this – ranging from Duchamp and Breton to Dali and Burroughs – bots are simply the next stage in the evolution of thought-provoking automated content creation. Bots that lack a semantic grounding often compensate with a statistical brio for superficial production that naturally yields outputs which can appear poetic and deeply metaphorical. In this presentation I shall consider how an AI bot can deliberately produce meaningful metaphors from a knowledge-base of everyday facts and beliefs. But we need not stop there: given an ability to produce metaphors on demand, a computational system can produce higher-level digital constructs such as stories and games that make the familiar seem strange and the strange seem oddly and meaningfully familiar. I will describe how metaphor and blending can play a pivotal role in these engaging new forms of digital content, to the extent that they truly do rest on metaphors all the way down.

Dr. Tony Veale is a Computer Scientist at University College Dublin, Ireland, where his research focuses on the computational modeling of creative linguistic phenomena, including metaphor, blending, simile, analogy and verbal irony. He leads the European Commission's PROSECCO network (PROSECCO-network.eu and @PROSECCO network), an international coordination action that aims to promote the scientific exploration of Computational Creativity. He is particularly interested in the generative creativity of metaphor, and builds generative models of metaphor, simile and blending which are made publically available as reusable Web services to promote the integration of figurative language processing capabilities in third-party applications. He is the author of the 2012 monograph on computational linguistic creativity titled

Exploding the Creativity Myth: The Computational Foundations of Linguistic Creativity from Bloomsbury press and co-author (with Ekaterina Shutova and Beata Beigman Klebanov) of Metaphor: A Computational Perspective from Morgan Claypool press. He is the creator of the metaphor-generating and story-telling Twitterbot @MetaphorMagnet and the founder of the educational Web-site RobotComix.com, which promotes the philosophy and practice of Computational Creativity to the general public.

Mozart's Laptop: Implications for Creativity in Multimedia Digital Libraries and Beyond

David Bainbridge

University of Waikato, and Director of the New Zealand Digital Library
Research Project, Hamilton, New Zealand

Abstract. If Mozart were alive today, what sorts of musical apps would such an innovative composer use on his laptop? In this keynote I will attempt to answer—at least in part—this question. We will metaphorically drop in on Wolfgang composing at home in the morning, at an orchestra rehearsal in the afternoon, and find him unwinding in the evening playing a spot of the new game Piano Hero which is (in my fictional narrative) all the rage in the Viennese coffee shops! From a pedagogical perspective, these three scenarios are chosen because they cover the main forms of digital music representation: audio, sheet music, and symbolic notation. In each case I will demonstrate software prototypes that combines digital music library and music information retrieval research to provide novel forms of access and management of musical digital content. I will then broaden the discussion and relate the work to other forms of media, and (going beyond this) contemplate whether the presented research fits the established definition of a digital library, or if it is perhaps time to repurpose traditional ideas about the structure and capabilities of digital libraries, or even revisit what we define as a digital library.

David Bainbridge is a Professor of Computer Science at the University of Waikato, and Director of the New Zealand Digital Library Research Project. He is an advocate of open source software, and an active coder on the Greenstone digital library project, and the spatial hypermedia system, Expeditee. His research interests include multimedia content analysis, and human computer interaction in addition to digital libraries. He has published widely in these areas, including the book How to Build a Digital Library, with colleagues Ian Witten and Dave Nichols, now into its second edition. David graduated with a Bachelor of Engineering in Computer Science from Edinburgh University, UK as the class medalist, and undertook his PhD in Computer Science at Canterbury University, New Zealand as a Commonwealth Scholar.

Contents

Web Archives

Semantics

Multimedia and Time Aspects

Digital Library Evaluation

Digital Humanities

e-Infrastructures

Short Papers

Posters and Demos

Tutorials

Digital Library Design

Realizing Inclusive Digital Library Environments: Opportunities and Challenges

Wondwossen M. Beyene[(✉)]

Oslo and Akershus University College of Applied Sciences, Oslo, Norway
Wondwossen.beyene@hioa.no

Abstract. Universal design, also known as inclusive design, envisions the design of products and services to be accessible and usable to all irrespective of their disability status, cultural background, age, etc. Libraries have been benefiting from the breakthroughs in accessibility research to design their environments as friendly as possible for all groups of uses. However, the present scenario of digital library environments characterized by different types of resources acquired or subscribed from different vendors operating with different rules, and who would maintain some form of control over the collections shows that adherence to guidelines by itself won't ensure inclusive digital library environments. The paper attempts to explore the matter taking the case of digital services run in selected libraries to identify trends that favor universal design and point challenges that need to be dealt with as part of further endeavors.

Keywords: Universal design · Inclusive design · Digital library · Accessibility · Digital services

1 Introduction

One of the features of the information society is the proliferation of digital technology in production of information in text, audio, video and graphics formats. As the result, we are witnessing growing volumes of electronic books, journals and other information resources as well as mass digitization of already existing print materials. However, increased digital presence of information resources did not translate well into accessibility of them by all users [1]. Therefore, ensuring the accessibility of born-digital and newly digitized materials to *all* users remains the inescapable and important challenge libraries face nowadays.

Expanding the scope of accessibility to include the needs of *all* puts the focus on universal design [2], also referred as inclusive design, which envisions the design of products and services to be accessible and usable to all regardless of their disability status, cultural background, age, etc. [3]. The ideals of universal design have been enshrined in international conventions such as Article 9 of the United Nations Convention on the Rights of Persons with Disabilities (UNCRPD), which requires state parties to promote "appropriate forms of assistance and support to persons with disabilities to ensure their access to information"[1]. They are also supported by

---

[1] http://www.un.org/disabilities/convention/conventionfull.shtml.

© Springer International Publishing Switzerland 2016
N. Fuhr et al. (Eds.): TPDL 2016, LNCS 9819, pp. 3–14, 2016.
DOI: 10.1007/978-3-319-43997-6_1

country-specific anti-discrimination laws such as the Americans with Disabilities Act (ADA)[2] and the Norwegian Anti-Discrimination and Accessibility Act[3]. These acts promote inclusion and equal participation of people with disabilities in education, government, entertainment, and other aspects of societal life.

Conventions and laws like those mentioned above help to frame accessibility as part of responsibilities of libraries. However, libraries credit themselves for being cognizant of the needs of people for disabilities long before the introduction of anti-discrimination laws, as can be exemplified by the reading rooms they setup for the blind or the effort they make to collect materials in alternative formats [4]. Therefore, the laws or conventions simply embolden their traditional responsibility. As the digital content and technology keep infiltrating the library world, libraries began to utilize guidelines and design specifications which mostly represent the state of knowledge in computer and information accessibility design [5]. As the result, library websites are being designed utilizing the Web Content Accessibility Guidelines (WCAG) and other accessibility standards and guidelines [6]. WCAG 2.0, approved as ISO/IEC 40500:2012, is part of guidelines produced by the World Wide Web Consortium (W3C)'s Web Accessibility Initiative (WAI) to guide development of web content including text, images, sounds and others so that they can be more accessible to people with disabilities [7].

There have been studies evaluating the conformance of library websites, databases, etc., to established accessibility criteria. The studies, like those discussed in the next section, employed a variety of methods including heuristic evaluation, automated accessibility testing, and user testing. However, the present scenario of digital library environments which is characterized by strong links with publishers, commercial indexing services, open access services, resource discovery tool providers and others makes accessibility a complex challenge that may not be achieved by merely following certain types of guidelines. Therefore, one way for ensuring accessibility of knowledge could be to address the technical as well as non-technical issues surrounding its creation, organization, dissemination, and the retrieval mechanisms put in place for its end users.

Therefore, instead of following the common product-based approach for evaluating accessibility of digital resources, this study chose to make an "environmental scan" of existing digital services that can be available through libraries. The purpose is to explore how the concerns of universal design are being addressed and to pinpoint the challenges that should be met. Its value could be interpreted as encouraging libraries to make inventory or self-assessment of their services and strategize on how to proceed in incorporating the needs of *all* of their users. To that end, this paper attempts to answer the following questions: "What digital services are available through libraries? What steps are being taken to make the services accessible for users with disabilities? What are the challenges for providing accessible digital services? What approaches could be considered to realize inclusive digital library environments? A case study of digital services at four libraries with relatively developed digital services was made to answer

---

[2] http://www.disabled-world.com/disability/ada/.

[3] http://app.uio.no/ub/ujur/oversatte-lover/data/lov-20130621-061-eng.pdf.

the questions. In-depth interviews involving eight people working in the digital services (with one of them communicated via email) were used to collect the required data.

The rest of this paper is organized as follows: review of related literature is presented next followed by explanation of the methodology used in the study. Then follow findings of the study, discussion on the findings, and recommendations for further research. Finally, the paper closes with the conclusion section.

2 Literature Review

Digital inclusion has been a concept used as a framework to ensure that individuals have access to digital technologies and the skills to use them [8]. Literature shows that the concept is hard to clearly define, but has been used in libraries to formulate policies and actions aimed at solving problems related to digital divide and digital literacy [8, 9]. The early conception of digital divide was used to connote unequal ownership of computational technology and access to Internet [8]. The proliferation of those technologies in people's daily lives introduced yet another layer of digital divide that alienated people with different types of disabilities from enjoying the benefits of the technologies [10]. Inaccessibility of technology creates digital exclusion, which also relates to social exclusion [11]. In this sense, inclusive/universal design in libraries can be considered as an outgrowth of their efforts in digital inclusion.

In the world of information services, we see two different models such as special access and universal access [12]. Users who have difficulties in reading printed text were traditionally referred to special libraries where they can be served with braille or talking book versions of the materials they want. However, the introduction of digital content in libraries has created an opportunity for including the print disabled users in any mainstream information service [13]. This coupled with advancements in assistive technologies and developments in universal design paves the way for digital services to be inclusively designed.

There have been studies made for evaluating accessibility of library websites for their compliance to established guidelines [14–16], accessibility and usability of online library databases [17], accessibility of digitized special collections [1], accessibility of academic eBook libraries [16], and accessibility of library content management systems [18]. These studies identified problems such as failure of library websites to fully comply with WCAG guidelines, inaccessibility of book pages to assistive devices, CAPTCHAs that complicate logon for users with visual impairment, thumbnails without alternative text, poorly described links, excessive navigation links, inaccuracy of some automatic testing tools, etc. Studies also showed that copyright protection measures that involve application of Digital Rights Management tools (DRMs) could cause accessibility problems as they may have the undesired effect of preventing assistive devices from rendering content to, users of screen reader technologies [19].

A content analysis of literature in library and information science (LIS) from 2000 to 2010 [20] shows that the "strongest theme" of accessibility research was related to web, database and software. The research added that the majority of those studies were accessibility testing. To sum up, this and the other papers reviewed in this study have presented results one can expect, i.e., mixed results showing some parts accessible while some other parts inaccessible, while evaluating accessibility of digital resources. Here, we tried to adopt a different approach that allows exploring the whole scenario in current digital library environments from a broader perspective. The following section explains the methodology used in the research.

3   Methodology

The study adopted exploratory research design as its main purpose was to understand the state of the art as much as possible, identify developments that favor universal design, and highlight challenges that need to be tackled. The exploratory design was favored because of the flexibility it provides for tackling problems which are not much explored in previous research [21]. The idea was to find libraries with well-developed digital services, which would provide a good case for identifying and discussing issues that determine inclusiveness in digital services. The scope of the study was limited to libraries in Oslo and four libraries with relatively developed digital services such as the University of Oslo (UiO) library, Oslo and Akershus University College (HiOA) library, the National Library of Norway, and Helsebiblioteket.no (which is an electronic health library) were included in the study. The selection of the libraries was judgmental primarily based on the scale of their digital services. However, the fact that they represent different types of libraries (academic, national, and special) could be taken as a plus. Semi-structured in-depth interviews were made with the personnel running the services. The questions were related to the research questions presented at the start of this paper. The study included a total of eight respondents. Four of them were interviewed in pairs, three individuals were interviewed individually and one other respondent chose to communicate via email. Demonstration of the digital services was made in some of the sessions. The responses were categorized into themes for analysis to finally provide answers for the research questions.

4   Results

4.1   Digital Services

The respondents were first asked to enumerate what they refer as digital services available through their libraries. Though it might not be an exhaustive list, the following were those raised during the discussions.

Access Services. Access to electronic resources including ebooks and subscribed electronic journals constitute the biggest share of digital services in the two academic (UiO and HiOA) and one special libraries. Access to institutional repositories was among the digital services mentioned. The UiO library added that they maintain special databases including a database of papyrus images, a collection of digitized old books, a database of Norwegian translation of literature in Spanish, a bibliography of Norwegian literary criticism, a database of Norwegian laws translated to other languages, and a database of Non-Norwegian literature before 1966. HiOA included ezproxy, which is an authentication software that provides remote access to licensed electronic resources, and link resolvers as part of their digital services.

Helsebiblioteket.no said they were setup to provide health information service to professionals and practitioners in healthcare, as part of evidence-based practice. As the result, their digital services include access to point of care tools, links to published and unpublished national and international guidelines, medical procedure catalogs, and patient leaflets translated to Norwegian from British Medical Journal (BMJ) best practices. They mentioned that they have access to ebooks and videos through the databases they have subscribed, though most of the videos are not relevant to the Norwegian practice. The other digital resources include meta collection of openly available resources relevant to Norwegian practice, commentaries on contents of their resources, and the digitized version of the Norwegian Handbook for Emergency Medicine available online and as a mobile app.

The Norwegian National Library maintains a digitized collection of books, newspapers, photo and radio. As the respondent from the library said, the legal deposit act grants them permission to digitize for preservation purposes. These resources are available for their users as part of digital services. Some items have been restricted to use within the national library, whereas others (for instance newer newspapers) are also available in public libraries. There are solutions for browsers and mobile devices alike.

Content Production. Currently, UiO and HiOA libraries publish 12 and 11 open access journals respectively. They also maintain institutional repositories. HiOA reproduce educational films that shall be published on Vimeo, YouTube, and on film service called film.hioa.no. They also produce and present compendiums with PDF/A format to students. Helsebiblioteket.no mentioned presentation of translated contents from English to Norwegian to suit the needs of their users.

Discovery Service. Oria, a discovery tool based on Ex Libris's Primo, is the principal tool used in the academic libraries. The UiO said they have developed additional tools. There is an app developed to locate and find a book in a shelf in their science library. This app shows map of the library with a pointer where the book is located on the shelf. The other application is called Book Motion and it is based on motion sensor Leap-Motion technology. Documentation on the library website[4] shows that this application enables users to browse through ebooks using hand gestures, without the need of using mouse or screen.

Other Services. Other services mentioned as part of digital services by the academic libraries include resource pages where a student can find databases and journals tailored to their field of studies; guides on reference styles, how to write a paper, etc. HiOA mentioned they run MOOC (Massive Online Open Course) page with information and instruction on how to use the library, how to search in the catalog etc. Facebook pages set for communicating with users were among those raised in the discussions.

---

[4] http://www.ub.uio.no/om/prosjekter/brukerdrevet-innovasjon/utviklingsprosjekter/bookmotion/.

4.2   Accessibility of Digital Services

The next question posed to the respondents was on what they have done to make their digital services inclusive of the needs of people with disabilities. The steps taken were largely described either as plans or measures taken to improve accessibility of the websites according to WCAG guidelines, in a bid to meet universal design requirements set by the Norwegian Agency for Public Management and eGovernment (Difi). Universal design has been legislated as requirement by the Norwegian Anti-Discrimination and Accessibility Act § 14[5] and Difi is responsible for monitoring whether the regulations are being met[6]. Difi states WCAG 2.0 level AA as a standard for universal design of websites with some exceptions regarding time-based media, audio description or media alternative (Prerecorded content), captions (live content) and audio description (prerecorded content).

The national library stated that their web pages comply with W3C CSS validation standards and are accessible for screen readers. They mentioned, though, the use of screen readers in connection to the text viewer used to view digitized materials like books has not found optimal solution partly due to the agreement between the National library and the rights holders. This agreement grants the National library the right to expose books within the framework of "bokhylla.no" (their digital library) for anyone in Norway, but as the respondent said, "the OCR text must not be exposed to the users, to prevent it from being downloaded, copied or modified. The book viewer therefore shows the scanned pages as they are, and suppresses the hidden text that is bundled with the photos of book pages". He also added that the library has prototyped solutions for activating the text behind paragraphs or pages, making it accessible for text readers or text-to-speech, but the solutions have not been deployed so far. However, it might be worth to add that the agreement between the library and representative of the right holders, which is found on the library's website[7], didn't exhibit a clause carrying provisions for users with disability.

The UiO library said the books they digitized (books whose copyright periods have expired) are OCR-treated, but added the difficulty posed by fonts used in the old books which were confused for pictures-which makes them difficult to be rendered by screen readers. They said these resources are made available for users but admitted that accessibility concerns were not well addressed. HiOA said they don't do digitization except producing compendiums. The compendiums are scanned, OCR treated and presented as PDF/A for students. The two academic libraries have discussed presentation of materials in alternative formats. For instance, HiOA publish one of their open access journals in PDF, HTML, mobile and EPUB formats. The UiO library said they are working to make their electronic resources, including the new discovery tool Oria, available through mobile devices.

Though it might not form part of their digital services, it is worth mentioning that both academic libraries have rooms equipped with tools such as adjustable tables,

---

[5] http://app.uio.no/ub/ujur/oversatte-lover/data/lov-20130621-061-eng.pdf.

[6] http://uu.difi.no/om-oss/english.

[7] http://www.nb.no/pressebilder/Contract_NationalLibraryandKopinor.pdf.

braille printer, braille keyboards, scanner mouse, screen magnifiers and other physical utilities for users with impairments. Users who require more help would be directed to the Norwegian Library of Braille and Talking Books (NLB) where they can have resources they want either audio or braille forms, depending on their request. A respondent acknowledged that could have affected the state of accessibility in their digital services saying, "Sometimes also we rely too much on this room that if people have to use we send them down there but it will be difficult if they want to use our services from home". The other respondent added, "… we don't offer more services in that regard because they (NLB) handle everything in Norway".

Helsebiblioteket.no said they have not been involved in digitizing documents except for the Norwegian Handbook for Emergency Medicine (Legevakt boka), a book very popular among students and anyone involved in emergency medicine. The digitization was done by the publisher and made available via browser and mobile platforms. When asked whether they have considered people with disability as potential consumers of their services, they said their mandate is to serve healthcare practitioners. However, they acknowledged their user could be anyone in Norway saying, "A lot of traffic comes from Google so we have a reason to believe that, especially out of the traffic on the patient leaflets, [it] comes from patients, their caregivers or family and friends". They said they are in the process of redesigning their entire system emphasizing the demands within the law regarding universal design citing Difi's WCAG recommendations as "our toolbox". During the discussion, they raised the possibility of presenting content in PDF, HTML and other alternatives but said they prefer HTML over PDF. The reason as they explained was HTML's mobile friendliness and the ease it provides for navigation through pages. They specifically mentioned ease of access to information for health professionals using mobile devices. Therefore, they have converted the national health guidelines from PDF to HTML and plan to do the same with their other local collections.

Accessibility of Subscribed Resources. As discussed in the first section, libraries are involved in reproduction as well as subscription of electronic resources. Respondents were asked if there are measures they have taken to ensure that materials they subscribe to are universally accessible to all of their users, including users with disabilities. Respondents from HiOA said that there had been a case where a user with screen reader couldn't access an ebook because of Digital Rights Management (DRM) tools restriction. The library had to buy the version of the ebook without DRM. They explained their policy of buying ebooks from vendors without DRM to the possible extent. However, they mentioned the difficulty of signing hundreds of agreements with small vendors, where there is a possibility of buying ebooks without DRM and the ease of signing agreements with big vendors but which usually use DRMs.

Helsebiblioteket.no said that most of their resources can be accessed by anyone in Norway without the need for logging in and that is part of their agreement with vendors. The whole point for this national IP access was the availability of the resources for their users and the ease of access. For instance as they said, "a doctor shouldn't be required to enter password every time he wants to access a resource".

All of the respondents acknowledged the agreement they make with publishers dictates the manner the resources are retrieved and used. For instance a respondent from one of the academic libraries said "in each agreement you have different kinds of solutions for access, some allow you only to view, the others allow download". However, they didn't specifically mention whether they insert in those agreements clauses that protect the right of access of users with disabilities.

Accessibility of Resource Discovery Tools. Resource discovery tools were among the list of digital services mentioned by the respondents from the academic libraries. Therefore, they were asked what they have done to make those tools accessible and usable by users with disabilities. As mentioned above, the academic libraries use Oria. The respondents said there is little they can change on Oria except some "font and small CSS changes". They need to contact the company if more changes are required. Oria is thought to provide a single point of search for all resources either owned or subscribed by libraries. The interface carries multiple search options by author, title, etc. and filters by media type, creation date, language, etc. The information on the company website states that the product meets international accessibility standards [22]. However, whether its feature-rich interface pauses problem for people with disabilities, for example for low vision and dyslectic users, would require further research.

Helsebibioteket.no explained that they didn't start up as a traditional library so at this time they don't have yet a unified catalog for all of their contents. They started up as a website, a service providing research information to healthcare professionals based on the paradigm of evidence-based healthcare, and evolving. There are two types of search systems their users can use: the one provided by journal databases and the other for local collections setup with the IBM Watson Data Explorer. It allows users to search by a term then filter by source, information type, language, and other facets.

Resource Description. The respondents were asked about the metadata schema they use and whether they have used elements in the schemas to describe resources by their accessibility qualities. This question was inspired by the study [23] that suggested Dublin Core's "Audience" and MARC 21's "Reading Level or Interest Grade level" could be used, for example, to describe resources suitable for dyslexic users or users with low English language skill. The academic libraries use both MARC 21 and Dublin Core but the respondents couldn't confirm whether they have used the metadata schemas to that extent. One of the reasons as two of the respondents said is that, mainstream libraries are not expected to have resources that require such special descriptions, as that is the task of NLB. The other reason is the increasing reliance on metadata generated by vendors of the electronic resources which makes it difficult to track the level and scale of metadata usage. The other respondent noted that perhaps annotation of resources by their formats as audio, video, etc. could be taken as describing them by their accessibility attributes.

5   Discussion

This paper aimed at making environmental scan of digital services that can be available through different types of libraries and asked, "What digital services are available through libraries? What steps are being taken to make them accessible for users with disabilities? What are the challenges for providing accessible digital services? What approaches could be considered to realize inclusive digital library environments?

The case of libraries included in this study showed that libraries could be involved in content production and presentation. List of digital services possible through today's libraries include access to locally produced resources, presentation of translated content, access to subscribed eResources, resource discovery tools, ezproxy, link resolvers, production of movies; maintenance of movie archives, digital repositories, open access journals; user-tailored resource pages, MOOC pages, digitization of resources, etc.

The case of libraries being involved in production of electronic resources gives them the opportunity to consider approaches that make the resources accessible to all. Their recognition of universal design at least as a requirement under the country's law; the startups to present textual content in PDF or PDF/A, EPUB, and HTML alternatives; and efforts on making digitized content accessible to screen readers could be taken as steps towards the right direction.

The study on the selected libraries, however, identified the tendency of relating universal design to fulfillment of WCAG guidelines and limiting the role of those guidelines to development of accessible websites. If utilized, WCAG 2.0 could help libraries in designing accessible digital products. For example, it can offer them guidelines for making text content readable and understandable to all users including users with assistive technologies. It may also help to present content in different ways, produce movies in a way that won't cause seizures on some users, and provide alternative text to videos or non-text formats so that the content can be available in forms such as large text, braille, symbols etc. However, the question is whether WCAG would be sufficient to serve as a framework to ensure inclusive digital library environments. As a matter of fact, the guidelines from WAI (including WCAG) has been critiqued for their emphasis on technical or design elements of accessibility, prompting others to fill the gap by putting more focus on organizational elements that impact digital inclusion [24]. A prominent example is the British Standard (BS) 8878, a process oriented standard aimed at helping organizations to embed inclusion strategically across key job-roles, policies, and decisions[8].

As shown in this paper, the state of digital services in libraries presents a complicated scenario. First, today's libraries host different types of resources acquired or subscribed from different vendors operating with different rules, and who would maintain some form of control over the collections. The fact that libraries negotiate access with vendors or copyrights holders shows that adherence to technical guidelines by itself won't ensure accessibility of the resources to all users. Digitization of books could be taken as a good trend that promotes digital services. However, the measures being taken to safeguard intellectual rights could have the effect of excluding users who depend on text-to-speech technologies, presenting a tug-of-war situation between protecting the right of access and the right of intellectual property protection. This underlines the need of including demands of universal design in the negotiations. Therefore, activities related to accessibility evaluation need to extend to evaluating the relationships between libraries and content providers. Second, the issue related to

---

[8] https://www.access8878.co.uk/bs8878-overview.aspx.

resource discovery tools could extend to the use of metadata to describe resources by their accessibility features so that they can be easily discovered by their users. Third, the fact that digital services are run on different technological platforms and managed by different groups of professionals such as librarians, IT Experts, and others adds up to the complexity of the environment. To sum up, the scenario of digital services in libraries calls for development of a framework that identifies all activities, processes, job roles and responsibilities, and provides a template for evaluating accessibility of the technical as well as nontechnical aspects of the services. Therefore, in addition to WCAG 2.0 guidelines, libraries need to be open for inspirations from other standards such as BS 8878.

This study showed the trend where libraries support users with special needs through setup of physical facilities and referring them to the NLB for further assistance. However, this may need to change, at least in digital services, for at least two reasons. First, the Norwegian Discrimination and Accessibility act §12 states that "Breach of the duty to ensure universal design pursuant to Sect. 13 or the duty to ensure individual accommodation in Sects. 16, 17 and 26 shall constitute discrimination."[9]. The law itself requires the services to be inclusive. Second, as a paper on higher education in Norway[10] showed, there is a model under development for production of talking books at a local higher education institute library, as part of the effort for making higher education accessible. This shows that the duties and responsibilities of NLB would be shared by academic libraries. Such developments entail the acceptance of the need of inclusive services at organizational level and strive to make services accessible and usable to all to the extent possible.

6   Directions for Further Research

Issues that might be interesting research directions for further studies have been identified during the course of this research. For instance, the utilization of LeapMotion or related technologies in gesture-based interactions can be extended as developments that help, for instance people with motor impairments, to browse through library catalogs. Use of apps to locate and find books has also the potential to help any user to navigate to the place where a book is shelved. Development of applications that allow users to access digitized content with or without assistive devices, but at the same time protect the intellectual right of the creator, could be perused as an endeavor that benefits both the user and the content creator. The issue of resource discovery and access in library systems is the other challenging area. The way resource discovery tools are being designed can simplify or complicate the process of information search and retrieval for users with different types of impairments. Research and development to help these groups of users to easily discover and access resource that fit their needs can be an important direction.

---

[9] http://app.uio.no/ub/ujur/oversatte-lover/data/lov-20130621-061-eng.pdf.
[10] http://www.universell.no/english/.

Beside the technical issues, the need for a holistic approach or framework that helps to ensure inclusive information resources has been highlighted in this paper. There has been a recommendation from past research that the Functional Requirements of Bibliographic Records (FRBR), which was first developed by International Federation of Library Associations, could be grown as a model for inclusive information environment [25]. Re-examination of this model or development of a better model could yet be another direction for further research.

7 Conclusion

This paper showed that there are potentials, opportunities, and also challenges to realize inclusive digital services. Startups and future considerations regarding production of local contents in alternative formats and utilization of WCAG guidelines to design accessible websites, as well as works on application of technology in libraries provide the basis for promoting the agenda of universal design in libraries. However as mentioned at the start of the paper, the best way for realizing inclusive digital library environments could be to address the technical and non-technical processes surrounding creation, acquisition, organization, and presentation of information resources to their potential users. The case of libraries discussed in this paper shows a trend of associating universal design with fulfillment of WCAG 2.0 guidelines which may not help in addressing organizational and policy related issues that need to be considered.

The scenario of current digital services is characterized by different types of resources managed on different technological platforms, controlled directly or indirectly by different actors involved in the production and organization of information, and run by diverse human resources. This calls for a holistic approach that helps to incorporate concerns of universal design in tasks, processes, activities, procedures, policies, rules and regulations involved in running digital services.

References

1. Southwell, K.L., Slater, J.: An evaluation of finding aid accessibility for screen readers. Inf. Technol. Libr. **32**(3), 34–46 (2013)
2. Bergman, E., Johnson, E.: Towards accessible human-computer interaction. Adv. Hum.-Comput. Interact. **5**(1), 87–114 (1995)
3. Persson, H., Åhman, H., Yngling, A.A., Gulliksen, J.: Universal design, inclusive design, accessible design, design for all: different concepts—one goal? On the concept of accessibility—historical, methodological and philosophical aspects. Univ. Access Inf. Soc. **14**(4), 505–526 (2015)
4. Bertot, J.C., Jaeger, P.T.: The ADA and Inclusion in Libraries (2015). https://american librariesmagazine.org/blogs/the-scoop/ada-inclusion-in-libraries/
5. Farb, S.: #16 Universal Design and Americans with Disabilities Act. http://www.ala.org/alcts/alcts_news/v11n3/gateway_(2003)
6. Colorado State University: Website Accessibility (2013). http://lib.colostate.edu/about/website-accessibility

7. Web Content Accessibility Guidelines (WCAG) Overview. https://www.w3.org/WAI/intro/wcag
8. Real, B., Bertot, J.C., Jaeger, P.T.: Rural public libraries and digital inclusion: issues and challenges. Inf. Technol. Libr. **33**(1), 6–24 (2013)
9. Jaeger, P.T., Bertot, J.C., Thompson, K.M., Katz, S.M., DeCoster, E.J.: The intersection of public policy and public access: digital divides, digital literacy, digital inclusion, and public libraries. Public Libr. Q. **31**(1), 1–20 (2012)
10. Burgstahler, S.: Equal Access: Universal Design of Libraries. http://www.washington.edu/doit/Brochures/Academics/equal_access_lib.html
11. Seale, J., Draffan, E.A., Wald, M.: Digital agility and digital decision-making: conceptualizing digital inclusion in the context of disabled learners in higher education. Stud. High. Educ. **35**(4), 445–461 (2010)
12. Bonnici, L.J., Maatta, S.L., Brodsky, J., Steele, J.E.: Second national accessibility survey: librarians, patrons, and disabilities. New Libr. World **116**(9/10), 503–516 (2015)
13. Lazar, J., Subramaniam, M., Jaeger, P., Bertot, J.: HCI public policy issues in US libraries. Interactions **21**(5), 78–81 (2014)
14. Comeaux, D., Schmetzke, A.: Web accessibility trends in university libraries and library schools. Libr. Hi Tech **25**(4), 457–477 (2007)
15. Comeaux, D., Schmetzke, A.: Accessibility of academic library web sites in North America: current status and trends (2002–2012). Libr. Hi Tech **31**(1), 8–33 (2013)
16. Harpur, P., Suzor, N.: The paradigm shift in realising the right to read: how ebook libraries are enabling in the university sector. Disabil. Soc. **29**(10), 1658–1671 (2014)
17. Stewart, R., Narendra, V., Schmetzke, A.: Accessibility and usability of online library databases. Libr. Hi Tech **23**(2), 265–286 (2005)
18. Walker, W., Keenan, T.: Do you hear what i see? Assessing accessibility of digital commons and CONTENTdm. J. Electron. Resour. Librarianship **27**(2), 69–87 (2015)
19. Ellis, K., Kent, M.: Disability and New Media. Routledge, London (2011)
20. Hill, H.: Disability and accessibility in the library and information science literature: a content analysis. Libr. Inf. Sci. Res. **35**(2), 137–142 (2013)
21. Brown, R.B.: Doing Your Dissertation in Business and Management: The Reality of Researching and Writing. SAGE, Thousand Oaks (2006)
22. Discovery and Delivery: Provide a Cutting-Edge Interface to Your Entire Collection. http://www.exlibrisgroup.com/?catid={14258C70-29B6-49B3-BF80-624A00EF89E6}#{C1D25BD0-1907-4715-BEF4-34451CF21C21}
23. Morozumi, A., Nevile, L., Sugimoto, S.: Enabling resource selection based on written english and intellectual competencies. In: Goh, D.H.-L., Cao, T.H., Sølvberg, I.T., Rasmussen, E. (eds.) ICADL 2007. LNCS, vol. 4822, pp. 127–130. Springer, Heidelberg (2007)
24. Hassell, J.: Including Your Missing 20% by Embedding Web and Mobile Accessibility, 1st edn. BSI British Standards Institution, London (2014)
25. Morozumi, A., Nagamori, M., Nevile, L., Sugimoto, S.: Using FRBR for the selection and adaptation of accessible resources. In: International Conference on Dublin Core and Metadata Applications (2006)

A Maturity Model for Information Governance

Diogo Proença$^{(\boxtimes)}$, Ricardo Vieira, and José Borbinha

IST/INESC-ID, Lisbon, Portugal
{diogo.proenca,rjcv,jlb}@tecnico.ulisboa.pt

Abstract. Information Governance (IG) as defined by Gartner is the "specification of decision rights and an accountability framework to encourage desirable behavior in the valuation, creation, storage, use, archival and deletion of information. Includes the processes, roles, standards and metrics that ensure the effective and efficient use of information in enabling an organization to achieve its goals".

Organizations that wish to comply with IG best practices, can seek support on the existing best practices, standards and other relevant references not only in the core domain but also in relevant peripheral domains. Thus, despite the existence of these references, organizations still are unable, in many scenarios, to determine in a straightforward manner two fundamental business-related concerns: (1) to which extent do their current processes comply with such standards; and, if not, (2) which goals do they need to achieve in order to be compliant.

In this paper, we present how to create an IG maturity model based on existing reference documents. The process is based on existing maturity model development methods that allow for a systematic approach to maturity model development backed up by a well-known and proved scientific research method called Design Science Research.

Keywords: Information governance · Maturity model

1 Introduction

A maturity model defines a pathway of improvement for organizational aspects and is classified by a maturity level. The maturity levels often range from zero to five, where zero consists on the lack of maturity and five consists of a fully mature and self-optimizing process. Maturity models can be used for assessing and/or achieving compliance since they allow the measurement of a maturity level and, by identifying the gap between the current and pursued level, allow the planning of efforts, priorities and objectives in order to achieve the goals proposed.

The use of maturity models is widely used and accepted, both in the industry and the academia [1]. There are numerous maturity models, virtually one for each of the most trending topics in such areas as Information Technology or Management. Maturity Models are widely used and accepted because of their simplicity and effectiveness. They depict the current maturity level of a specific aspect of the organization, for example IT, Outsourcing or Project Management, in a meaningful way, so that stakeholders can clearly identify strengths and improvement points and prioritize what

© Springer International Publishing Switzerland 2016
N. Fuhr et al. (Eds.): TPDL 2016, LNCS 9819, pp. 15–26, 2016.
DOI: 10.1007/978-3-319-43997-6_2

they can do in order to reach higher maturity levels, showing the outcomes that will result from that effort which enables stakeholders to decide if the outcomes justify the effort needed to go to higher levels and results in a better business and budget planning.

The remaining of this paper is structured as follows: Sect. 2 presents the related work that can influence the development of the maturity model, Sect. 3 presents the development strategy for the maturity model as well as a first example of the maturity model based on the ISO16363 and ISO20652 and based on the levels from SEI CMMI [2]. Section 4 presents the assessment strategy for the maturity model. Section 5 presents the conclusions of this paper. The maturity model presented here is being developed in the context of the E-ARK project.

2 Related Work

This section details the related work relevant for this paper, namely the maturity model fundamentals and maturity assessment methods. These are essential to understand the remaining of this paper.

2.1 Maturity Model Fundamentals

To evaluate maturity, organizational assessment models are used, which are also known as stages-of-growth models, stage models, or stage theories [12].

The concept of maturity is a state in which, when optimized to a particular organizational context, is not advisable to proceed with any further action. It is not an end, because it is a mobile and dynamic goal [3]. It is rather a state in which, given certain conditions, it is agreed not to continue any further action. Several authors have defined maturity, however many of the current definitions fit into the context in which each a particular maturity model was developed.

In [4] maturity is defined as a specific process to explicitly define, manage, measure and control the evolutionary growth of an entity. In turn, in [5] maturity is defined as a state in which an organization is perfectly able to achieve the goals it sets itself. In [6] it is suggested that maturity is associated with an evaluation criterion or the state of being complete, perfect and ready and in [7] as being a concept which progresses from an initial state to a final state (which is more advanced), that is, higher levels of maturity. Similarly, in [8] maturity is related with the evolutionary progress in demonstrating a particular capacity or the pursuit of a certain goal, from an initial state to a final desirable state. Still, in [9] it is emphasized the fact that this state of perfection can be achieved in various ways. The distinction between organizations with more or less mature systems relates not only to the results of the indicators used, but also with the fact that mature organizations measure different indicators when comparing to organizations which are less mature [10]. While the concept of maturity relates to one or more items identified as relevant [11], the concept of capability is concerned only with each of these items. In [12] maturity models are defined as a series of sequential levels, which together form an anticipated or desired logical path from an initial state to a final state of maturity. These models have their origin in the area of quality [13, 14]. The Organizational Project

Management Maturity Model (OPM3) defines a maturity model as a structured set of elements that describe the characteristics of a process or product [15, 16]. In [17] maturity models are defined as tools used to evaluate the maturity capabilities of certain elements and select the appropriate actions to bring the elements to a higher level of maturity. Conceptually, these represent stages of growth of a capability at qualitative or quantitative level of the element in growth, in order to evaluate their progress relative to the defined maturity levels.

Some definitions found involve organizational concepts commonly used, such as the definition of [18] in which the authors consider a maturity model as a "... a framework of evaluation that allows an organization to compare their projects and against the best practices or the practices of their competitors, while defining a structured path for improvement." This definition is deeply embedded in the concept of benchmarking. In other definitions, such as in the presented by [19] there appears the concern of associating a maturity model to the concept of continuous improvement.

In [20], the maturity models are particularly important for identifying strengths and weaknesses of the organizational context to which they are applied, and the collection of information through methodologies associated with benchmarking. In [21] it was concluded that the great advantage of maturity models is that they show that maturity must evolve through different dimensions and, once reached a maturity level, sometime is needed for it to be actually sustained. In [22] it was concluded that project performance in organizations with higher maturity levels was significantly increased. Currently, the lack of a generic and global standards for maturity models has been identified as the cause of poor dissemination of this concept.

2.2 Maturity Assessment

An assessment is a systematic method for obtaining feedback on the performance of an organization and identify issues that affect performance. Assessments are of extreme importance as organizations are constantly trying to adapt, survive, perform and influence despite not being always successful. To better understand what they can or should change to improve the way they conduct their business, organizations can perform organizational assessments. This technique can help organizations obtain data on their performance, identify important factors that help or inhibit the achievement of the desired outcomes of a process, and benchmark them in respect to other organizations. In the last decade, the demand for organizational assessment are gaining ground with the implementation of legislation that mandate good governance in organizations, such as, the Sarbanes-Oxley Act [23] and the BASEL accords in financial organizations [24]. Moreover, funding agencies are using the results of these assessments to understand the performance of organizations which they fund (e.g., Not for profit organizations, European Commission, Banks, Research institutes) as a means to determine how well organizations are developing the desired outcomes, and also to better understand the capabilities these organizations have in place to support the achievement of the desired outcome.

The result of an assessment effort will be a set of guidelines which will allow for process improvement. Process improvement is a way of improving the approach taken for organizing and managing business processes and can involve also executing improvements to existing systems. There are several examples of process improvement such as compliance with existing legislation. Process improvement often results in process redesign which involves understanding the requirements of a stakeholder and developing processes which meet the stakeholders' expectations. This often means that the existing processes supporting a specific part of business need to be adapted, or even made from scratch to meet the stakeholders' expectations. When the processes need to be made from scratch we are dealing with process reengineering which is a way to introduce radical changes in the business processes of an organization and changes the way a business operates. In this way, process reengineering starts from scratch by determining how the key business activities need to be reengineered to meet stakeholders' expectations. One well known example, is the transition from traditional banking services to on-line banking services.

The ISO/IEC 15504, describes a method that can be used to guide the assessment of organizational processes, which is depicted in Fig. 1. The ISO15504 assessment method is composed of seven main steps which are then further detailed in atomic tasks.

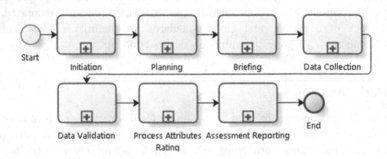

Fig. 1. ISO15504 assessment process overview

2.3 Maturity Models in Related Areas

This section presents the several maturity models from the Information Management, Records Management and Information Governance domains that can influence the development of the maturity model proposed in this paper. Each Maturity Model is presented starting with a small description of the model, the aim of the model, scope, attributes and levels. These attributes further detail the maturity model by decomposing certain aspects of the maturity model domain. Some of the attributes being used are sections or principles. Although there are other attributes being used, such as, dimensions. The synthesis of the analyzed maturity models is presented in Table 1.

Table 1. Synthesis of the analyzed maturity models

Maturity model	Attributes		Maturity levels
	Name	Number	
Asset management maturity model [25]	Dimensions/category	4	1 (Initial); 2 (repeatable); 3 (defined); 4 (managed) and 5 (optimizing)
Digital asset management (DAM) maturity model [26]	Categories/dimensions	4/15	Level 1 (Ad-Hoc); Level 2 (incipient); level 3 (formative); level 4 (operational) and level 5 (optimal)
Information governance maturity model [27]	Principles	8	Level 1 (sub-standard); level 2 (in development); level 3 (essential); level 4 (proactive) and level 5 (transformational)

3   Development of a Maturity Model for Information Governance

One recurrent criticism of maturity models is that they lack empirical foundation and traceability [28]. The main reason for the criticism is that existing maturity models typically do not follow a theoretical framework or methodology for their development [28]. In fact, there is an absence on literature regarding methods and practices for the design and development of maturity models [28].

One of the most known development model for maturity models is the one from Becker in [29], a procedure based on a scientific research method called Design Science Research (DSR). The well-argued claim of the design procedure [29] is that these fundamental requirements should drive the development of every maturity model. Apart from evaluating well-known models according to these dimensions, the article also delineates a set of steps to correctly develop a maturity model. It depicts which documentation should result from each step, and includes an iterative maturity model development method that proposes that each iteration of the maturity model should be implemented and validated before going to a new iteration. The procedure delineates eight requirements [29], (1) Comparison with existing maturity models is presented and clearly argues for the need of a new model or the adaptation of an existing one; (2) Iterative Procedures are followed to ensure a feedback loop and refinement; (3) The principles, quality and effectiveness behind the design and development effort of a maturity model should pass through an iterative Evaluation step; (4) The design and development of maturity models should follow a Multi-methodological Procedure which use must be well founded; (5) During the development of a maturity model there should be a clear Identification of Problem Relevance so that the problem solution can be relevant to practitioners and researchers; (6) Problem Definition should include the application domain for the maturity model and also detail the intended benefits and constraints of application; (7) There should be a Targeted Presentation of Results regarding the users' needs and application constraints and, (8) The design of a maturity model must include Scientific Documentation, which details the whole process design for each step of the process, as well as, the methods applied, people involved and the obtained results.

One limitation of existing maturity models is that it is not typically not clear which requirements were used for the design and development of the model. In other words, there is a weak or inexistent traceability between the maturity model and the requirements that are used as reference. Consequently, stakeholders that wish to use the maturity model are unable to understand if the model is aligned with current best practices. To address the aforementioned traceability problem the maturity model described in this paper is based in well-known references of IG. Due to the fact that IG is a multi-disciplinary fields that covers several disciplines the range of standards and references documents is vast and include references, such as, the ISO 16363, ISO 20652, ISO 14721, MoREQ 2010, ISO 16175, ISO 23081, ISO 30301, ISO 27001, among others.

The maturity model for information governance, depicted further on in this section, consists of three dimensions:

- **Management:** "The term management refers to all the activities that are used to coordinate, direct, and control an organization." [30]
- **Processes:** "A process is a set of activities that are interrelated or that interact with one another. Processes use resources to transform inputs into outputs." [30]
- **Infrastructure:** "The term infrastructure refers to the entire system of facilities, equipment, and services that an organization needs in order to function." [30]

These dimensions provide different viewpoints of information governance which help to decompose the maturity model and enable easy understanding.

For each dimension we have a set of levels, from one to five, where one show the initial phase of maturity of a dimension and level five shows that the dimension is fully mature, self-aware and optimizing. These levels and their meaning were adapted from the levels defined for SEI CMMI [2].

Management

Level 1 (Initial): Management is unpredictable; the business is weakly controlled and reactive. The required skills for staff are neither defined nor identified. There is no planned training of the staff.

Level 2 (Managed): There is awareness of the need for effective management within the archive. However, there are no policies defined. The required skills are identified only for critical business areas. There is no training plan, however training is provided when the necessity arises.

Level 3 (Defined): The documentation, policies and procedures that allows for effective management are defined. There is documentation of skill requirements for all job positions within the organization. There is a formal training plan defined; however it is not enforced.

Level 4 (Quantitatively Managed): The organization monitors its organizational environment to determine when to execute its policies and procedures. Skill requirements are routinely assessed to guarantee that the required skills are present in the organization. There are procedures in place to guarantee that a skill is not lost when staff leaves the archive. There is a policy for knowledge sharing of information within the organization that is described in the training plan. The training plan is also assessed routinely.

Level 5 (Optimizing): Standards and best practices are applied. There is an effort for the organization to undergo assessment for certification of standards. The organization is seen as an example of effective management among its communities and there is continuous improvement of all management procedures. There is encouragement of continuous improvement of skills, based both on personal and organizational goals. Knowledge sharing is formally recognized in the organization. The organization staff contributes to external best practice.

Processes

Level 1 (Initial): Ingest, Archival and Dissemination of content are not done in a coherent way. Procedures are ad-hoc and undefined, the archive may not even be prepared to ingest, archive and disseminate content.

Level 2 (Managed): There is evidence of procedures being applied in an inconsistent manner and based on individual initiative. Due to fact that the processes are not defined, most of the times the applied procedures cannot be repeated.

Level 3 (Defined): The Ingest, archival and dissemination processes are defined and in place. For ingest, is defined which content the archive accepts and how to communicate with producers, the creation of the Archival Information Package is defined as well as the Preservation Description Information necessary for ingesting the object into the archive. For archival, preservation planning procedures are defined and the preservation strategies are documented. For dissemination, the requirements that allow the designated community to discover and identify relevant materials arc in place, and access policies are defined.

Level 4 (Quantitatively Managed): The Ingest, Archival and Dissemination processes are actively managed for their performance and adequacy. There are mechanisms to measure the satisfaction of the designated community. There are procedures in place that measure the efficiency of the ingest, archival and dissemination processes and identify bottlenecks in these processes.

Level 5 (Optimizing): There is an information system that allows for process performance monitoring in a proactive way so that the performance data can be systematically used to improve and optimize the processes.

Infrastructure

Level 1 (Initial): The infrastructure is not managed effectively. Changes in the infrastructure are performed in a reactive basis, when there is hardware/software malfunction or it becomes obsolete. There are no security procedures in place. The organization reacts to threats when they occur.

Level 2 (Managed): There is evidence of procedures being applied to manage the infrastructure. There is awareness of the need to properly define the procedures that allow for effective management of the infrastructure that supports the critical areas of the business. There are security procedures in place. However, individuals perform these procedures in different ways and there is no common procedures defined.

Level 3 (Defined): Infrastructure procedures are defined and in place. There are technology watches/monitoring, there are procedures to evaluate when changes to software and hardware are needed, there is software and hardware available for performing backups and there are mechanisms to detect bit corruption and reporting it. Security procedures are defined and being applied in the organization. The security risk are analyzed, the controls for these risks are identified and there is disaster preparedness and recovery plans.

Level 4 (Quantitatively Managed): There are procedures in place that actively monitor the environment to detect when hardware and software technology changes are needed. The hardware and software that support the services are monitored so that the organization can provide appropriate services to the designated community. There are procedures in place to record and report data corruption that identify the steps needed to replace or repair corrupt data. The security risks are analyzed periodically and new controls are derived from these risk factors. There are procedures to measure the efficiency of these controls to treat the security risks identified. Disaster preparedness and recovery plans are tested and measured for their efficacy.

Level 5 (Optimizing): There is an information system that monitors the technological environment and detects when changes to hardware and software are needed and reacts to it by proposing plans to replace hardware and software. There is also a system that detects data corruption and identifies the necessary steps to repair the data and acts without human intervention. To allow for continuous improvement there are also mechanisms to act upon when the hardware and software available no longer meets the designated community requirements. There is an information system that manages security and policy procedures and the disaster and recovery plans which allows for continual improvement. There is a security officer that is a recognized expert in data security.

4   Assessment of a Reality Using the Maturity Model for Information Governance

This section details the assessment strategy used in the development of the maturity model proposed in this paper. According to [8] there are three possible types of maturity assessment methods, (1) self-assessment where there are guidelines and forms for an organization to perform self-assessment (2) third-party assisted where a third-party help the organization in applying the assessment method and, (3) certified professionals where organizations can apply for certification, in which case a group of competent assessors will perform the assessment.

For the purpose of this maturity model we opted for the self-assessment method as it provides a way for organizations to assess their IG practice while maintaining a low cost to the organizations. Moreover, we defined three main scopes for the assessment depending on the nature of the organization being assessed. The first is the Information Governance Maturity Assessment (IGMA) for Producers (IGMA P) which is used to assess organization that create content. The second is the IGMA for Archives (IGMA A) which is used to assess organizations that archive content, such as, archives. Finally, the IGMA for producers and archives (IGMA PA) which is used to assess organization that are both producers of content and also archive content. An example of such organization would be a national archive that digitizes content and then archives it using an in-house infrastructure. For each of these scopes the assessment questionnaire will be different and only appropriate to the organizations that meet a certain scope. Despite this fact, the self-assessment questionnaire, for each of the three scopes is comprised of a set of questions.

The questionnaire starts by providing an introduction. This introduction provides details on the purpose of the questionnaire, how it will be analysed, and clarifies concepts being constantly used throughout the questionnaire. This questionnaire consists of a set of questions that will be used to determine the maturity levels of an organization for each of the three dimensions of the IG Maturity Model.

The answers provided will then be analysed by the Information Governance Maturity Model development team and a report will be issued detailing all the findings of the assessment. For each question there is a field respondents can use to provide additional comments, clarifications or a justification to the answer. These comments will be considered by the assessment team when evaluating the answers.

For the first assessment, as a proof of concept, we asked seven organizations that are comprised in the IGMA P to fill the self-assessment questionnaire for the processes dimension of the maturity model which was comprised of 35 questions. The results from this first assessment are presented in Fig. 2.

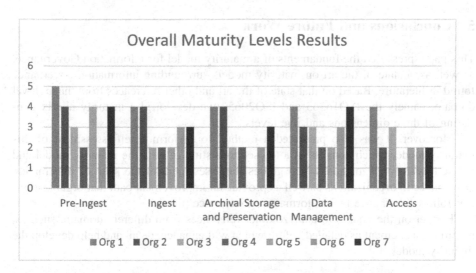

Fig. 2. Results of the first assessment using the maturity model for information governance

Organization 1 is the one which achieved the best overall results, especially in pre-ingest and access it achieved the best results. Organization 2 achieved the second best results. However there are still some enhancements to perform in the access capability where it achieved maturity level 2. Despite this fact, the access capability is not the focus in Organization 2. Organization 7 also shows a high level maturity across the capabilities measured in the assessment. However, as in Organization 2, there are still some important enhancements to perform to the access capability. In organization 7, the importance of the access capability is considerable due to it being one of the focuses of the organization.

The other four organizations showed similar results among the capabilities. With some exceptions for organization 3, where it shows higher maturity levels for pre-ingest and the access capabilities. Another exception is organization 6 which shows higher maturity levels for ingest and data management capabilities. Organization 5 did not answer to the questions for the archival storage and preservation and as the result no maturity level was calculated. As this is not the focus capability of the organization there is no major problem with this fact.

There are still several capabilities at maturity level 1 or 2 for all organizations except organization 1. These should be addressed as soon as possible to reach at least maturity level. This is due to the fact that maturity level 3 is considered an intermediate level between lack of definition of consistency of mechanism and procedures typical of maturity level 1 and 2; and the documentation and assessment of mechanism and procedures typical of maturity level 4 and 5. Maturity level 3 depicts aspects that are consistent and defined throughout the organizational context and shows a state of change in this context from no definition to improvement. The results of the E-ARK project will help the organizations to reach this maturity level and will also assist other organizations to reach higher levels of maturity and as result improve archival practice.

5   Conclusions and Future Work

This paper presented the fundaments of a maturity model for information Governance, as well as, a state of the art on maturity models surrounding information governance found in literature. Based on that state of the art and other references from the archival domain, namely the ISO16363 and ISO20652 we developed a maturity matrix consisting of three dimensions and five levels.

Moreover, it was also presented a method to perform a self-assessment of this maturity model which consists of a toolset consisting of both the maturity model and the self-assessment method which guides the assessment of the state of information governance in organizations as well as provide an improvement path that organizations can follow to enhance their information governance practice.

Further on the goal is to analyze other references from different domain, such as, records management as detailed before which will enhance, detail and help develop the maturity model.

Acknowledgements. This research was co-funded by FCT – Fundação para a Ciência e Tecnologia, under project PEst-OE/EEI/LA0021/2013 and by the European Commission under the Competitiveness and Innovation Programme 2007-2013, E-ARK – Grant Agreement no. 620998 under the Policy Support Programme. The authors are solely responsible for the content of this paper.

References

1. Shang, S., Lin. S.: Understanding the effectiveness of capability maturity model integration by examining the knowledge management of software development process. Total Qual. Manag. Bus. Excellence **20**(5), 509–521 (2009)
2. CMMI Product Team, CMMI for services, version 1.3. Software Engineering Institute. Carnegie Mellon University, Technical report CMU/SEI-2010-TR-034 (2010)
3. Tonini, A., Carvalho, M., Spínola, M.: Contribuição dos modelos de qualidade e maturidade na melhoria dos processos de software. Produção **18**(2), 275–286 (2008)
4. Paulk, M., Curtis, B., Chrissis, M., Weber, C.: Capability Maturity Model for software, Version 1.1 CMU/SEI-93-TR-24, Pittsburgh, Pennsylvania, USA, Carnegie Melon University (1993)
5. Anderson, E., Jessen, S.: Project maturity in organizations. Int. J. Proj. Manag. Account. **21**, 457–461 (2003)
6. Fitterer, R., Rohner, P.: Towards assessing the networkability of health care providers: a maturity model approach. Inf. Syst. E-Bus. Manag. **8**, 309–333 (2010)
7. Sen, A., Ramamurthy, K., Sinha, A.P.: A model of data warehousing process maturity. IEEE Trans. Softw. Eng. **38**(2), 336–353 (2012). doi:10.1109/TSE.2011.2. http://ieeexplore.ieee.org/stamp/stamp.jsptp=&arnumber=5680911&isnumber=6173074
8. Mettler, T.: A Design Science Research Perspective on Maturity Models in Information Systems. Institute of Information Management, University of St. Gallen, St. Gallen (2009)
9. Amaral, A., Araújo, M.: The organizational maturity as a conductive field for germinating business sustainability. In: Proceedings of Business Sustainability I Conference, Póvoa do Varzim, Portugal (2008)
10. Cooke-Davies, T., Arzymanowc, A.: The maturity of project management in different industries: an investigation into variations between project management models. Int. J. Proj. Manag. **21**(6), 471–478 (2003)
11. Hillson, D.: Maturity - good or bad? Project Manager Today, p. 14 (2008)
12. Röglinger, M., Pöppelbuß, J.: What makes a useful maturity model? A framework for general design principles for maturity models and its demonstration in business process management. In: Proceedings of the 19th European Conference on Information Systems, Helsinki, Finland, June 2011
13. Brookes, N., Clark, R.: Using maturity models to improve project management practice. In: Proceedings of the POMS 20th Annual Conference, Florida, USA, 1–4 May 2009
14. Demir, C., Kocabaş, I.: Project management maturity model (PMMM) in educational organizations. Procedia Soc. Behav. Sci. **9**, 1641–1645 (2010)
15. OPM3: Organizational Project Management Maturity Model. Project Management Institute, Newtown Square (2003)
16. Hersey-Miller, L.: Organizational Project Management Maturity Model (OPM3). Office of Major Projects, Quarter 3, pp. 1–8, September 2005
17. Kohlegger, M., Maier, R., Thalmann, S.: Understanding maturity models: results of a structured content analysis. In: Proceedings of the I-KNOW 2009 and I-SEMANTICS 2009, Graz, Austria, 2–4 September 2009
18. Korbel, A., Benedict, R.: Application of the project management maturity model to drive organisational improvement in a state owned corporation. In: Proceedings of 2007 AIPM Conference, Tasmania, Australia, 7–10 October 2007
19. Jia, G., Chen, Y., Xue, X., Chen, J., Cao, J., Tang, K.: Program management organization maturity integrated model for mega construction programs in China. Int. J. Proj. Manag. **29**, 834–845 (2011)

20. Koshgoftar, M., Osman, O.: Comparison between maturity models. In: Proceedings of the 2nd IEEE International Conference on Computer Science and Information Technology, vol. 5, pp. 297–301 (2009)
21. Prado, D.: Gerenciamento de Programas e Projetos nas Organizações. Minas Gerais: Nova Lima (2004)
22. Jamaluddin, R., Chin, C., Lee, C.: Understanding the requirements for project management maturity models: awareness of the ICT industry in Malaysia. In: Proceedings of the 2010 IEEE IEEM, pp. 1573–1577 (2010)
23. Tarantino, A.: Governance, Risk and Compliance Handbook. Wiley, Hoboken (2008)
24. IT Governance Institute, IT Control Objectives for BASEL II – The Importance of Governance and Risk Management for Compliance (2006)
25. Lei, T., Ligtvoet, A., Volker, L., Herder, P.: Evaluating asset management maturity in the netherlands: a compact benchmark of eight different asset management organizations. In: Proceedings of the 6th World Congress of Engineering Asset Management (2011)
26. Real Story Group, DAM Foundation: The DAM Maturity Model. http://dammaturitymodel.org/
27. ARMA International: Generally Accepted Recordkeeping Principles - Information Governance Maturity Model. http://www.arma.org/principles
28. Röglinger, M., Pöppelbuß, J., Becker, J.: Maturity models in business process management. Bus. Process Manag. J. 18(2), 328–346 (2012)
29. Becker, J., Knackstedt, R., Pöppelbuß, J.: Developing maturity models for IT management – a procedure model and its application. Bus. Inf. Syst. Eng. 1(3), 212–222 (2009)
30. ISO 9001:2008: Quality management systems – Requirements (2008)

Usage-Driven Dublin Core Descriptor Selection

A Case Study Using the Dendro Platform for Research Dataset Description

João Rocha da Silva[1](✉), Cristina Ribeiro[2], and João Correia Lopes[2]

[1] Faculdade de Engenharia da Universidade do Porto/INESC TEC, Porto, Portugal
joaorosilva@gmail.com
[2] DEI—Faculdade de Engenharia da Universidade do Porto/INESC TEC,
Porto, Portugal
{mcr,jlopes}@fe.up.pt

Abstract. Dublin Core schemas are the core metadata models of most repositories, and this includes recent repositories dedicated to datasets. DC descriptors are generic and are being adapted to the needs of different communities with the so-called Dublin Core Application Profiles. DCAPs rely on the agreement within user communities, in a process mainly driven by their evolving needs. In this paper, we propose a complementary automated process, designed to help curators and users discover the descriptors that better suit the needs of a specific research group. We target the description of datasets, and test our approach using Dendro, a prototype research data management platform, where an experimental method is used to rank and present DC Terms descriptors to the users based on their usage patterns. In a controlled experiment, we gathered the interactions of two groups as they used Dendro to describe datasets from selected sources. One of the groups had descriptor ranking on, while the other had the same list of descriptors throughout the whole experiment. Preliminary results show that 1. some DC Terms are filled in more often than others, with different distribution in the two groups, 2. selected descriptors were increasingly accepted by users in detriment of manual selection and 3. users were satisfied with the performance of the platform, as demonstrated by a post-study survey.

Keywords: Research data management · Ontologies · Linked Data · Ranking · User feedback

1 Introduction

Data is becoming an important research output. Studies prove that papers with associated data have higher citation rates over time, providing additional recognition of the author's work [1,2]. With informal data sharing taking place in spite of the lack of adequate resources for data curation [3], it is clear that many researchers want to share data, and also that there is a great demand for datasets with high-quality metadata as a complement of publications. On the other hand, research data description is recognized as an expensive process

© Springer International Publishing Switzerland 2016
N. Fuhr et al. (Eds.): TPDL 2016, LNCS 9819, pp. 27–38, 2016.
DOI: 10.1007/978-3-319-43997-6_3

in time and resources. Informal metadata is created as researchers gather their datasets [4,5] and used to share data within research groups in an ad-hoc manner [6]. However, producing metadata good enough for sharing with the community requires much more effort and a higher degree of knowledge of metadata practices [7].

High-quality metadata is important for the discovery, retrieval and interpretation of research datasets [8]. Good results would be expected from researchers working together with data curators to adequately capture the production context of a dataset [9]. Currently, however, curators are unavailable in many research groups and there are few financial resources for data curation—especially on the so-called long-tail of science [10]. As a result, researchers themselves have to describe their data, by their own initiative or to comply with funding requirements [11].

Repositories often adopt the Dublin Core Metadata Element Set (DCMES) [12], since it is generic and widely supported in repository software. However, the quality of metadata records can be improved by introducing more descriptors and helping researchers decide which to include in the metadata records of their resources.

In this paper we propose and test a usage-based descriptor ranking approach for qualified DC, designed to select sets of descriptors that depend on the data's production context, from the more extensive DC Metadata Terms Schema [13]. Descriptors are selected and ordered depending on factors such as the resource being described or the past usage of descriptors by the users and their research groups. The descriptor lists emerge from the actual usage of descriptors in the different research groups, and can contribute to the formalization of an Application Profile (AP) [14], or to improve an existing one. The automated generation of the profile cannot replace the broad analysis and community-wide understanding that forms the basis of a Dublin Core Application Profile (DCAP) [15,16]. However, it may provide additional quantitative information regarding the changing needs of a community—which has been an important driving force for the improvement of existing DCAPs [17].

2   Dendro

In the long tail of science, data management is mostly performed "a posteriori"— that is, at the end of the research workflow, and after publishing results. This places research data at risk, since researchers often seek new projects as funding ends, making it hard to obtain timely and comprehensive metadata for these large numbers of research products. To prevent this, the research data management process should start as early as possible [18,19], ideally as researchers have knowledge of their data production context and are actively producing their datasets [20]. This need has been identified in the past by the ADMIRAL project [21], and more recently by the EUDAT infrastructure, which provides separate solutions designed for the storage of data inside research groups (B2DROP) and for description and deposit into prominent repositories (B2SHARE) [22].

Dendro[1] brings dataset deposit and description closer to an earlier moment in the research workflow. It is a data management platform under development at FEUP, which acts as a file storage, description and sharing platform. It combines a "Dropbox"-like interface with some description features usually found in a semantic Wiki [23]. Users start by creating a project, which is a shared storage area among project contributors. Afterwards, they can upload files and create folders, while collaboratively producing metadata records for every element in the directory structure of the project [24].

Its extensible data model is built on ontologies, allowing users to easily "mix-and-match" descriptors from different schemas in their metadata records. The simplicity and expressiveness of its data model also facilitate the platform's decommissioning when it becomes necessary to do so, increasing the survivability of the data and metadata contained inside. When compared to existing solutions supported on a relational model, this graph data model combined with full Linked Open Data representation allows the representation of metadata in a more flexible, interoperable and arguably, simpler way when compared to most repository platforms [25]. These alternatives often implement protocols such as OAI-PMH or specific REST APIs to expose their metadata records to external systems [26]. Instead, we propose that these systems should rely on URI de-referentiation as proposed by the Linked Data guidelines [27] and SPARQL, which allows for much more sophisticated querying.

2.1   User Interface

In this work, we take advantage of Dendro's ability to combine arbitrary sets of descriptors from different ontologies in metadata records, helping research groups discover and select the sets of descriptors that might be more adequate to the description of their datasets.

Figure 1 shows the user interface of Dendro, with its descriptor selection features highlighted. The layout is divided into three main sections, from the left to the right: the file browser, which allows users to navigate in their files and folders; the metadata editor, which shows all the descriptors and controls for users to enter their corresponding values (text boxes, date pickers or maps, depending on the nature of the descriptor) and finally, on the rightmost position there is the descriptor selection area where users can pick descriptors to be included in the current metadata record. For each descriptor, two additional buttons are present: one to promote a descriptor to "Favorite", and another to "Hide" that descriptor. When the user selects the "Favorite" button for the first time, the descriptor will be marked as "Project Favorite" for all the project collaborators, moving it up in the list of suggested descriptors within that project. A second press will promote the descriptor to "User Favorite", making the descriptor go up in the list for the current user only, without influencing the lists presented to the other project collaborators.

---

[1] http://dendro.fe.up.pt/blog/index.php/dendro/.

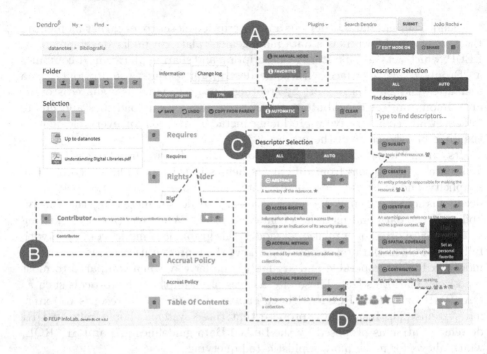

Fig. 1. The main user interface of Dendro (Color figure online)

A shows the interface modes that users can switch between. When the "Manual mode" is active, no descriptor suggestions are automatically added to the metadata editor; when the user switches to the "Automatic" mode, a set of descriptors selected by the platform are automatically added to the metadata editor at the centre, if they are not already filled in for the current record. In "Favorites" mode, descriptors that were marked as "Project Favorites" or "User Favorites" are automatically added. The interface highlights descriptors suggested via the "Automatic" mode in yellow and descriptors added via the "Favorites" mode in green (**B**).

Dendro also provides feedback to the user as to why descriptors are included in the list, in order to improve system *transparency* [28,29]. For each descriptor in (**C**), a set of icons can appear to indicate the reasons why a descriptor is included in the list (e.g. "Frequently used in project", "Frequently used in the entire platform", "Used in textually similar resources", etc.).

2.2   Descriptor Ranking

Different types of interactions are recorded as users perform tasks in Dendro, such as selecting a descriptor or filling it in. Afterwards, they are counted and used as part of calculation formulas to yield n descriptor *features* $f_{1..n}$. Summing these features produces the final score S_d for descriptor d (Eq. 1). Table 1 details the features and their calculation formulas.

Table 1. Descriptor features and scoring formulas.

	Feature	Description	Formula
f_1	Frequently used overall	The descriptor was filled in k_1 times over all records in the Dendro instance	$f_1 = \text{Max}(+1.0 * k_1; +80.0)$
f_2	Recently used	The descriptor has been filled in k_2 times by the user in the last 30 days	$f_2 = \text{Max}(+2.0 * k_2; +80.0)$
f_3	Used in active project	The descriptor has been filled in k_3 times in the project that contains the resource being described	$f_3 = \text{Max}(+2.0 * k_3; +80.0)$
f_4	Auto accepted in editor	The user has filled in this descriptor after it was automatically added to the metadata editor area	$f_4 = +80.0$
f_5	From textually similar	The descriptor is present in other k_5 Dendro resources that are deemed to be textually similar to the one being described	$f_5 = \text{Max}(+20.0 * k_5; +80.0)$
f_6	Auto rejected in editor	The user has removed this descriptor from the metadata editor after it was automatically added	$f_6 = -80.0$
f_7	Favorite accepted in editor	The user has filled in this descriptor after it was automatically added to the metadata editor, in "Favorites" selection mode (see Area **A** of Fig. 1)	$f_7 = +80.0$
f_8	Project favorite	The descriptor has been marked as a project favorite by the user or another collaborator of the current project	$f_8 = +80.0$
f_9	User favorite	The descriptor has been marked as a personal favorite by the user	$f_9 = +80.0$
f_{10}	Hidden in project	The descriptor has been hidden in the project that contains the resource being annotated	N/A (Forces hiding)
f_{11}	Hidden for user	The descriptor has been hidden by the user, regardless of the active project	N/A (Forces hiding)

$$S_d = \sum_{i=1}^{n} f_i \tag{1}$$

The first 4 rows represent implicit feedback features; for example, filling in a descriptor is implicit feedback because it expresses an implicit acceptance of that descriptor without disrupting the description process; setting a descriptor as a favorite, on the other hand, is explicit feedback because it expresses an intention of the user to change the ranking of the descriptor in the list, and is not a necessary part of the description work. The value of each descriptor feature depends on the number of interactions of that type (k_i), which is multiplied by an empirically determined weight that expresses their relative importance.

In some cases, a maximum value for the component is also set. Row 5 shows a component that is calculated from a similarity measure given by the search index used in Dendro (ElasticSearch). If there are many resources with similar text contents (only valid for certain file types from which text can be extracted), the descriptors used to describe those similar resources receive a score bonus. Rows 6 through 9 indicate score components that are calculated from explicit feedback over the suggestions. Setting a descriptor as a "Favorite" or rejecting it when it is suggested, for example, require the user to stop the description tasks to work towards changing the descriptor ranking, so they have to be more strongly weighted than those in the first 4 rows. Finally, some descriptors will also not be presented, regardless of their score, if the user marks them as hidden for himself or for the current project (rows 10 and 11).

3   User Study

To test our approach, we had the collaboration of 23 students of the Digital Archives and Libraries course at FEUP. This course is part of a Masters course of Information Science, so all students were already aware of the concepts that are relevant for creating quality metadata, such as the notions of metadata schema, descriptor and Dublin Core. These students are expected to carry out curatorial tasks in their work environment, which might be a library, a digital repository, an archive, or another system that requires the support of information management experts—in fact, some are already data management professionals. While the size of the user group is not large enough to be considered a large scale study, it is in line with similar user studies in IR and recommendation [30–32].

We decided to split our user group into separate subgroups for A–B testing to reduce learning bias. The two distinct user groups also allowed us to observe their learning behaviors separately as they interacted with the system. To improve the realism of this experiment, users were kept unaware of the real goal of the experiment (evaluating descriptor usage); instead, we instructed them to provide the best possible descriptions for the datasets.

The groups were named U_{Rec} (users of the Dendro with descriptor ranking functionality) and U_{Alpha} (users of the Dendro that presented descriptors ordered alphabetically). Each of the two was further divided into subgroups of 3 students, and randomly attributed to U_{Rec} or U_{Alpha}. Each of the subgroups was tasked, over three weeks, with the description of several datasets from different online sources from distinct research domains and natures.

The interface elements in areas **A**, **B** and **D** (see Fig. 1) were only available in the D_{Rec} instance. The **C** elements was available on both Dendro instances, but the "Auto" button was not visible, only the list of descriptors being shown.

4   Results Analysis

We define *effort interactions* as those interactions that users have to perform to build a metadata sheet in addition to filling in the descriptors themselves.

An example of an effort interaction is adding a descriptor from the list **C** to the metadata editor **B**, by clicking on the corresponding button (Fig. 1). Conversely, those interactions that do not require any effort towards the descriptor lists are *non-effort interactions*. For example, Fig. 1 shows several descriptors in yellow (which means automatically added) in the metadata editor at the center. Since the user did not have to manually select them, we record a non-effort interaction for every one of those descriptors, but only after it is saved.

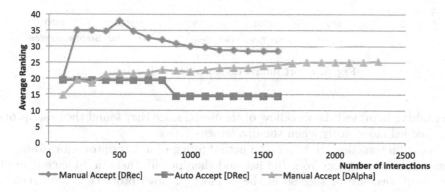

Fig. 2. Average position of the selected descriptors in the selection lists

Figure 2 shows a comparison of the average position of the descriptors as they were selected from the list. Lower values indicate an average position higher up in the list; the lower the value the better the ranking performs, since users find what they need in the top positions and do not need to look further down. For every value of the x axis (an interaction that occurred at an instant T), the corresponding y is the result of an average of the positions of the selected descriptors of all interactions that occurred before T.

From this chart we can see that the average ranking of the manually accepted descriptors in D_{Alpha} tends to assume an average of 25, which is almost half of 54. Since there are 54 descriptors in DCTERMS and D_{Alpha} shows all the descriptors ordered alphabetically, it seems natural that after a sufficiently large number of interactions the average position tends to assume the middle position of the list. The "Manual Accept [DRec]" series shows the average position of the descriptors selected in D_{Rec} while it is in manual mode ("ALL" option is selected in **C**, Fig. 1). In this mode, D_{Rec} shows the same alphabetically-ordered list as present in D_{Alpha} so the long-term behavior is understandably similar.

The chart also shows the "Auto Accept [DRec]" series, which is the average position of the selected descriptors when the "AUTO" option is selected in D_{Rec} (Fig. 1C). When this option is active, the order of the descriptors is given by the ranking algorithm. Since the average ranking of the selected descriptors is consistently lower than in the other two series, we can conclude that in this mode users consistently selected descriptors higher up on the ranked list when compared to the alphabetical order. This is a positive outcome that indicates that

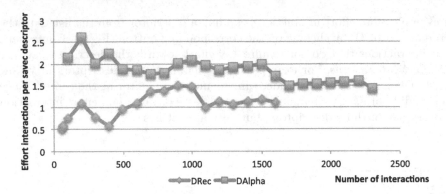

Fig. 3. Effort interactions per saved descriptor

the ranking improved the workflow of the users, since they found the descriptors they needed more easily when the option was active.

Not all interactions result in the actual filling of a descriptor—for example, users might select them from the list and then not fill them in. Since our most important measure of success is the number of descriptors that are actually filled in, we decided to plot the number of effort interactions for every descriptor that is filled in. Figure 3 shows the evolution of these ratios for the two Dendro instances. In the x axis are the effort interactions x_i recorded in the system, and the y axis values are calculated by dividing the number of effort interactions recorded at the time of an interaction x_i by the number of descriptors present in the metadata records at that moment.

The D_{Rec} version achieves a significantly lower number of interactions for each descriptor that is saved. In the case of D_{Alpha}, this number is consistently higher than 1; since it has no automatic descriptor selection capabilities, users have to manually pick every descriptor (an effort interaction) from the list before they can fill it in. In contrast, the average number of interactions per descriptor in D_{Rec} is sometimes lower than 1. This happens because users fill in descriptors that are automatically added to the metadata editor instead of rejecting them (which is also an effort interaction), hence lowering the number of effort interactions required to fill in the same number of descriptors. This is another indication of an improvement introduced by the ranking approach versus the alphabetical ordering.

All the charts show that there is a significantly lower number of interactions overall registered in D_{Rec} when compared to D_{Alpha}—we believe this to be a consequence of a difference in performance and attention to detail between the user groups, although they were randomly split. We can also observe some learning behavior in our users. The first interactions over the system originate large variations in the number of interactions carried out per saved descriptor. As the number of interactions grows, the values stabilize and become lower, indicating an increase in efficiency by the users.

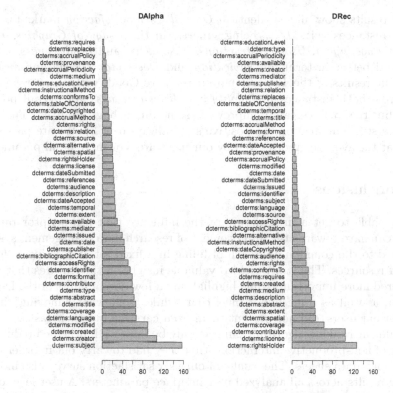

Fig. 4. Distribution of descriptor instances, per descriptor, at the end of the user study

Figure 4 shows the distribution of descriptor instances by their DCTERMS element at the end of the user study. When comparing the two charts, it is apparent that the D_{Alpha} distribution has a larger "tail" when compared to D_{Rec}. Moreover, there is a higher frequency of descriptor usage for the top-n descriptors in D_{Rec} than in D_{Alpha}. This concentration of descriptor usage in the top-n descriptors can be explained by the way that the lists are produced, because they can highly benefit descriptors that have been somehow favored in the past.

This behavior can be positive or not, as at first it may seem that users are liking the descriptor lists and that those are actually automating repetitive work. However, it can also be due to other reasons, such as the similarity between the datasets themselves—which we tried to counteract by requiring users to retrieve data from three different sources—or simply the inability of our algorithm to introduce previously unused descriptors in the lists.

4.1   User Satisfaction Survey

Our user survey was anonymous; 18 users from the initial 23 that participated, 9 from each Dendro instance. Several user interface and user experience parameters were evaluated by the survey respondents, who graded each parameter according to a 1("Poor")-5("Excellent") scale.

The results show almost identical *Overall User Satisfaction* marks for both Dendro instances, with D_{Rec} being superior in the *Usage of Graphics, Usage of color, Page layout, Usage instructions, Ease of use* and *Intuitiveness*. D_{Alpha} performed better in terms of *Navigation* and *Descriptor display and explanations*. The results of the *Features* parameter were Good, with an average score of 4 across both instances. *Reliability* and *Responsiveness* got overall positive scores, but they will certainly improve as some minor bugs are fixed. The average score still remained on the positive side (always over 3). We are pleased to note that the average grades given by our users are 3 or more in all parameters.

5   Conclusions

We were able to get a first picture of the influence that a descriptor ranking system can have towards the improvement of research data management system, compared to the common approach of filling in a fixed set of descriptors for all types of resources. The study yielded valuable insight on the features that users considered more important, while highlighting a few design faults of the Dendro solution, as well as some minor bugs that, while "somewhat annoying" in the words or our users, did not prevent them from carrying out their tasks.

While our ranking algorithm can certainly be improved, it helped reduce the number of non-productive interactions that they had to carry out in order to fill in their metadata sheets. The results of our user satisfaction survey also indicate positive results across all analyzed user interface parameters. A user-level quantitative analysis will provide more valuable insight as to why, for example, the U_{Rec} subgroup ended up saving a significantly lower number of descriptors than the U_{Alpha} counterpart. Our current conjecture is that one or more of the work groups in the U_{Alpha} group have decided to perform a much more comprehensive description of their selected datasets.

Dendro has the ability to record the full history of metadata changes in any record. This makes it an adequate testing ground for studying the evolution of metadata records, so it can provide valuable insight on how much effort curators need to put in to improve and maintain metadata records after their creation by researchers. We believe that such results might add to existing work that reports on the quality of submitted datasets containing DC-compliant metadata [33].

We will continue to improve the ranking algorithm, giving it new features. For example, we want to make it possible for it to suggest descriptors that were never used before by neither the active user nor the collaborators of the active project. We are also planning to include domain-specific metadata descriptors from ontologies other than DC. This approach will provide a novel solution for the usage-driven construction of application profiles, which has long been a purely manual process carried out by research data curators. While not completely replacing the curator's validation and oversight roles, these automatic data description capabilities will provide researchers greater autonomy in the description of their datasets, freeing curators from metadata creation to validation roles.

The base data of these studies is available in a demonstration project at our Dendro demo instance[2].

Acknowledgements. This work is financed by the ERDF – European Regional Development Fund through the Operational Programme for Competitiveness and Internationalisation - COMPETE 2020 Programme and by National Funds through the Portuguese funding agency, FCT - Fundação para a Ciência e a Tecnologia within project POCI-01-0145-FEDER-016736.

References

1. Piwowar, H., Vision, T.: Data reuse and the open data citation advantage. PeerJ **1**, e175 (2013)
2. Piwowar, H., Day, R., Fridsma, D.: Sharing detailed research data is associated with increased citation rate. PLoS ONE **2**(3), e308 (2007)
3. Science Magazine: Dealing with data. Challenges and opportunities. Introduction. Science **331**, pp. 692–693 (2011). New York, NY
4. Jahnke, L., Asher, A., Keralis, S.D.C.: The Problem of Data. Council on Library and Information Resources, Washington, D.C. (2012)
5. Rocha, J., Ribeiro, C., Correia Lopes, J.: Managing research data at U.Porto: requirements, technologies and services. In: Innovations in XML Applications and Metadata Management: Advancing Technologies. IGI Global (2012)
6. Martinez-Uribe, L.: Using the data audit framework: an Oxford case study. Technical report, Oxford Digital Repositories Steering Group, JISC (2009)
7. Borgman, C.: The conundrum of sharing research data. J. Am.Soc. Inf. Sci. Technol. **63**(6), 1059–1078 (2012)
8. Lord, P., Macdonald, A.: Data curation for e-Science in the UK: an audit to establish requirements for future curation and provision. Technical report, JISC (2003)
9. Lyon, L.: Dealing with Data: Roles, Rights, Responsibilities and Relationships (2007)
10. Heidorn, P.B.: Shedding light on the dark data in the long tail of science. Libr. Trends **57**(2), 280–299 (2008)
11. Leonelli, S., Spichtinger, D., Prainsack, B.: Sticks and carrots: encouraging open science at its source. Geo: Geogr. Environ. **2**(1), 12–16 (2015)
12. Dublin Core Metadata Initiative: Dublin Core Metadata Element Set, Version 1.1 (2012)
13. Dublin Core Metadata Initiative: DCMI Metadata Terms (2012)
14. Heery, R., Patel, M.: Application profiles: mixing and matching metadata schemas. Ariadne, no. 25 (2000)
15. Malta, M.C., Baptista, A.A.: State of the art on methodologies for the development of a metadata application profile. In: Dodero, J.M., Palomo-Duarte, M., Karampiperis, P. (eds.) MTSR 2012. CCIS, vol. 343, pp. 61–73. Springer, Heidelberg (2012)
16. Malta, M., Baptista, A.: A panoramic view on metadata application profiles of the last decade. Int. J. Metadata Semant. Ontol. **9**(1), 58–73 (2014)

---

[2] See http://dendro.fe.up.pt/demo.

17. Krause, E.M., Clary, E., Greenberg, J., Ogletree, A.: Evolution of an application profile: advancing metadata best practices through the dryad data repository. In: Proceedings of the International Conference on Dublin Core and Metadata Applications 2015, pp. 63–75 (2015)
18. Martinez-Uribe, L., Macdonald, S.: User engagement in research data curation. In: Agosti, M., Borbinha, J., Kapidakis, S., Papatheodorou, C., Tsakonas, G. (eds.) ECDL 2009. LNCS, vol. 5714, pp. 309–314. Springer, Heidelberg (2009)
19. Eynden, V.V.D., Corti, L., Bishop, L., Horton, L.: Managing and Sharing Data: A Guide to Good Practice, 3rd edn. UK Data Archive University of Essex, Colchester (2011)
20. Ball, A.: Scientific data application profile scoping study report. Technical report, UKOLN, University of Bath, Bath, UK (2009)
21. Hodson, S.: ADMIRAL: a data management infrastructure for research activities in the life sciences. Technical report, University of Oxford (2011)
22. Hanahoe, H., Baxter, R., Carter, A., Reetz, J., Riedel, M., Ritz, R., van de Sanden, M., Wittenberg, P.: Second EUDAT Conference Report (2014)
23. Rocha, J., Castro, J., Ribeiro, C., Correia Lopes, J.: The Dendro research data management platform: applying ontologies to long-term preservation in a collaborative environment. In: iPres 2014 Conference Proceedings (2014)
24. Rocha, J., Castro, J., Ribeiro, C., Correia Lopes, J.: Dendro: collaborative research data management built on linked open data. In: Proceedings of the 11th European Semantic Web Conference (2014)
25. Rocha, J., Ribeiro, C., Correia Lopes, J.: Ontology-based multi-domain metadata for research data management using triple stores. In: Proceedings of the 18th International Database Engineering and Applications Symposium (2014)
26. Amorim, R., Castro, J., Rocha, J., Ribeiro, C.: A comparative study of platforms for research data management: interoperability, metadata capabilities. In: Rocha, A., Correia, A.M., Costanzo, S., Reis, L.P. (eds.) New Contributions in Information Systems and Technologies. Advances in Intelligent Systems and Computing, vol. 353, pp. 101–111. Springer, Berlin (2015)
27. Berners-Lee, T.: Linked Data—Design Issues (2006)
28. Sinha, R., Swearingen, K.: The role of transparency in recommender systems. In: Extended Abstracts on Human Factors in Computing Systems (CHI 2002), p. 830 (2002)
29. Swearingen, K., Sinha, R.: Beyond algorithms: an HCI perspective on recommender systems. In: 2001 ACM SIGIR 2001 Workshop on Recommender Systems , pp. 1–11 (2001)
30. Joachims, T., Granka, L., Pan, B.: Accurately interpreting clickthrough data as implicit feedback. In: Proceedings of the 28th Annual International ACM SIGIR Conference on Research and Development in Information Retrieval, pp. 154–161 (2005)
31. Strickroth, S., Pinkwart, N.: High quality recommendations for small communities: the case of a regional parent network. In: Proceedings of the Sixth ACM conference on Recommender Systems, pp. 107–114 (2012)
32. Goy, A., Magro, D., Petrone, G., Picardi, C., Segnan, M.: Ontology-driven collaborative annotation in shared workspaces. Future Gener. Comput. Syst. 54, 435–449 (2015)
33. Greenberg, J., Swauger, S., Feinstein, E.M.: Metadata capital in a data repository. In: Proceedings of the International Conference on Dublin Core and Metadata Applications 2013, Lisbon, pp. 140–150. Data Dryad Repository (2013)

Search

Retrieving and Ranking Similar Questions from Question-Answer Archives Using Topic Modelling and Topic Distribution Regression

Pedro Chahuara[1,2], Thomas Lampert[1,3(✉)], and Pierre Gançarski[1]

[1] Laboratoire ICube, Université de Strasbourg, Strasbourg, France
{pedro.chahuara,lampert,pierre.gancarski}@unistra.fr
[2] Xerox Research Centre Europe, Meylan, France
[3] Laboratoire Quantup, Strasbourg, France

Abstract. Presented herein is a novel model for similar question ranking within collaborative question answer platforms. The presented approach integrates a regression stage to relate topics derived from questions to those derived from question-answer pairs. This helps to avoid problems caused by the differences in vocabulary used within questions and answers, and the tendency for questions to be shorter than answers. The performance of the model is shown to outperform translation methods and topic modelling (without regression) on several real-world datasets.

Keywords: Collaborative Question Answering · Question and answer retrieval · LDA · Neural network · Topic modelling · Regression

1 Introduction

During the last decade internet based Collaborative Question Answering (CQA) platforms have increased in popularity. These platforms offer a social environment for people to seek answers to questions, and where the answers are offered by other community members. Users pose questions in natural language, as opposed to queries in web search engines, and community members propose answers in addition to voting and rating the information posted on the platform. Some of the most popular CQA sites are Yahoo! Questions, Quora, and StackExchange. Besides public CQA websites, similar systems can be found in industry, for example in retail and business websites where users can pose questions about a company's product and a group of specialists can give support.

This content has attracted the attention of researchers from a number of domains [1–7] who aim to automatically return existing, relevant information from the CQA database when a novel question is submitted. Proposed approaches fall into two categories: determining the most relevant answers to a question [8,9]; and determining similar questions [1,5,8]. The latter is the problem that is covered in the present work. As such, the system helps to remove the delay needed for other community members to answer; and the list of related questions provides material for users to acquire more knowledge on the topic of their question.

© Springer International Publishing Switzerland 2016
N. Fuhr et al. (Eds.): TPDL 2016, LNCS 9819, pp. 41–53, 2016.
DOI: 10.1007/978-3-319-43997-6_4

Solving this problem is not a trivial matter as semantically similar questions and answers can be lexically dissimilar [1,2], referred to as the 'lexical chasm' [9]. For instance the questions "Where can I watch movies on the internet for free?" and "Are there any sites for streaming films?" are semantically related but lexically different. The opposite case is also possible—questions having words in common may have different semantic meanings. Besides the need for accurately identifying a question's semantics, a solution to the problem must deal with noisy information such as: misspelt words, polysemy, and short questions.

Similar questions are typically found by comparing the query question to the content of existing questions as it has been shown that finding similar questions based solely on their answers does not perform well [1,5]. Nevertheless Xue et al. demonstrated that combining information derived from existing questions and their answers outperforms the other strategies [5]. In recent years topic modelling has been applied to this problem [2,10,11] as it reduces the dimensionality of textual information when compared to classical methods such as bag-of words and efficiently handles polysemy and synonymy. These approaches, however, have thus far only been used to model the questions in the archives. As such, the contribution of the present work is twofold: firstly, the application of Latent Dirichlet Allocation (LDA) to model topics among the questions and answers in the archive; and secondly, the use of a regression step to estimate the appropriate QA topic distribution from that of a novel question.

This paper is organized as follows: Sect. 2 presents the state of the art, Sect. 3 the study's methodology, Sect. 4 the experimental setup and results, which are discussed in Sect. 5, and conclusions are presented in Sect. 6.

2   Related Work

The principal challenge when retrieving related questions and answers in a QA database given a new question is the lexical gap that may exist between two semantically similar questions. In general, a method that intends to solve the problem of question retrieval should be composed at least of two main parts: a document representation that can properly express the semantics and context of QAs in the database; and a mechanism for comparing the similarity of documents given their representations. The most widespread document representation methods in the literature are those based on bag-of-words (BOW), which explicitly represents each of the document's words. Comparison is achieved by computing the number of matching words between two BOW representations. There exist several variations of this class of methods, each weighting words that have specific properties in the dataset, such as tf-idf and BM25 [12]. This class of methods is able to measure two documents' lexical similarity but it does not capture information regarding their semantics and context.

In QA databases, questions and answers are often short and contain many word variations resulting from grammatical inflection, misspelling, and informal abbreviations. As a consequence, BOW representations in QA corpora produce a vector representation that can be too sparse. Besides sparsity, BOW representations do not provide a measure of co-occurrence or shared contextual information, which can increase the similarity of related documents.

An approach that overcomes these limitations is the translation model, first proposed for use in this context by Jeon et al. [1]. Their method consists of two stages: first a set of semantically similar questions are found by matching their answers using a query-likelihood language model; and subsequently, word translation probabilities are estimated using the IBM translation model 1 [13]. Several extensions have been proposed [5–7] including the use of external corpora [14,15] such as Wikipedia. Xue et al. [5] propose an extension that combines the IBM translation model (applied to the questions) with a query likelihood language model (applied to the answers). Translation-based models have become the state-of-the-art in query retrieval [10,16] but they suffer from some limitations: they do not capture word co-occurrences nor word distributions in the corpora.

In the last decade Topic Modeling has become an important method for text analysis. Since the topics that characterise a document can be considered a semantic representation, it is possible to use topic distributions inferred using a method such as Latent Dirichlet Allocation (LDA) [17] to measure the semantic similarity between documents in a corpora. Consequently, several approaches for applying topic modelling to QA archives have been proposed: Zhang et al. [2] retrieve similar questions by measuring lexical and topical similarities [2]; Cai et al. [10] combine the result of LDA and translation models; Vasiljević et al. [11] explore combining a document's word count and topic model similarity into one measure; and Yang et al. [8] form a generative probabilistic method to jointly model QA topic distributions and user expertise. In all of the above-mentioned topic modelling approaches similarity is calculated using the questions that exist in the database.

This work explores the possibility of deriving topic distributions from existing questions and answers, and proposes a method to relate these to the topic distribution of a novel question. Some work has been done in this direction; Zolaktaf et al. [18] model the question topics and then use them to condition the answer topics. This work proposes to model question and question-answer topics independently and then to learn a mapping between them. Furthermore, it extends topic modeling to include distributed word representations.

3   Methodology

A corpora C of size $L = |C|$ consists of many question-answer pairs: $C = \{(q_1, a_1), (q_2, a_2), \ldots, (q_L, a_L)\}$, where $Q = \{q_1, q_2, \ldots, q_L\}$ and $A = \{a_1, a_2, \ldots, a_L\}$, $\forall (q_i, a_i) \in C : q_i \in Q, a_i \in A$, are question and answer sets (respectively).

Questions in a CQA corpora tend to be shorter than answers and may contain few relevant words, which limits a model's ability to discover underlying trends. An approach to mitigate this is to assume that each question q_i contains its text, and possibly keywords, a title, and a description. This assumption is not a requirement as meta-data may not always be available; however its absence may

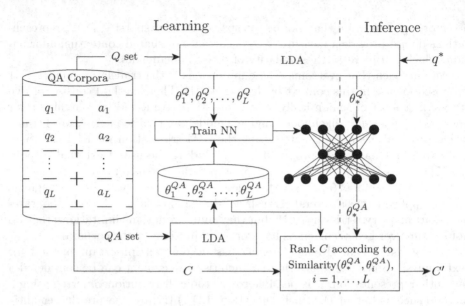

Fig. 1. An overview of the system's training and inference (see text for details).

limit the ability of a model to represent the questions. We discuss this further in this section and propose methods to overcome the problem of question sparsity. Furthermore, each question in a QA corpora may have multiple answers and these are concatenated to form each element a_i as they all provide contextual information that can be exploited to determine the question's relevance.

Figure 1 presents the proposed methodology. The task of similar question retrieval implies ranking the pairs contained in the QA Corpora (C) according to their similarity to a query question q^*, producing a partially ordered set C' such that its first element has the highest similarity (the top, say, ten elements of which can then be returned as suggestions). In the learning phase of the proposed methodology, the QA corpora is used to train two topic models (Sect. 3.1): LDA on the set Q, and LDA on the set QA, in which each pair $(q_i, a_i) \in C$ is concatenated to form a single document. This results in topic distributions associated with the sets Q and QA and each element contained therein (θ_i^Q and θ_i^{QA} respectively). A regression model is trained using the samples θ_i^Q and θ_i^{QA} (Train NN) to learn the translation function between the Q and QA topic distributions (Sect. 3.2). During inference the Q set LDA model is used to determine the topic distribution of a query question (θ_*^Q) which is translated to a QA topic distribution (θ_*^{QA}) using the regression model. Finally, a similarity measure (Sect. 3.2) is used to rank the QA Corpora (QA) according the similarity between each pair's topic distribution (θ_i^{QA}) and the query question's QA topic distribution (θ_*^{QA}) obtained from the regression model. The LDA and regression models are discussed in more detail in the following subsections.

3.1   Latent Dirichlet Allocation and Distributed Word Representation

In this work we assert that topic modeling provides a representation of the elements in C that facilitates the discovery of semantically similar questions; particularly when these similar questions do not have words in common.

Latent Dirichlet Allocation (LDA) [17] is a generative probabilistic model that enables us to describe a collection of discrete observations in terms of latent variables. The plate notation representing LDA is presented in Fig. 2. When applied to a corpora, LDA models the generation of each document by means of two stochastic independent processes and can be summarised as follows

1. For each document d in the collection D, randomly choose a distribution over topics $\theta_d \sim Dir(\alpha)$, where α is the Dirichlet prior.
2. For each word w_n in document d:
    (a) choose a topic from the distribution over topics in Step 1. $z_{d,n} \sim \text{Mult}(\theta_d)$;
    (b) choose a word from the vocabulary distribution $w_{d,n} \sim \text{Mult}(\phi_{z_{d,n}})$.

After learning a corpora's latent variables a topic is represented as a multinomial distribution of words, and a document by a multinomial distribution of topics.

The LDA algorithm described above treats words as explicit constraints, which inhibits its effectiveness when words are rare. A solution is to treat words as features [19] and the method used to calculate a word's features then influences its topic membership. This allows us to exploit a word's semantic similarity to augment information in short questions by giving similar topic membership probabilities to semantically equivalent words. For example, the words "educator", "education", "educational", and "instruction" should have similar probabilities within a certain topic, even if some of these words appear rarely in the corpus.

Mikolov et al. [20] introduced the continuous bag-of-words and Skip-gram neural network models that produce a continuous-valued vectorial word representation by exploiting the content of large textual databases. Distances between these vectors are proportional to the semantic difference of the words they represent, and thus these vectors can be used as features in many NLP tasks. In this work, the Word2vec vector representation is used to group semantically related words; its use for this application was first proposed by Petterson et al. [19].

In the original LDA algorithm, a word is generated by the process $w_{d,n} \sim \text{Mult}(\phi_{z_{d,n}})$ where $\phi_{z_{d,n}}$ is the multinomial distribution (Mult) of word probabilities in topic $z_{d,n}$ over the whole vocabulary. In order to introduce the distributed representation of words, we define a function $v : \mathbb{R} \to \mathbb{R}^r$ that maps

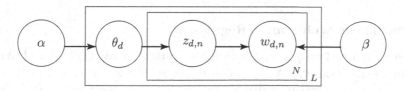

Fig. 2. LDA plate notation, β is the Dirichlet prior on the per-topic word distribution.

a word to its vectorial representation learnt by Word2vec, where r the number of latent features used for the distributed word representation (in practice this function is represented by a matrix $\omega \in \mathbb{R}^{N \times r}$, where N is the vocabulary size). Given two words w and w' their semantic similarity can be found by applying the cosine similarity function, see Eq. (4), to their vectorial representations, i.e. Similarity$(v(w), v(w'))$. A set of words that are similar to w, Ω_w, can be obtained by defining a threshold τ such that $\Omega_w = \{w' \mid \text{Similarity}(v(w), v(w')) > \tau\}$. This set can be used to define an alternative distribution of word probabilities $\phi'_{z_{d,n}}$ for topic $z_{d,n}$, in which the probability of a word w is given by

$$\phi'_{z_{d,n}}(w) = \frac{1}{c} \sum_{w' \in \Omega_w} \exp\left(\phi_{z_{d,n}}(w') \, \text{Similarity}\,(v(w), v(w'))\right), \tag{1}$$

where c is a normalisation factor. This modified distribution gives a high probability to semantically related words. Finally we consider each word w to be sampled from a linear combination of the original and modified distributions

$$w_{d,n} \sim \lambda \, \text{Mult}(\phi_{z_{d,n}}) + (1 - \lambda) \, \text{Mult}(\phi'_{z_{d,n}}), \quad 0 \le \lambda \le 1. \tag{2}$$

We fixed λ to 0.9 so that the results of standard LDA are not excessively altered.

In order to implement this modification, the Gibbs Sampling-based algorithm proposed by [21] was adapted so that at each step the probability of topic t being present in document d given word w is estimated as follows:

$$p(z = t \mid w) = \frac{\alpha \beta}{\beta V + n_{.|t}} + \frac{n_{t|d} \beta}{\beta V + n_{.|t}} + \frac{(\alpha + n_{t|d}) \lambda n_{w|t}}{\beta V + n_{.|t}}$$
$$+ \frac{1 - \lambda}{c} \sum_{w' \in \Omega_w} \exp\left(\frac{n_{w'|t} \, \text{Similarity}\,(v(w), v(w'))}{n_{.|t}}\right), \tag{3}$$

where $n_{w|t}$ is the number of words w assigned to topic t, $n_{t|d}$ is the total number of words in document d assigned to topic t, $n_{.|t} = \sum_w n_{w|t}$, $\alpha = 35/T$ is the Dirichlet prior of the per document topic distribution (for number of topics T), and $\beta = 0.01$ [21,22]. Small values of α and β result in a fine-grained decomposition into topics that address specific areas [21,22].

This method is applied to two document collections (Fig. 1), Q and QA, which results in two topic models: $T_Q = \{1, \ldots, K_Q\}$ in which each question q_i is represented by the distribution of topics θ_i^Q; and $T_{QA} = \{1, \ldots, K_{QA}\}$ in which each pair (q_i, a_i) is represented by the distribution of topics θ_i^{QA}.

3.2   Nonlinear Multinomial Regression

When a query question q^* is entered the left-to-right method [23] is used to infer its topic distribution, θ_*^Q. A regression model is therefore needed to obtain an estimate of θ_*^Q mapped to a distribution of topics in the QA set, θ_*^{QA}. Mapping the distribution of question topics to the distribution of question-answer topics avoids problems that occur when limited vocabularies are used in a question. This information is augmented with that derived from the set of answer terms, thus by mapping a query question to the space of question-answers it is possible to calculate its similarity using words that do not exist in the question vocabulary (and therefore are not represented in the topic distribution T_Q). Performing this mapping also provides a means to model the relationship between question semantics and existing question-answer semantics (which will be discussed further in Sect. 5): given a query question q^* the model estimates a topic distribution in the space of concatenated questions and answers, which can be compared to the distributions of existing QA pairs.

Determining the topic distribution in the space of documents comprising questions and answers, given the topic distribution of a new question is a problem of multinomial regression. For which we use a multilayer perceptron neural network (NN), which are nonlinear multinomial regression models [24,25]. The NN is trained using the set of topic distributions for each document in Q and QA, θ_i^Q and θ_i^{QA} (respectively) where $i = 1, \ldots, L$, and therefore the input and output layers have as many nodes as the number of topics used to model these sets, K_Q and K_{QA} (respectively). Sigmoid activation functions are used in the hidden layer and softmax in the output layer to ensure that outputs sum to one.

In application the input of the NN is the topic distribution of the query question according to latent topic model of the existing questions, represented by θ_*^Q, and its output is an estimate of its distribution in the QA latent topic model, θ_*^{QA}. The cosine similarity measure allows us to rank existing questions q_i according to their similarity to θ_*^{QA}, i.e.

$$\text{Similarity}\left(\theta_*^{QA}, \theta_i^{QA}\right) = \frac{\theta_*^{QA} \cdot \theta_i^{QA}}{\|\theta_*^{QA}\|\|\theta_i^{QA}\|}, \tag{4}$$

where $\|x\|$ is the length of vector x, and therefore the most similar existing questions appear at the top of the ranked list that is output by the system.

4   Evaluation

This section describes the data, experimental setup, and comparison algorithms used to evaluate the proposed approach.

4.1   Data

Four categories, derived from two different CQA sources, are used for the evaluation. The first two are the *Health* and *Computers & Internet* (referred to herein as Computers) categories in the publicly available Yahoo! Questions L6 (Yahoo! Answers Comprehensive Questions and Answers version 1.0) dataset[1]. The second two are the *Physics* and *Geographic Information Systems* (GIS) categories

---

[1] Available from http://webscope.sandbox.yahoo.com/catalog.php?datatype=l.

Table 1. Summary of the datasets

	Y! Health	Y! Computers	SE Physics	SE GIS
# of Questions	40050	40050	41201	36520
# of Answers	133747	116946	77767	59873
# of Answers per Question	3.34	2.92	1.89	1.64
Average Question Length (# of words)	22.26	27.33	122.31	145.31
Average Answer Length (# of words)	27.33	67.91	195.77	108.32

taken from the publicly available StackExchange (SE) dataset[2]. The question sets extracted from the Yahoo! dataset were created by concatenating the question text and description (when available), and the question sets extracted from the SE dataset were created by concatenating the question title, tags, and text. The answer sets were created by concatenating all the answers provided by different users for a particular question. Table 1 summarises these datasets.

Preprocessing was performed before data use: stop words were removed using Mallet's standard English list (543 words), non-English characters were removed, and lemmatization was performed to reduce the number of inflected word forms.

Fifty randomly selected questions from each category were used for testing and the remaining pairs were used as training data. Therefore four models were calculated using each algorithm, one for each category. The output of each model (the top ten most similar results for each test question) were manually labelled as relevant or not and this was used to calculate the evaluation statistics.

The Word2vec model requires training in order to learn the word embedding space, and this was realised using an additional corpus of Google news and Yahoo! Questions QA pairs (from categories other than those presented previously). The reason for including documents form Yahoo! Questions in this corpus is that it enables words that are specific to the dataset—such as abbreviations, misspellings, and technical jargon—to be learnt.

A modified version of Mallet, which implements the Gibbs sampling method proposed by Yao et al. [21], was used for Topic Modeling. The number of topics were empirically set to 140 and 160 for the Q and QA sets (respectively) and the size of the neural network's hidden layer was empirically set to 180 using 100 questions-answer pairs (these were subsequently removed from the corpus).

4.2   Results

The proposed method, referred to henceforth as LDA$^+$, was compared to four state-of-the-art algorithms: Translation1, the IBM translation approach proposed by Jeon et al. [1]; Translation2, the combined translation and query-likelihood language model proposed by Xue et al. [5]; an autoencoder based method proposed by Socher et al. [26]; to establish the benefit of word2vec, LDA* (as described within Sect. 3 excluding word2vec); and to establish the benefit of the regression stage, LDA† (as described within Sect. 3 excluding the regression step).

---

[2] Available from https://archive.org/details/stackexchange.

Table 2. Performance of each method using two Yahoo! Questions categories and two StackExchange categories (to two significant figures). The highest results for each measure and category are in bold and italics indicate statistical significance when compared to LDA$^+$ using a paired two-sample t-test with an alpha level of 0.05.

Method	Category	MAP	P@1	P@2	P@4	P@7	P@10
Translation1	Y! Health	0.38	0.64	0.60	0.51	0.53	0.51
Translation2		0.36	0.50	0.52	0.49	0.48	0.45
LDA†		*0.28*	0.50	0.52	0.45	*0.42*	*0.38*
LDA*		*0.29*	*0.48*	*0.51*	0.48	*0.44*	*0.41*
LDA$^+$		**0.43**	**0.66**	**0.64**	**0.57**	**0.55**	**0.53**
Translation1	Y! Computers	*0.21*	0.50	0.41	0.36	*0.33*	*0.31*
Translation2		0.28	0.46	0.44	0.42	0.38	0.38
LDA†		0.26	0.50	0.43	0.39	0.37	0.34
LDA*		0.31	0.50	0.48	0.45	0.43	0.394
LDA$^+$		**0.35**	**0.52**	**0.55**	**0.49**	**0.48**	**0.46**
Translation1	SE Physics	*0.34*	*0.36*	*0.41*	*0.45*	*0.43*	*0.42*
Translation2		*0.30*	*0.38*	*0.35*	*0.40*	*0.40*	*0.39*
LDA†		0.67	0.82	0.78	0.76	0.75	0.74
LDA*		*0.62*	*0.72*	*0.78*	0.80	0.75	*0.71*
LDA$^+$		**0.71**	**0.86**	**0.89**	**0.84**	**0.79**	**0.77**
Translation1	SE GIS	*0.19*	*0.32*	*0.30*	*0.31*	*0.33*	*0.30*
Translation2		*0.14*	*0.30*	*0.27*	*0.27*	*0.23*	*0.21*
LDA†		*0.59*	*0.76*	*0.77*	*0.73*	*0.70*	*0.69*
LDA*		*0.59*	0.92	0.86	*0.77*	*0.70*	*0.66*
LDA$^+$		**0.75**	**0.98**	**0.92**	**0.85**	**0.83**	**0.82**

Mean Average Precision (MAP) and Precision at N (P@N) are used to summarise retrieval performance within each category. The autoencoder was found to be computationally infeasible when applied to the described datasets and therefore its retrieval performance is not presented. The results obtained using the remaining methods are presented in Table 2. A cursory validation of these results was performed by comparing the translation methods' figures to those presented in the literature using the same method and data source (but not the same partitioning) and they fall within the observed range [6,10,14,15].

The results show that in all of the datasets, LDA$^+$ outperforms all other methods. However, the difference is much more pronounced when the length of the question and answers increase (as is the case in the SE datasets). In this situation, the translation methods fail to find relevant documents whereas all of the LDA methods do (due to the increase in information). It is difficult to separate the performances of LDA with Word2vec and LDA with regression (LDA† and LDA*), but when combined (LDA$^+$) a performance increase is observed.

5   Discussion

Within the translation based approaches [1,5] the translation probabilities of equal source and target words are fixed to 1. This forces questions that share words in common with the query question to be highly ranked. Conversely, LDA[†], LDA[*], and LDA[+] perform matching based upon shared topics, and inherently accounts for words that represent multiple concepts by decreasing their probabilities in the topics that they appear. To illustrate this, Table 3 presents an example of retrieved questions using LDA[+] and the two translation based approaches[3] (the points discussed in this section were observed in all of the categories but to save space we present examples from the Health category). In the first example, presented in the top half of the table, the QA pairs retrieved by LDA[+] do not contain the words "lift" and "weight" even though they are relevant to the query. The excessive contribution from the word "weight" causes the translation models to retrieve questions that are related to body weight instead of weight lifting. The second example illustrates a query in which all the retrieved QA pairs are relevant. As before, the translation methods result in questions that have words in common with the query question (as does LDA[+]); in this case Translation2 associates a high translation probability between "hair" and "mustache" (sic).

Table 3. Examples of the top two retrieved QA pairs for each method given a query question (using the Yahoo! Health category).

Query question	Retrieved QA pairs		
	LDA[+]	Translation1	Translation2
How many days a week should you lift weight?	When are you to old to build muscle mass from work out?	How do I gain body weight?	My weight is 90 lb how could I gain more weight?
	What is a 1 set rep?	My weight is 90 lb how could I gain more weight?	Will lose weight faster than average if I workout?
Can you make your hair grow faster	Is there any way to get rid of razor bump?	What important function do our body hair play?	How can I make my mustache grow faster?
	What's the best herbal remindie for hair loss?	What can be do to prevent hair loss?	What is hair?

Table 4 demonstrates the benefit of performing the multinomial regression. It presents the representative words (those that have high probability in the topic's word distribution) of three of the topics derived from the question (Q) set and the question + answers (QA) set. It demonstrates that the topics derived from the QA set better represent the themes that appear in health documents, whilst the topics of derived from the Q set are less distinguishable. For example, the words in Topic 3 appear to represent depression, however, the words derived

---

[3] Mistakes in the questions are original to the data.

Table 4. Words that comprise the topics derived from the questions (Q) and question-answer (QA) sets of the Yahoo! Health category.

Topic	Q
1	treatment, effective, method, suggest, special, option, undergo, acupuncture, indian, prognosis, bad, acupuncture
2	test, result, show, urine, blood, negative, positive, pap, pass, testing, smear, screen, lab, tuberculosis, perform
3	depression, suffer, depress, deal, solution, seek, advise, cost, remain, viagra, clinical, dysfunction, overcome, admit

Topic	QA
1	pill, product, work, market, effective, company, fda, safe, ingredient, call, wont, sell, approve, brand, generic
2	study, research, show, find, percent, report, health, american, accord, result, evidence, national, researcher
3	depression, depress, feel, mood, medication, talk, anxiety, anti, therapy, psychiatrist, antidepressant

from the QA set are more coherent. This is because of the limited vocabulary used in questions and their typically short length.

Furthermore, the topics derived from the Q set tend to represent the semantics of expressions commonly used in questions (and not in answers), for example the phrases "an effective method" and "effective treatment". The word "effective" in the topics derived from the QA set is associated with the topic representing medical products. Consequently, when a question such as "What is an effective sleeping aid?" is posed to a model trained on the QA set, topics in which the words "method" and "treatment" have high probability would not be considered. The model trained on the Q set, however, results in a high probability of Topic 1, and the regression stage of LDA$^+$ causes this to be mapped to the distribution in which the words "treatment" and "method" have higher probabilities. Another example is provided by Topic 2, here the word "result" is often mentioned in questions posed by those who have performed medical tests, while in answers the word usually refers to the results of health research studies.

6   Conclusions

This paper has presented a novel model that fuses topic modelling with Word2vec and a regression stage for ranking relevant questions-answer pairs within Collaborative Question Answering platforms. The performance of the proposed method has been evaluated using several real-world datasets, and it has been shown to outperform translation based methods and LDA with each innovation separately in all cases. Most notably when the dataset contains long questions and answers. It achieves this by allowing the model to overcome the differences in vocabulary used in questions and answers, helps to deal with the sparsity often encountered in questions (due to their relatively short length), and allows the method to exploit all available information.

References

1. Jeon, J., Croft, B.W., Ho Lee, J.: Finding similar questions in large question and answer archives. In: CIKM, pp. 84–90 (2005)
2. Zhang, W.N., et al.: A topic clustering approach to finding similar questions from large question and answer archives. PLoS ONE **9**, e71511 (2014)
3. Wang, K., Ming, Z., Chua, T.S.: A syntactic tree matching approach to finding similar questions in community-based QA services. In: SIGIR, pp. 187–194 (2009)
4. Cui, H., Sun, R., Li, K., Kan, M.Y., Chua, T.S.: Question answering passage retrieval using dependency relations. In: SIGIR, pp. 400–407 (2005)
5. Xue, X., Jeon, J., Croft, W.B.: Retrieval models for question and answer archives. In: SIGIR, pp. 475–482 (2008)
6. Lee, J.T., et al.: Bridging lexical gaps between queries and questions on large online Q&A collections with compact translation models. In: EMNLP, pp. 410–418 (2008)
7. Bernhard, D., Gurevych, I.: Combining lexical semantic resources with question & answer archives for translation-based answer finding. In: ACL-IJCNLP, vol. 2, pp. 728–736 (2009)
8. Yang, L., et al.: CQArank: jointly model topics and expertise in community question answering. In: CIKM, pp. 99–108 (2013)
9. Berger, A., Caruana, R., Cohn, D., Freitag, D., Mittal, V.: Bridging the lexical chasm: statistical approaches to answer-finding. In: SIGIR, pp. 192–199 (2000)
10. Cai, L., Zhou, G., Liu, K., Zhao, J.: Learning the latent topics for question retrieval in community QA. In: IJCNLP, pp. 273–281 (2011)
11. Vasiljević, J., Ivanović, M., Lampert, T.: The application of the topic modeling to question answer retrieval. In: ICIST, pp. 241–246 (2016)
12. Robertson, S.E., Walker, S.: Some simple effective approximations to the 2-poisson model for probabilistic weighted retrieval. In: SIGIR, pp. 232–241 (1994)
13. Brown, P., et al.: The mathematics of statistical machine translation: paramter estimation. Comput. Linguist. **19**, 263–311 (1993)
14. Zhou, G., et al.: Improving question retrieval in community question answering using world knowledge. In: IJCAI, pp. 2239–2245 (2013)
15. Singh, A.: Entity based Q&A retrieval. In: EMNLP, pp. 1266–1277 (2012)
16. Zhou, G., et al.: Statistical machine translation improves question retrieval in community question answering via matrix factorization. In: ACL, pp. 852–861 (2013)
17. Blei, D.M., et al.: Latent dirichlet allocation. J. Mach. Learn. Res. **3**, 993–1022 (2003)
18. Zolaktaf, Z., Riahi, F., Shafiei, M., Milios, E.: Modeling community question-answering archives. In: Proceedings of the Workshop on Computational Social Science and the Wisdom of Crowds at NIPS (2011)
19. Petterson, J., et al.: Word features for latent dirichlet allocation. In: NIPS, vol. 23, pp. 1921–1929 (2010)
20. Mikolov, T., Chen, K., Corrado, G., Dean, J.: Efficient estimation of word representations in vector space. In: ICLR Workshop (2013)
21. Yao, L., Mimno, D., McCallum, A.: Efficient methods for topic model inference on streaming document collections. In: SIGKDD, pp. 937–946 (2009)
22. Griffiths, T., Steyvers, M.: Finding scientific topics. PNAS **101**, 5228–5235 (2004)

23. Wallach, H., Murray, I., Salakhutdinov, R., Mimno, D.: Evaluation methods for topic models. In: ICML, pp. 1105–1112 (2009)
24. Ripley, B.: Pattern Recognition and Neural Networks. Cambridge University Press, London (1996)
25. Bentz, Y., Merunka, D.: Neural networks and the multinomial logit for brand choice modelling: a hybrid approach. J. Forecast. **19**, 177–200 (2000)
26. Socher, R., et al.: Dynamic pooling and unfolding recursive autoencoders for paraphrase detection. In: NIPS (2011)

Survey on High-level Search Activities
Based on the Stratagem Level
in Digital Libraries

Zeljko Carevic$^{(\boxtimes)}$ and Philipp Mayr

GESIS - Leibniz Institute for the Social Sciences,
Unter Sachsenhausen 6-8, 50667 Cologne, Germany
{zeljko.carevic,philipp.mayr}@gesis.org

Abstract. High-level search activities for Digital Libraries (DLs) introduced by Fuhr et al. [8] go beyond basic query searches because they
include targeted and structured searches like e.g. a journal run or citation searching. In this paper, we investigate if and how typical high-
level search activities are really used in current DLs. We conducted an
online survey with 129 participating researchers from different fields of
study that aims at getting a quantitative view on the usage of high level
search activities in DLs. Although our results indicate the usefulness
of high-level search activities, they are not well supported by modern
DLs with regards to the users' state of search, e.g. looking at a relevant
or not relevant document. Furthermore, we identified differences in the
information seeking behavior across the respondents. Respondents with
a higher academic degree significantly considered journals and conference proceedings as more useful than respondents with a lower academic
degree.

Keywords: Information filtering · Search process · Stratagems ·
Interactive IR · Survey · Digital Libraries

1 Introduction

Digital Libraries (DLs) offer direct access to a vast number of bibliographic
records. As more publications are made available in electronic format the amount
of material a user needs to assess becomes difficult to manage. This leads to
highly fragmented interactive sessions in which users perform various types of
search activities [14]. During the past, different models have been proposed that
aim to model the information seeking behaviour, e.g. the information seeking
behaviour model proposed by Bates [2]. Based on empirical studies of the information seeking behaviour of experienced library users she identified four levels
of search activities that, amongst other, differ in their complexity: *moves, tactics, stratagems* and *strategies*. A move is the lowest unit of search activities like
entering a query term or selecting a certain document. Tactics are described as a

© Springer International Publishing Switzerland 2016
N. Fuhr et al. (Eds.): TPDL 2016, LNCS 9819, pp. 54–66, 2016.
DOI: 10.1007/978-3-319-43997-6_5

combination of many moves like the selection of a broader search term or breaking down complex search queries into subproblems. Bates defines a stratagem as follows: "a stratagem is a complex of a number of moves and/or tactics, and generally involves both a particular identified information search domain anticipated to be productive by the searcher, and a mode of tackling the particular file organization of that domain" [2]. Hence, a stratagem could be for instance a "journal run" where a user identifies a journal to be productive for his or her research and browses the latest publications in that journal. Another example for a stratagem is to follow references in a certain document that might lead to potentially relevant material. Finally, strategies are combinations of moves, tactics and stratagems, thus, forming the highest search activity as they cover the whole information seeking process.

Moves are considered lower-level search activities whereas tactics, stratagems and strategies are considered high-level search activities [8]. Moves and tactics are commonly used during an information seeking episode and have been subject in research. Popular examples are search term recommender or thesauri that support the user in choosing appropriate query terms [10]. Stratagems and strategies, on the contrary, have attracted less attention although they are undoubtedly reasonable search activities. Fuhr et al. [8] for instance have developed a federated DL aimed at providing strategic support to the user.

In this paper, we present the results of an online survey that empirically evaluates high-level search activities based on the stratagem level. To the best of our knowledge such a survey has not been conducted yet, but is able to provide us with a deeper understanding of the users' information need when performing high-level search activities. The online survey aims at getting a quantitative view on the usage of high level search activities in DLs with respect to the users' state of search. We aim at answering the following research question:

RQ: What kind of stratagems do users perform when looking for relevant documents?
We look at different types of stratagems that were derived from the examples proposed in [2]. We investigate the usefulness of certain stratagems with regards to the users' state of search. We distinguish the users' state of search by: (a) the user has found a relevant document and wants to find similar documents and (b) the user performs a stratagem search without a preceeding document.

The paper is structured as follows: first we briefly discuss related work regarding high-level search activities and their empirical evaluation (cf. Sect. 2). In Sect. 3 we describe the setup of our survey and general demographic information about our respondents. In Sect. 4 we show the results of our survey as well as significant differences between certain groups of respondents by field of study and academic degree. A discussion of the results and possible implications on the design of DLs is presented in Sect. 5.

2   Related Work

Based on the work of Bates [2], there have been many attempts to support users with high-level search activities. In the literature, the two most notable directions in search or retrieval support are: (a) the implementation of high-level search activities as direct system functionality with proactive (e.g. spelling corrections) or automatic (e.g. normalization of spelling variants) user functions, and (b) support systems for user guidance. Theses systems often try to adapt the search situation and the capability of the user and then recommend suitable search tactics. Xie [17] and Joo and Xie [12] e.g. investigated the relationships between users' search tactic selections and search outputs while conducting exploratory searches in digital libraries.

While high-level search functions can be found in practice in DLs, support systems for user guidance have been rarely accepted. Most probably this has to do with the lacking consideration of the search situation and history and the deficient inclusion of the capabilities of the user in the search process. A qualitative study observing the usage of a DL was presented in [7]. The study starts with the observation that experienced users of DLs are more effective than non-experts. The purpose of the study was then to investigate the nature of experienced DL users in more detail in order to design interfaces that support unexperienced users. Bates concepts describe the mechanisms of search activities and tasks in a very generalized way, as an information seeking model. These concepts of specific search tactics in an evolving search have been implemented in an academic Web environment by Fuhr et al., in the project Daffodil [8].

Today many other state-of-the-art DLs support the search tactics outlined in Bates. Carevic and Mayr [5] have recently introduced simple bibliometric-enhanced search facilities which are derived from Bates' stratagems and could be easily integrated to DLs. They outlined the idea of extended versions of journal run or citation search for interactive information retrieval. Hienert et al. have studied the user acceptance of various search term recommender systems which have been integrated and evaluated in a live DL [10]. They found that the combined recommendation service which interconnects a thesaurus service with additional statistical relations outperformed all other services. Their findings contradict the typical observation that users are ignorant of advanced search features. In experimental prototypes Brajnik et al. [4] and Bhavnani et al. [3] focused on the integration of strategic tool support in the user interface to help the users to find more comprehensive information when searching. In their position paper, Wilson and Schraefel [16] highlighted the importance of exploratory search interfaces which could profit from hybrid IR/HCI approaches.

3   Online Survey

The survey was available for two weeks during August 2015 and was primarily designed for researchers and postgraduate students but not limited to a particular field of study. The respondents were recruited via collaborating universities and institutes, mailing lists and social media (Twitter, Facebook). To keep the survey maintainable for the respondents, we focused on two of the six stratagems proposed by Bates: (a) journal and conference run and (b) citation and reference search. The survey consisted of 28 questions that were divided into four parts. The first two parts concerned the general usage of journal and conference runs as well as citations and references. In the third part of the survey the respondents were given a scenario in which they were looking for related material to a given search task in a journal named *Addiction*. Alongside the scenario, we defined six randomly arranged options on how the articles from that particular journal could be ranked (e.g. by title, by authors, etc.). The respondents were then asked to order these six options according to their preference. The fourth part of the survey concerned the general usage of stratagems in DLs. To this end, we presented the respondents all six stratagems derived from [2] (see Table 3) and asked them to rank these activities by their usage when searching for relevant documents. The survey concluded with 9 demographic questions and two optional questions regarding feedback and contact information. The quality of the survey was ensured in a two fold way: first our department *Survey Design and Methodology* evaluated the Likert scales of the survey. Second we have performed two pretests with colleagues.

Demographics: In total there were 204 respondents of which 129 completed the survey. We report on all available responses even if the survey was not completed by the respondent. 62.6 % of the respondents were male. The ages ranged from 23 to 79 years (mean $= 40.3$, sd $= 12.2$, N $= 128$). The respondents were asked to choose their field of work from a set of 26 options. In total 12 fields were chosen with the majority of the respondents coming from the field of Computer and Information Science (50.4 %) and Social Sciences (28.2 %). Regarding the academic degree of the respondents 54.2 % replied to have a master's, diploma or bachelor's degree, 32.1 % obtained a doctoral degree, 12.2 % were professors and 1.5 % of the respondents were undergraduate. When asked to rate their experience in searching DLs 24.4 % considered their experience as expert, 42.7 % as high, 21.4 % as moderate, 10.7 % as little and 0.8 % as none at all. Alongside their experience we asked the respondents about their usage of DLs and Google Scholar using a 5-point Likert scale ranging from never to very often. 59 % of the respondents use DLs "often" or "very often" (median $= 4$, mode $= 4$, N $= 131$) and 71.8 % use Google Scholar "often" or "very often" (median $= 4$, mode $= 5$, N $= 131$)[1].

4   Results

In the following (Sects. 4.1–4.4) we present the results of our online survey on high-level search activities. In Sect. 4.5 we look for significant differences between respondents clusters.

---

[1] The central tendency of the Likert scales is presented by using median and mode values throughout the paper due to the ordinality of the scales [11].

4.1 Journal and Conference Run

In the first part of the survey we asked some general questions concerning journal and conference runs on a five point Likert scale using different item labels for each question (example for an item label regarding a question about usefulness ranging from: not at all useful (1), rather not useful (2), neither useful nor not useful (3), rather useful (4), very useful (5)). For each of the questions the negative item was left aligned. The results are displayed in Table 1 (each item-label is highlighted).

Table 1. General questions about journal and conference run (N = 156).

Task	Mdn	Mode	M	SD
How **useful** are conference proceedings or journals as a source for relevant documents during your search task?	5 (very useful)	5	4.31	0.89
How **satisfied** are you with the support of current Digital Libraries (e.g. ACM DL, Web of Science) browsing through conference proceedings or journals?	3 (neither satisfied nor unsatisfied)	4	3.27	0.9
How **important** is the quality of a conference (ranking) or a journal (e.g. the impact factor) for your confidence in the source?	4 (rather important)	4	3.44	1.06

Furthermore, we investigate how frequently the respondents use journal or conference runs. To this end we asked two questions with items ranging from "never" to "very often": (a) "How often do you browse through conference proceedings or journals to find relevant documents?" We then asked the same question but from a different perspective (b) "After finding a document (e.g. ACM DL, Web of Science) that is relevant for your current search task: How often do you browse through the conference proceedings or journals the document was published in?".

A journal run without preceding document (a) was selected "often" or "very often" by 54.9 % (median = 4, mode = 4, N = 142) of the respondents. Regarding a journal run with a preceding document (b), 35.2 % replied to use this search activity often to very often (median = 3, mode = 3, N = 142). If the respondent replied that he/she never or rarely browses through conference proceedings or journals, he/she was asked to justify his/her decision using a free-text form. A frequently stated reason was a preference for searching instead of browsing.

We Summarize: Journals and conference proceedings are considered a very useful source. 54.9 % of the respondents browse through a journal or conference proceedings often to very often in order to find relevant documents but only 35.2 % use this stratagem as a follow-up search activity starting from a relevant document.

4.2   References and Citations

In the second part of the survey we asked some general questions concerning the usage of citations and references on a five point Likert scale using different item labels. The results are displayed in Table 2.

Table 2. General questions about citations and references (N = 140).

Task	Mdn	Mode	M	SD
How **important** is the number of citations a document has received to you?	3 (neither important nor unimportant)	4	3.33	0.91
How would you rate the **usefulness** of citation rankings (e.g. h-index) where documents are ranked by the number of received citations?	3 (neither useful nor not useful)	4	3.23	0.99
Assuming there is a key document in a particular field. How **important** is it to you to find central authors citing that particular document?	4 (rather useful)	4	3.60	1.07

Furthermore, we examined how frequently the respondents use references or citations after finding a relevant document. To this end we asked two questions with items ranging from never to very, often: (a) "Starting from a relevant document: How often do you use references to find other relevant documents for your search task?" and (b) "Starting from a relevant document: How often do you use citations to find ..?". Regarding citations, the options "often" to "very often" were selected by 65.7 % (median = 4, mode = 5, N = 140) of the respondents and by 82.1 % (median = 4, mode = 4, N = 140) regarding references. If the respondent replied that he/she never or rarely used citations or references he/she was asked to provide some additional information why. For references only one response was available. He/she replied to prefer semantic tools. With respect to using citations there was an overall agreement that they are more difficult to find and therefore not that commonly used.

We Summarize: Citation and reference search are frequently used search activities. Citations and references are as well commonly used. 65 % of the respondents used citations and 82 % used references often to very often. The central tendency regarding features like the h-index or the general citation count ranges between the mid-point and a rather positive tendency.

4.3   Stratagem Usage in DLs

In [2], six example stratagems are proposed. We want to investigate the usefulness of these search activities with respect to the following search scenario that was presented to the respondents:

Table 3. Stratagem usage for the given scenario. Mean values range from lowest rank (6) to highest rank (1). (N ≥ 125)

Ranking option	M	SD	Mdn	Mode
Follow references in the current document	2.38	1.24	2	2
Inspect the list of documents that cite the current document	2.79	1.50	2	2
Keywords that describe the current document as search terms	2.82	1.63	3	1
Look for papers the authors of the current document has/have published	3.46	1.21	3	3
Browse the conference/journal the current document was published in	4.10	1.53	4	5
Browse a thesaurus to find classification terms related to the current document	5.21	1.30	6	6

"Please consider the following scenario. You want to find out about the current state of the art in a particular field. You have already found one document that is useful to your current work task."

Alongside this scenario, we gave the respondents the six example stratagems derived from [2]. We then asked to order all these activities from best to worst regarding the given scenario using a drag and drop user interface. The list of stratagems as well as the results are illustrated in Table 3. Additionally, the respondents were given a free-text field where they were asked to provide other search activities they use to find related material. Amongst others, the respondents mentioned: using recommender systems, asking colleagues, using some kind of bibliometric measure/feature like co-author search.

We summarize that references, citations and keywords are the most commonly used search activities for finding relevant documents. The importance of citations and references was evident throughout the entire survey. Previous tasks showed that journals and conference proceedings are a useful source as well. Compared to the other stratagems these search activities do not appear to be used very often.

4.4   Organizing Journal Articles

Articles in a journal or in conference proceedings can be sorted in various ways (i.e. by date, by title, etc.). The ranking of the articles strongly depends on the current search task of the user. Using a search task scenario, we compare different ranking options for a journal run. The respondents were given a scenario displayed in Fig. 1 where they are looking for related material in a journal named *Addiction*. Alongside the scenario, we defined six randomly arranged options on how to rank articles from that particular journal. Four of the six options (issue date, title, author, and citation count) are well known and widely implemented

Please consider the following situation: You are about to write an essay about 'Alcohol Consumption in Germany and its Demographic Distribution'. You start your search by entering the search terms 'alcohol consumption germany'. You find a relevant document (see illustration) that was published in a journal named 'Addiction'. After reading the document you want to see more material from that particular journal.

Developments in alcohol consumption in reunited Germany

by Bloomfield, Kim; Grittner, Ulrike; Kramer, Stephanie

In: Addiction, 100 (2005), 12, p. 1770-1778 : table(s)

Cited by: 3

Fig. 1. Scenario for the task on organizing journal articles

in today's DLs. The two remaining options rank the articles based on previous search activities that were described in the scenario. One option ranks the articles by the previously entered query term (alcohol consumption Germany) and the other option ranks the articles by similarity to the current relevant article the user was inspecting based on title ("Developments in alcohol consumption.."). The respondents were then asked to order all these options from best to worst using a drag and drop user interface. The results for this task are displayed in Table 4.

Table 4. Task on organizing journal articles. Mean values range from lowest rank (6) to highest rank (1). ($N \geq 128$)

Ranking option	M	SD	Mdn	Mode
By the entered query terms (alcohol consumption Germany)	2.08	1.34	2	1
By similarity to the current document based on title (Developments in alcohol consumption..)	2.23	1.32	2	2
By title	3.95	1.49	4	5
By issue and date	3.95	1.66	4	6
By number of citations	4.08	1.42	4	4
By author	4.42	1.31	5	6

We Summarize that respondents assess the ranking options based on previous search activities noticeably higher. A ranking based on the previously entered query term was slightly more often chosen (mean $= 2.08$, sd $= 1.34$) than the ranking option based on similarity to the current relevant article (mean $= 2.23$, sd = 1.32). Both search activity based options clearly outperform the other four ranking options that are well known and commonly used in DLs.

4.5 Diversity in Respondents

The survey was primarily designed for researchers but not limited to a particular field of study. This leads to a variety of respondents from diverse fields of study and academic degree. Utilizing a non-parametric Mann-Whitney test ($\alpha \leq 0.05$) we seek for significant differences in respondents' groups. In particular we separated the respondents by field of study and by academic degree. The decision for utilizing a non-parametric Mann-Whitney test is due to highly skewed distributions in the responses.

By Field of Study: The majority of respondents are either Computer and Information Scientists (50.4 %) or Social Scientists (28.2 %). A significant difference between the two groups was found in the task on organizing journal articles (see Sect. 4.4). Computer and Information Scientists preferred a ranking by citations more frequently than Social Scientists (u = 1989, p = 0.047), whereas Social Scientists preferred a ranking based on title (u = 2031, p = 0.042). A further difference was found for the ranking option based on authors which was preferred by Social Scientists (u = 2002, p = 0.045). However, these differences have only marginal influence on the overall result of the ranking task because both groups agreed to perform a ranking based on preceding search activities (by similarity to the preceding article or by the entered query term).

By Academic Degree: To look for significant differences by academic degree we created two groups of respondents: The group of *junior researchers* consists of researchers having a bachelor's, master's or diploma degree (55 %) and the group of *senior researchers* consists of researchers having a doctoral degree or a professorship (45 %). Significant differences between these two groups could be found in various responses. It shows that *senior researchers* consider conference proceedings and journals as more useful for their search task (u = 2396, p = 0.017), that they are more satisfied with the current support of DLs (u = 1971, p = 0.049) and consider the number of citations as more important (u = 2112, p = 0.035). Furthermore, we found differences in the usage of Dls. *Senior researchers* utilise DLs (u = 2647, p = 0.008) and Google Scholar (u = 2142, p = 0.032) more frequently. Both systems are less frequently used by *junior researchers*.

In Sect. 4.3 we asked the respondents to rank different stratagems by their usage when looking for related material. *Senior researchers* ranked "inspecting the list of citations" on the first position (u = 2661, p = 0.017). The usage of "keywords that describe the current document as search terms" was ranked on position two by *junior researchers* and on position three by *senior researchers* (u = 2043, p = 0.042).

We Summarize that there are significant differences in the usage of high-level search activities depending on the academic degree. Respondents with a higher academic degree significantly considered journals and conference proceedings as more useful than respondents with a lower academic degree. Furthermore, they utilise DLs and Google Scholar significantly more often.

5 Discussion

In the following we discuss the results with respect to the research question introduced in Sect. 1.

RQ: What Kind of Stratagems Do Users Perform When Looking for Relevant Documents? The survey showed that stratagems are commonly used search activities across a wide range of respondents. Journals and conference proceedings are considered very useful sources by the majority of the respondents. Citations and references are as well commonly used. 65 % of the respondents used citations and 88 % used references often to very often. However, the higher usage of references over citations does not necessarily reflect the usefulness of these search activities. Several respondents stated that various DLs do not provide access to citations. Therefore, citation search is not utilized that often. A better access to citation information in DLs could balance this difference. When looking for related material to a given relevant document the respondents preferred to use citations, references and keywords.

We showed that the usage of stratagems depends on the users' state of search. The journal run for instance is less often utilized when looking for related material to a given relevant document compared to a journal run that is performed without viewing a preceding document. Using a free text form some respondents pointed out that the content of journals or conference proceedings is topically too broad to discover something similar and therefore not that often utilized. A similar argumentation is discussed in the well-known *berrypicking* modell by Bates [1] where she argues that:" .. a journal with a broad subject is unlikely to fulfil a users information need but more useful to monitor a certain area of research whereas very specific journals are likely to meet a researchers interest."

By clustering the respondents into different groups according to their academic degree and by field of study we showed significant differences in the usage of certain stratagems. *Senior researchers* for instance asses "inspecting the list of citations" as a more valuable stratagem than *junior researchers* who prefer to use keywords that describe a certain document. Although we cannot make any assumptions which of the groups are more effective in satisfying their information needs we can assume that more experienced users are as well more effective. This was also observed in [7]. We assume that senior researchers are more interested in highly specific publications than junior researchers who try to get a broader overview about a certain topic.

Implications for the Design of DLs: The results of the survey showed significant differences in the usage of certain stratagems between *senior researchers* and *junior researchers*. The former, more experienced group utilise DLs and Google Scholar significantly more often. To better support unexperienced users in certain situations DLs could be designed to suggest search activities that experienced users (e.g. *senior researchers*) utilise when solving a search task. Comparable approaches can be found in the literature. In [4] a coaching approach is developed that provides the user with strategic help on potentially useful search activities that were derived from [2] using a rule based mechanism. A similar approach was presented in [13] where an *Adaptive Support for Digital Libraries (ASDL)* was developed that covers sixteen predefined search activity suggestions. A more user oriented approach would be to identify search activities that expert users perform and use their search behaviour as a search strategy suggestion. This was also the main observation in [7].

Current DLs treat each activity on the level of stratagems as a basic unit ignoring search activities that have been performed in previous steps. The task on organizing journal articles for instance (see Sect. 4.4) shows a need for session-context sensitive browsing. The results indicate that the ranking of documents during a journal run should stronger relate to the users' search activity (e.g. entered query term or the inspected document). To the best of our knowledge this would be a novel feature in DLs that users could benefit from during a journal/conference run. This approach has already been proposed in [5] and is further underpinned by the results of our survey. However, the scenario that was used to assess the users' opinion on organizing journal content was composed using a relevant document as a starting point. It would be interesting to see whether the outcome of the task significantly changes when using a negative scenario in which they start from a non-relevant article. A negative scenario could possibly lead to a lower performance of search activity based ranking options. This will be investigated as part of a larger user study we are conducting. Whether search activity based ranking options are applicable during other stratagem search activities, like citation search, needs to be evaluated as well.

6   Outlook

For this survey we focused on stratagems introduced by Bates 25 years ago. Future work can be to identify further high-level search activities that are suitable to solve todays information seeking problems (compare [15]). To this end it is necessary to get a more qualitative view of high-level search activities to fully understand the users' task and goal. We are therefore conducting a user study including interviews and search diaries in which we expect to get a more detailed view on the users' information need when performing high-level search activities.

The results obtained in the survey provide a general overview about the usage of high-level search activities across a broad range of discipline and academic degree. However, it is challenging to generalize the results due to the artificial situation of the online survey. This could be overcome by looking at high-level search activities from their real usage in transaction logs. This would provide us with quantitative behavioural usage data about the frequency of certain search activities. We are therefore conducting a large scale transaction log study to look at usage data of certain search activities in a DL for the Social Sciences Sowiport². Besides their usage frequency we measure the usefulness [6] of certain search activities using pseudo relevance feedback (e.g. bookmark or cite a certain document found in Sowiport) as proposed in [9]. The codebook of the survey and the anonymised data is available at http://dx.doi.org/10.7802/1257.

Acknowledgment. We thank the Interactive Information Retrieval group at GESIS for discussions concerning previous versions of this paper. This work was partly funded by the DFG, grant no. MA 3964/5-1; the AMUR project at GESIS.

References

1. Bates, M.J.: The design of browsing and berrypicking techniques for the online search interface. Online Rev. **13**(5), 407–424 (1989)
2. Bates, M.J.: Where should the person stop and the information search interface start? Inf. Process. Manag. **26**(5), 575–591 (1990)
3. Bhavnani, S.K., Bichakjian, C.K., Johnson, T.M., Little, R.J., Peck, F.A., Schwartz, J.L., Strecher, V.J.: Strategy hubs: domain portals to help find comprehensive information. JASIST **57**(1), 4–24 (2006)
4. Brajnik, G., Mizzaro, S., Tasso, C., Venuti, F.: Strategic help in user interfaces for information retrieval. JASIST **53**(5), 343–358 (2002)
5. Carevic, Z., Mayr, P.: Extending search facilities via bibliometric-enhanced stratagems. In: Proceedings of the 2nd Workshop on Bibliometric-enhanced Information Retrieval, pp. 40–46. CEUR-WS.org, Vienna (2015)
6. Cole, M., Liu, J., Belkin, N., Bierig, R., Gwizdka, J., Liu, C., Zhang, J., Zhang, X.: Usefulness as the criterion for evaluation of interactive information retrieval. In: Proceedings of the Third Human Computer Information Retrieval Workshop, Washington, DC (2009)
7. Fields, B., Keith, S., Blandford, A.: Designing for expert information finding strategies. In: Fincher, S., Markopoulos, P., Moore, D., Ruddle, R. (eds.) People and Computers XVIII Design for Life, pp. 89–102. Springer, London (2005)
8. Fuhr, N., Klas, C.-P., Schaefer, A., Mutschke, P.: Daffodil: an integrated desktop for supporting high-level search activities in federated digital libraries. In: Agosti, M., Thanos, C. (eds.) ECDL 2002. LNCS, vol. 2458, pp. 597–612. Springer, Heidelberg (2002)
9. Hienert, D., Mutschke, P.: A usefulness-based approach for measuring the local and global effect of IIR services. In: Proceedings of the 2016 ACM on Conference on Human Information Interaction and Retrieval, CHIIR 2016, pp. 153–162. ACM, New York (2016)
10. Hienert, D., Schaer, P., Schaible, J., Mayr, P.: A novel combined term suggestion service for domain-specific digital libraries. In: Gradmann, S., Borri, F., Meghini, C., Schuldt, H. (eds.) TPDL 2011. LNCS, vol. 6966, pp. 192–203. Springer, Heidelberg (2011)
11. Jamieson, S.: Likert scales: how to (ab) use them. Med. Educ. **38**(12), 1217–1218 (2004)

---

² sowiport.gesis.org.

12. Joo, S., Xie, I.: How do users' search tactic selections influence search outputs in exploratory search tasks? In: Proceedings of the 13th ACM/IEEE-CS Joint Conference on Digital Libraries, pp. 397–398. ACM, New York (2013)
13. Kriewel, S., Fuhr, N.: Adaptive search suggestions for digital libraries. In: Goh, D.H.-L., Cao, T.H., Sølvberg, I.T., Rasmussen, E. (eds.) ICADL 2007. LNCS, vol. 4822, pp. 220–229. Springer, Heidelberg (2007)
14. O'Day, V.L., Jeffries, R.: Orienting in an information landscape: how information seekers get from here to there. In: Proceedings of the INTERCHI 1993, pp. 438–445. IOS Press (1993)
15. Stelmaszewska, H., Blandford, A., Buchanan, G.: Designing to change users' information seeking behaviour: a case study. In: Adaptable and Adaptive Hypermedia Systems. Idea Group (2005)
16. Wilson, M.L., Schraefel, M.C.: Bridging the gap: using IR models for evaluating exploratory search interfaces. In: SIGCHI 2007 Workshop on Exploratory Search and HCI (2007), 28 April 2007
17. Xie, H.I.: Patterns between interactive intentions and information-seeking strategies. Inf. Process. Manag. **38**(1), 55–77 (2002)

Content-Based Video Retrieval in Historical Collections of the German Broadcasting Archive

Markus Mühling[1]([⊠]), Manja Meister[4], Nikolaus Korfhage[1], Jörg Wehling[4],
Angelika Hörth[4], Ralph Ewerth[2,3], and Bernd Freisleben[1]

[1] Department of Mathematics and Computer Science, University of Marburg,
Hans-Meerwein-Str. 6, 35032 Marburg, Germany
{muehling,korfhage,freisleb}@informatik.uni-marburg.de
[2] German National Library of Science and Technology (TIB), Welfengarten 1B,
30167 Hannover, Germany
ralph.ewerth@tib.eu
[3] Faculty of Electrical Engineering and Computer Science,
Leibniz Universität Hannover, Appelstr. 4, 30167 Hannover, Germany
[4] German Broadcasting Archive, Marlene-Dietrich-Allee 20,
14482 Potsdam, Germany
{manja.meister,joerg.wehling,angelika.hoerth}@dra.de

Abstract. The German Broadcasting Archive (DRA) maintains the cultural heritage of radio and television broadcasts of the former German Democratic Republic (GDR). The uniqueness and importance of the video material stimulates a large scientific interest in the video content. In this paper, we present an automatic video analysis and retrieval system for searching in historical collections of GDR television recordings. It consists of video analysis algorithms for shot boundary detection, concept classification, person recognition, text recognition and similarity search. The performance of the system is evaluated from a technical and an archival perspective on 2,500 h of GDR television recordings.

Keywords: German Broadcasting Archive · Automatic content-based video analysis · Content-based video retrieval

1 Introduction

Digital video libraries become more and more important due to the new potentials in accessing, searching and browsing the data [2,6,17]. In particular, content-based analysis and retrieval in large collections of scientific videos is an interesting field of research. Examples are Yovisto[1], ScienceCinema[2] and the TIB|AV-portal[3] of the German National Library of Science and Technology (TIB). The latter provides access to scientific videos based on speech recognition,

---

[1] http://www.yovisto.com.
[2] http://www.osti.gov/sciencecinema.
[3] http://av.tib.eu.

© Springer International Publishing Switzerland 2016
N. Fuhr et al. (Eds.): TPDL 2016, LNCS 9819, pp. 67–78, 2016.
DOI: 10.1007/978-3-319-43997-6_6

visual concept classification, and video OCR (optical character recognition). The videos of this portal stem from the fields of architecture, chemistry, computer science, mathematics, physics, and technology/engineering.

The German Broadcasting Archive (DRA) in Potsdam-Babelsberg provides access to another valuable collection of scientifically relevant videos. It encompasses significant parts of the audio-visual tradition in Germany and reflects the development of German broadcasting before 1945 as well as radio and television of the former German Democratic Republic (GDR). The DRA was founded in 1952 as a charitable foundation and joint institution of the Association of Public Broadcasting Corporations in the Federal Republic of Germany (ARD). In 1994, the former GDR's radio and broadcasting archive was established. The archive consists of the film documents of former GDR television productions from the first broadcast in 1952 until its cessation in 1991, including a total of around 100,000 broadcasts, such as contributions and recordings of the daily news program "Aktuelle Kamera", political magazines such as "Prisma" or "Der schwarze Kanal", broadcaster's own TV productions including numerous films, film adaptations and TV series productions such as "Polizeiruf 110", entertainment programs ("Ein Kessel Buntes"), children's and youth programs (fairy tales, "Elf 99") as well as advice and sports programs. Access to the archive is granted to scientific, educational and cultural institutions, to public service broadcasting companies and, to a limited extent, to commercial organizations and private persons. The video footage is often used in film and multimedia productions. Furthermore, there is a considerable international research interest in GDR and German-German history. Due to the uniqueness and importance of the video collection, the DRA is the starting point for many scientific studies.

To facilitate searching in videos, the DRA aims at digitizing and indexing the entire video collection. Due to the time-consuming task of manual video labeling, human annotations focus on larger video sequences and contexts. Furthermore, finding similar images in large multimedia archives is manually infeasible.

In this paper, an automatic video analysis and retrieval system for searching in historical collections of GDR television recordings is presented. It consists of novel algorithms for visual concept classification, similarity search, person and text recognition to complement human annotations and to support users in finding relevant video shots. In contrast to manual annotations, content-based video analysis algorithms provide a more fine-grained analysis, typically based on video shots. A GDR specific lexicon of 91 concepts including, for example, "applause", "optical industry", "Trabant", "military parade", "GDR emblem" or "community policeman" is used for automatic annotation. An extension of deep convolutional neural networks (CNN) for multi-label concept classification, a comparison of a Bag-of-Visual Words (BoVW) approach with CNNs in the field of concept classification and a novel, fast similarity search approach are presented. The results of automatically annotating 2,500 h of GDR television recordings are evaluated from a technical and an archival perspective.

The paper is organized as follows: Sect. 2 describes the video retrieval system and the video analysis algorithms. In Sect. 3, experimental results on the GDR television recordings are presented. Section 4 concludes the paper and outlines areas for future research.

2   Content-Based Video Retrieval

The aim of content-based video retrieval is to automatically assign semantic tags to video shots for the purpose of facilitating content-based search and navigation. Figure 1 shows an overview of our developed video retrieval system. First, the videos are preprocessed. This step mainly consists of shot boundary detection [9] in conjunction with thumb generation for the keyframes, which are required for later visualization purposes. The aim of shot boundary detection is the temporal segmentation of a video sequence into its fundamental units, the shots. Based on video segmentation, the following automatic content-based video analysis algorithms are applied: concept classification, similarity search, person and text recognition. The resulting metadata are written to a database and serve as an intermediate description to bridge the "semantic gap" between the data representation and the human interpretation. Given the semantic index, arbitrary search queries can be processed efficiently, and the query result is returned as a list of video shots ranked according to the probability that the desired content is present. In the following, the content based video analysis algorithms are described in more detail.

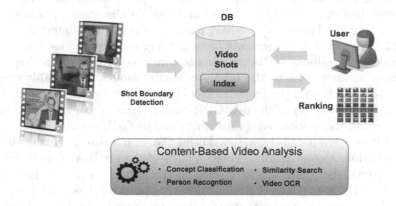

Fig. 1. Video retrieval system.

2.1   Visual Concept Classification

The classification of visual concepts is a challenging task due to the large complexity and variability of their appearance. Visual concepts can be, for example, objects, sites, scenes, personalities, events, or activities. The definition of the GDR specific concept lexicon is based on the analysis of user search queries with a focus on queries that were experienced as difficult and time-consuming to answer manually. A lexicon of 91 concepts was defined after analyzing more than 36,000 user queries received within a five-year period from 2008 to 2013. Recent user queries, assumed to be of future research interest, were summarized thematically and ordered by frequency. The concept lexicon comprises events such as "border control" and "concert", scenes such as "railroad station" and "optical industry", objects like "Trabant" or activities such as "applauding". Manually annotated training data was used to build the concept models. For this purpose, a client-server based annotation tool was built to facilitate the process of training data acquisition and to select a sufficient quantity of representative training examples for each concept. We defined a minimum number of 100 positive samples per concept.

Recently, deep learning algorithms fostered a renaissance of artificial neural networks, enabled by the massive parallel processing power of modern graphics cards. Deep learning approaches, especially deep CNNs, facilitated breakthroughs in many computer vision fields [4,11,14,23]. Instead of using hand-crafted features such as SIFT descriptors [16], CNNs learn the features automatically during the training process. A CNN consists of several alternating convolution and max-pooling layers with increasingly complex feature representations and typically has several fully connected final layers.

State-of-the-art network architectures for image recognition [14,22] as well as the current datasets [8,27] consider only a single concept per image ("single-label"). In contrast, real world concept classification scenarios are multi-label problems. Several concepts, such as "summer", "playground" and "teenager", may occur simultaneously in an image or scene. While some approaches use special ranking loss layers [10], we have extended the CNN architecture using a sigmoid layer instead of the softmax layer and a cross entropy loss function.

Furthermore, millions of training images are needed to build a deep CNN model from scratch. Due to the relatively small amount of available training data, we have adapted a pre-trained CNN classification model (GoogleNet [22] trained on ImageNet [8]) to the new GDR concept lexicon using our multi-label CNN extension and performed a fine-tuning on the GDR television recordings. The models were trained and fine-tuned using the deep learning framework Caffe [12]. In addition, a comparison between a state-of-the-art BoVW approach and our deep multi-label CNN was performed on the publicly available, fully annotated NUSWIDE scene dataset [7]. The BoVW approach is based on a combination of different feature representations relying on optimized SIFT variants [19]. The different feature representations are combined in a support vector machine (SVM)-based classifier using multiple kernel learning [20]. Our deep multi-label CNN approach significantly outperformed the BoVW approach by a relative performance improvement of almost 20 %. In contrast to binary SVM classifiers, deep neural networks are inherently multi-class capable so that only a single compact model has to be built for all concept classes. While the runtime of the BoVW approach takes already 1.97 s for feature extraction and the classification runtime depends linearly on the number of concepts, our deep multi-label CNN takes only 0.35 s on the CPU (Intel Core i5) and just 0.078 s on the GPU

(Nvidia Titan X). Although deep neural networks are computationally expensive in the training phase, they are very efficient at the classification stage. Altogether, deep multi-label CNNs provide better recognition quality, compact models and faster classification runtimes.

2.2 Similarity Search

Since the DRA offers researchers a large number of video recordings containing several millions of video shots, the need for a system that helps to rapidly find desired video shots emerges. While scanning through the whole video archive is practically infeasible for humans, a possible solution is to index the videos via concepts as described in Sect. 2.1. However, this approach requires manually annotated training images for learning the concept models. Additionally, search queries are restricted to the vocabulary of predefined concepts, and new concept models have to be developed on demand. In contrast to textual concept-based queries, image-based queries provide users more flexibility and new ways of searching.

While query-by-content based on low-level features turned out to be insufficient to search successfully in large-scale multimedia databases, image representations learned by deep neural networks greatly improved the performance of content-based image retrieval systems [25]. They are less dependent on pixel intensities and are clearly better suited for searching semantic content. However, high-dimensional CNN features are not well suited for searching efficiently in large video collections. Fast search in large databases is an essential requirement for practical use. For this purpose, proposals for learning binary image codes for compact representations and fast matching of images have been made. Krizhevsky and Hinton [13], for example, used very deep autoencoders, and Lin et al. [15] extended a CNN to learn binary hash codes for fast image retrieval.

In this section, an approach for fast content-based similarity search in large video databases is presented. For efficient storage and matching of images, binary codes are learned by deep CNNs. This mapping of images to binary codes is often referred to as "semantic hashing" [21]. The idea is to learn a "semantic hash function" that maps similar images to similar binary codes. To learn the hash function, a method similar to the approach described by Lin et al. [15] is used. Based on a pre-trained CNN classification model, we devised a coding layer and an appropriate loss layer for error propagation for the network architecture. The best results were obtained using Chatfield et al.'s network [5] trained on the PLACES dataset [27]. An advantage of using pre-trained classification models is the speed-up in training time. To obtain high-level features, we built the coding layer on top of the last fully-connected layer. Furthermore, the hash function can be adapted to unlabeled datasets by using the predictions of the pre-trained classification model for error propagation.

The overall video retrieval system is based on the analysis of keyframes, i.e., representative images. In our approach, five frames (the first, the last, and three in between) per video shot are used as keyframes for indexing. Given the hash function, the keyframes of the video collection are fed into the deep CNN and the

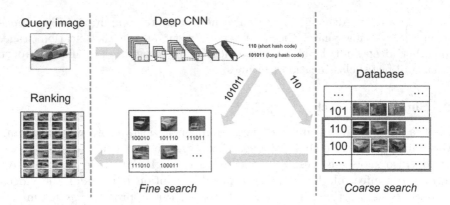

Fig. 2. Content based similarity search.

mapped binary codes are stored in the database. Based on the resulting index, queries-by-image can be answered by matching the binary code of the given image to the database. The overall retrieval process is shown in Fig. 2. Given the query image, the binary codes are extracted using the learned deep CNN. We use a two-stage approach based on semantic hashing. First, a coarse search is performed using 64-bit binary codes resulting in a "short" list of potential results. The hamming distance is applied to compare the binary codes, and a vantage point tree [26] is used as an additional index structure to accelerate the search process. The resulting short list consists of the 10,000 nearest neighbors. The longer the binary codes, the more accurate the image representations are. Therefore, in the second stage, a refined search using 256-bit binary codes is performed on the short list, and the images are ranked according to the distance to the query image. Differently from Lin et al. [15], hash codes are used for the refined search as well. In our two-stage approach, both coding layers, for 64-bit and 256-bit binary codes, are integrated into the same architecture resulting in a concurrent computation at training and testing time. Finally, the resulting images are mapped to video shots.

2.3 Person Recognition

Based on the analysis of user search queries, the GDR specific concept lexicon has been extended by 9 personalities, including "Erich Honecker", "Walter Ulbricht", "Hilde Benjamin", "Siegmund Jähn", "Hermann Henselmann", "Christa Wolf", "Werner Tübke", "Stephan Hermlin" and "Fritz Cremer". Instead of using concept classification, personalities are handled using a face recognition approach. Therefore, feature representations of known persons are stored in a face database, and a face recognition system was built that scans the video shots and recognizes the identity of a detected face image by comparing it to the face database. Finally, the resulting index of person occurrences can be used in search queries. The face processing pipeline consists of several components: face detection, face alignment and face recognition. For the face recognition component, we evaluated the following approaches: Fisherfaces [3], Local Binary Pattern Histograms [1] and a commercial library, called FaceVACs[4]. Furthermore, we evaluated whether a preprocessing step using grayscale histogram equalization, training data augmentation using Google search queries, or a face tracking component improve the recognition accuracy. Based on the results of our evaluations, we finally used the method of Viola and Jones [24] for face detection and FaceVACs for face alignment and recognition.

2.4 Text Recognition (Video OCR)

Superimposed text often hints at the content of a video shot. In news videos, for example, the text is closely related to the current report, and in silent movies it is used to complement the screen action with cross headings. Involved algorithms can be distinguished by their objective, whether it is text detection, also called localization, text segmentation, or optical character recognition (see Fig. 3).

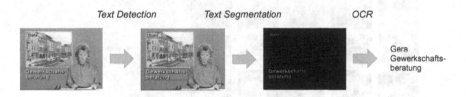

Fig. 3. Text recognition pipeline.

We developed a text recognition system that allows users to search for in-scene and overlaid text within the video archive. For this purpose, the I-frames of the videos are analyzed and the recognized ASCII text is stored in the database. Based on the resulting index, OCR search queries can be answered by a list of video shots ranked according to the similarity to the query term. Due to poor technical video quality and low contrast of text appearances, the similarities between the query term and the words in the database are calculated using the Levenshtein distance.

For text detection, localization and segmentation in video frames, a method based on Maximally Stable Extremal Regions (MSER) [18] is used. It is able to detect both overlaid text, as well as text within the scene, for example on banners. Experimental results revealed that the text segmentation component plays an important role in the case of poor quality videos. Text segmentation crops localized text out of the image to yield black letters on a white background. This step is necessary to feed the result into an OCR algorithm that transforms the image into machine-readable text. A non-uniform background would normally impair this process. For OCR, we finally used the open source library Tesseract[5].

---

[4] http://www.cognitec.com.
[5] http://code.google.com/p/tesseract-ocr/.

3   Experimental Results

A web-based GUI has been developed to automatically respond to user queries related to concepts, persons, similar images and text. The retrieval results are presented to the user in the form of a ranked list of video shots (see Fig. 4) where each video shot is represented by five video frames and a probability score indicating the relevance. Furthermore, a video player allows to visually inspect the video shots.

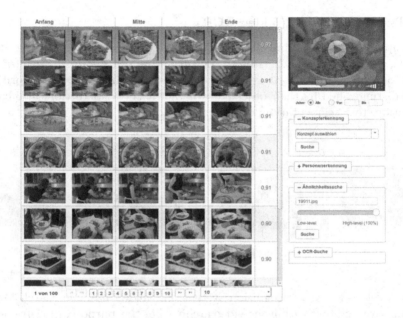

Fig. 4. Retrieval results for a query image showing a meal.

3.1   Data

In total, more than 3,000 h of historical GDR television recordings have been digitized by now. The video footage is technically very challenging. Many recordings are in grayscale and of low technical quality; the older the recordings, the poorer the video quality. The temporal segmentation of the videos resulted in approximately 2 million video shots. From these shots, 416,249 have been used for the training process and 1,545,600 video shots, corresponding to about 2,500 h of video data, for testing.

3.2   Results

In the following, the results for concept classification, person recognition, similarity search and video OCR are presented. The results are evaluated using the average precision (AP) score that is the most commonly used quality measure in video retrieval. The AP score is calculated from the list of video shots as follows:

$$AP(\rho) = \frac{1}{|R \cap \rho^N|} \sum_{k=1}^{N} \frac{|R \cap \rho^k|}{k} \psi(i_k) \quad \text{with} \quad \psi(i_k) = \begin{cases} 1 & \text{if } i_k \in R \\ 0 & \text{otherwise} \end{cases} \quad (1)$$

where N is the length of the ranked shot list, $\rho^k = \{i_1, i_2, \ldots, i_k\}$ is the ranked shot list up to rank k, R is the set of relevant documents, $|R \cap \rho^k|$ is the number of relevant video shots in the top-k of ρ and $\psi(i_k)$ is the relevance function. Generally speaking, AP is the average of the precisions at each relevant video shot. To evaluate the overall performance, the mean AP score is calculated by taking the mean value of the AP scores from different queries.

Concept Classification and Person Recognition. In total, 86 concepts, comprising 77 concepts and 9 persons, were evaluated. Of the originally 91 concepts 14 were dismissed due to an insufficient number of training images. However, another 14 of the 77 evaluated concepts have less than 100 training images. Altogether, 118,020 positive training examples were gathered for learning the concept models. The retrieval results for concepts and persons were evaluated based on the top-100 and top-200 ranked video shots. Although 14 concepts have less than 100 training images and despite poor video quality, we obtained a mean AP of 62.4 % and 58.0 %, respectively. Even concepts occurring predominantly in grayscale shots of low video quality yielded good results, such as "daylight mining" with 84.3 % AP. These results reveal the high robustness of the proposed multi-label deep CNN approach with respect to low quality historical video data.

For person recognition, we achieved a very good result of 83.3 % mean AP on the top-100 and 81.1 % mean AP on the top-200 video shots and even 100 % AP for distinctive and frequently occurring personalities, such as Erich Honecker and Walter Ulbricht. In summary, we achieved a mean AP of 64.6 % and 60.5 % on the top-100 and top-200, respectively, for both concepts and persons.

Similarity Search. The interpretation whether two images are similar is subjective and context specific. The definition of similarity ranges from pixel-based similarity to image similarity based on the semantic content. How much low-level and semantic similarity contribute to the retrieval results can be individually adjusted in the GUI. Furthermore, two use cases have been implemented: searching by video frames selected from within the corpus and searching by external images, e.g., downloaded from the Internet. In our evaluation, we focus on the more difficult task of semantic similarity using 50 external query images from the Internet chosen collaboratively by computer scientists and archivists. An example result for a query image showing a meal is presented in Fig. 4. Each retrieval result was evaluated up to the first 100 video shots obtaining a mean AP of 57.5 %. The image similarity search relies on representative images from the video shots. Based on an image corpus of more than 7 million images, we achieved a very fast response time of less than 2 s.

Video OCR. For the task of text retrieval, 46 query terms according to previously observed search query preferences of DRA users have been evaluated. Based on these 46 search queries, like "Abschaffung der Todesstrafe", "Kinder- und Jugendspartakiade", "Öffnungszeiten", "Protestbewegung", "Rauchen verboten", "Warschauer Vertrag", "Planerfüllung", "Nationale Front", "Gleichberechtigung", "Mikroelektronik" or "Staatshaushalt" a very satisfying retrieval performance of 92.9 % mean AP has been obtained. As expected, the results for overlaid text are significantly better than for text within the scene.

3.3   Archivist's Perspective

In this section, the presented content-based video retrieval system is evaluated from an archivist's perspective with a focus on usability and usefulness for archivists and DRA users. Since users of the archive are often looking for everyday scenes in the former GDR, concepts such as "pedestrian", "supermarket", "kitchen", "camping site", "allotment" or "production hall", in conjunction with the high search quality for most of these concepts, are a valuable contribution to help researchers finding appropriate scenes. Concepts with an AP score of more than approximately 50 % turned out to be practically very useful. 66 % of the concepts achieved an AP score of more than 50 %.

Searching manually for persons in videos is a quite time consuming task, particularly for less known members of the Politbüro and Ministers of the GDR. Thus, the high quality of the provided automatic person indexing algorithms is a great benefit for archivists as well as for users of the archive.

The implemented similarity search system tremendously extends the accessibility to the data in a flexible and precise way. It provides complementary search queries that are often hard to verbalize. In addition, it allows an incremental search. Previous results may serve as a source of inspiration for new similarity search queries for refining search intentions.

Another useful search option is offered by video OCR. OCR search results are very helpful since overlaid text is often closely related to the video content. The system recognizes the majority of slogans, locations and other terms precisely.

Altogether, the fine-grained automatic annotation is a very valuable supplement to human-generated meta-data. Due to the variability of the content-based video retrieval system, different user needs are taken into account. The combination of different search modalities allows to answer a wide range of user queries leading to more precise results in significantly less time.

4   Conclusion

In this paper, we have presented a content-based video retrieval system for searching in historical collections of GDR television recordings. Novel algorithms for visual concept classification, similarity search, person recognition and video OCR have been developed to complement human annotations and to support users in finding relevant video shots. Experimental results on about 2,500 h of

GDR television recordings have indicated the excellent video retrieval quality in terms of AP as well as in terms of usability and usefulness from an archivist's perspective.

There are several areas for future work. First, the concept-based approach requires manually annotated training images for learning the concept models. New concepts have to be developed on demand. Therefore, it is important to reduce the effort for training data acquisition. Second, the audio modality can be used to improve the detection performance of several audio related concepts. Finally, a similarity search query can be very subjective depending on a specific user in given situation. New strategies have to be developed to predict the user's intention.

Acknowledgements. This work is financially supported by the German Research Foundation (DFG-Programm "Förderung herausragender Forschungsbibliotheken", "Bild- und Szenenrecherche in historischen Beständen des DDR-Fernsehens im Deutschen Rundfunkarchiv durch automatische inhaltsbasierte Videoanalyse"; CR 456/1-1, EW 134/1-1, FR 791/12-1).

References

1. Ahonen, T., Hadid, A., Pietikäinen, M.: Face recognition with local binary patterns. In: Pajdla, T., Matas, J.G. (eds.) ECCV 2004. LNCS, vol. 3021, pp. 469–481. Springer, Heidelberg (2004)
2. Albertson, D., Ju, B.: Design criteria for video digital libraries: categories of important features emerging from users' responses. Online Inf. Rev. **39**(2), 214–228 (2015)
3. Belhumeur, P.N., Kriegman, D.J.: Eigenfaces vs. fisherfaces: recognition using class specific linear projection. IEEE Trans. Pattern Anal. Mach. Intell. **19**(7), 711–720 (1997)
4. Breuel, T.M., Ul-Hasan, A., Al-Azawi, M.A., Shafait, F.: High-performance OCR for printed English and Fraktur using LSTM networks. In: Proceedings of the International Conference on Document Analysis and Recognition, pp. 683–687 (2013)
5. Chatfield, K., Simonyan, K., Vedaldi, A., Zisserman, A.: Return of the devil in the details: delving deep into convolutional nets. In: Proceedings of the British Machine Vision Conference, pp. 1–11 (2014)
6. Christel, M., Kanade, T., Mauldin, M., Reddy, R., Sirbu, M., Stevens, S.M., Wactlar, H.D.: Informedia digital video library. Commun. ACM **38**(4), 57–58 (1995)
7. Chua, T.S., Tang, J., Hong, R., Li, H., Luo, Z., Zheng, Y.: NUS-WIDE: a real-world web image database from National University of Singapore. In: Proceedings of the ACM International Conference on Image and Video Retrieval, pp. 48:1–48:9 (2009)
8. Deng, J., Dong, W., Socher, R., Li, L.-J., Li, K., Fei-Fei, L.: ImageNet: a large-scale hierarchical image database. In: IEEE Conference on Computer Vision and Pattern Recognition (CVPR 2009), pp. 2–9 (2009)
9. Ewerth, R., Freisleben, B.: Video cut detection without thresholds. In: Proceedings of the 11th International Workshop on Signals, Systems and Image Processing (IWSSIP 2004), Poznan, Poland, pp. 227–230 (2004)

10. Gong, Y., Jia, Y., Leung, T., Toshev, A., Ioffe, S.: Deep convolutional ranking for multilabel image annotation. arXiv preprint arXiv:1312.4894 (2013)
11. Graves, A., Mohamed, A., Hinton, G.: Speech recognition with deep recurrent neural networks. In: IEEE International Conference on Acoustics, Speech and Signal Processing (ICASSP 2013), pp. 6645–6649 (2013)
12. Jia, Y., Shelhamer, E., Donahue, J., Karayev, S., Long, J., Girshick, R., Guadarrama, S., Darrell, T.: Caffe: convolutional architecture for fast feature embedding. In: Proceedings of the ACM International Conference on Multimedia, pp. 675–678 (2014)
13. Krizhevsky, A., Hinton, G.: Using very deep autoencoders for content-based image retrieval. In: Proceedings of the European Symposium on Artificial Neural Networks, pp. 1–7 (2011)
14. Krizhevsky, A., Sutskever, I., Hinton, G.E.: ImageNet classification with deep convolutional neural networks. In: Advances in Neural Information Processing Systems, pp. 1–9 (2012)
15. Lin, K., Yang, H.F., Hsiao, J.H., Chen, C.S.: Deep learning of binary hash codes for fast image retrieval. In: Proceedings of the IEEE Conference on Computer Vision and Pattern Recognition (CVPR), pp. 27–35 (2015)
16. Lowe, D.G.: Object recognition from local scale-invariant features. In: Proceedings of the 7th IEEE International Conference on Computer Vision, vol. 2, pp. 1150–1157 (1999)
17. Marchionini, G., Geisler, G.: The open video digital library. D-Lib Mag. 8(12), 1082–9873 (2002)
18. Matas, J., Chum, O., Urban, M., Pajdla, T.: Robust wide-baseline stereo from maximally stable extremal regions. Image Vis. Comput. 22(10), 761–767 (2004)
19. Mühling, M.: Visual concept detection in images and videos. Ph.D. thesis, University of Marburg (2014)
20. Mühling, M., Ewerth, R., Zhou, J., Freisleben, B.: Multimodal video concept detection via bag of auditory words and multiple kernel learning. In: Schoeffmann, K., Merialdo, B., Hauptmann, A.G., Ngo, C.-W., Andreopoulos, Y., Breiteneder, C. (eds.) MMM 2012. LNCS, vol. 7131, pp. 40–50. Springer, Heidelberg (2012)
21. Salakhutdinov, R., Hinton, G.: Semantic hashing. Int. J. Approximate Reasoning 50(7), 969–978 (2009)
22. Szegedy, C., Liu, W., Jia, Y., Sermanet, P., Reed, S., Anguelov, D., Erhan, D., Vanhoucke, V., Rabinovich, A.: Going deeper with convolutions. In: IEEE Conference on Computer Vision and Pattern Recognition (CVPR), pp. 1–9 (2015)
23. Taigman, Y., Yang, M., Ranzato, M., Wolf, L.: DeepFace: closing the gap to human-level performance in face verification. In: Proceedings of the IEEE Conference on Computer Vision and Pattern Recognition (CVPR), pp. 1–8 (2014)
24. Viola, P., Jones, M.: Rapid object detection using a boosted cascade of simple features. In: Proceedings of the 2001 IEEE Computer Society Conference on Computer Vision and Pattern Recognition, vol. 1, pp. 511–518 (2001)
25. Wan, J., Wang, D., Hoi, S.C.H., Wu, P.: Deep learning for content-based image retrieval: a comprehensive study. In: Proceedings of the ACM International Conference on Multimedia (MM), pp. 157–166 (2014)
26. Yianilos, P.N.: Data structures and algorithms for nearest neighbor search in general metric spaces. In: Annual ACM-SIAM Symposium on Discrete Algorithms, pp. 311–321 (1993)
27. Zhou, B., Lapedriza, A., Xiao, J., Torralba, A., Oliva, A.: Learning deep features for scene recognition using places database. Adv. Neural Inf. Process. Syst. 27, 487–495 (2014)

User Aspects

Profile-Based Selection of Expert Groups

Georgios A. Sfyris, Nikolaos Fragkos, and Christos Doulkeridis[✉]

Department of Digital Systems,
School of Information and Communication Technologies,
University of Piraeus, 18534 Piraeus, Greece
george.sfiris@gmail.com, nifragos@gmail.com, cdoulk@unipi.gr

Abstract. In a wide variety of daily activities, the need of selecting a group of k experts from a larger pool of n candidates ($k < n$) based on some criteria often arises. Indicative examples, among many others, include the selection of program committee members for a research conference, staffing an organization's board with competent members, forming a subject-specific task force, or building a group of project evaluators. Unfortunately, the process of *expert group selection* is typically carried out manually by a certain individual, which poses two significant shortcomings: (a) the task is particularly cumbersome, and (b) the selection process is largely subjective thus leading to results of doubtful quality. To address these challenges, in this paper, we propose an automatic profile-based expert group selection mechanism that is supported by digital libraries. To this end, we build textual profiles of candidates and propose algorithms that follow an IR-based approach to perform the expert group selection. Our approach is generic and independent of the actual expert group selection problem, as long as the candidate profiles have been generated. To evaluate the effectiveness of our approach, we demonstrate its applicability on the scenario of automatically building a program committee for a research conference.

1 Introduction

Expert group selection is an significant task that naturally arises in plenty aspects of our everyday life. Indicative examples include the selection of program committee (PC) members for a research conference, staffing an organization's board with competent members, forming a subject-specific task force, or building a group of project evaluators. Practically in all cases, the importance of selecting competent groups of experts is paramount, and determines the successful completion of the task at hand.

Typically, the process of expert group selection is performed manually by some individuals. For example, the PC chairs of a research conference are assigned the role of forming the program committee, by selecting and inviting prominent researchers who have expertise in the topics of interest of the conference. Unfortunately, this process has various shortcomings, First, it is very cumbersome and requires great effort from the PC chairs. Second, it is highly subjective and this may undermine the quality of the reviewing process.

© Springer International Publishing Switzerland 2016
N. Fuhr et al. (Eds.): TPDL 2016, LNCS 9819, pp. 81–93, 2016.
DOI: 10.1007/978-3-319-43997-6_7

Motivated by such shortcomings, in this paper, we study the problem of automatic selection of expert groups based on a profile repository, which can be created from digital libraries. To this end, we propose a generic framework for expert group selection, termed $\mathcal{EGS}$, which takes as input a set of query terms and retrieves a group of k experts that satisfy the input requirements. In fact, the framework can be parameterized by an optimization goal and a suitable algorithm that solves the respective optimization problem. We focus on a specific instantiation $\mathcal{EGS}^{cov}$ of our framework, where the optimization goal is *coverage* of the input query terms, which is directly applicable to the PC selection problem when the aim is to choose PC members whose combined expertise cover the areas of interest of the conference.

In summary, the contributions of this paper include:

- We propose the *Expert Group Selection* ($\mathcal{EGS}$) framework, which relies on a profile database to automatically select a group of k experts that satisfy a given set of query terms and an optimization goal (Sect. 3.1).
- We formalize and study a specific problem statement (denoted $\mathcal{EGS}^{cov}$) with optimization goal the coverage of the input set of query terms by the group of k selected experts (Sect. 3.2).
- We show how our framework can be employed in order to generate profiles for experts and perform profile-based retrieval of experts given a query term (Sect. 4).
- We propose two algorithms that return results of high quality, completely automatically without requiring any user intervention (Sects. 5 and 6).
- We demonstrate the quality of returned groups of experts, by means of empirical evaluation on real data of research conferences, where our algorithms suggest a list of k program committee members that cover the topics/areas of interest of the respective conference (Sect. 7).

In addition, we review related work in Sect. 2, and we present our conclusions and sketch future research directions in Sect. 8.

2   Related Work

According to Balog [1], there are two principal approaches to expert finding, termed *Candidate* and *Document*. Using the probabilistic ranking principle [12], Fang and Zhai attempt to create a general model for expert finding [5]. In addition, there have been proposals for graph-based ranking schemes [6]. The advantage of this concept is that it naturally utilizes hyperlinks between documents and professional connections between people [13]. A complete system for combined full text and ontology search is the work of Bast et al. [3]. However, the main focus of the paper is on how to make the given semantic information searchable fast and conveniently, which is totally different from our approach. In [9], it is argued that taking into account contextual as well as content-based factors (such as the recency of an expert's knowledge) can improve the task of expert finding. While we do not explicitly take into account temporal factors, it is fairly easy to modify the scoring function that we employ in order to boost the scores of experts based on temporal information.

Another model proposed for ranking entities [4] builds on the well-known vector space model. We employ a very similar method as baseline in our evaluation, and we demonstrate that our approach achieves results of higher quality. Related to this work, as demonstrated in [2], the model can also be extended to accommodate additional knowledge independent of the relevance to the documents, such as time, the link structure, and other local or non-local evidence. A specialized work on the topic of expert finding is [7], where a greedy and an evolutionary algorithm are combined to assign papers submitted to a conference to reviewers. Again, this is a different problem than the one targeted in our work, since they focus on a matching problem, whereas we perform the task of expert group selection aiming to optimize the result with respect to coverage.

Probably the closest approach to our problem has been tackled by Karimzadehgan and Zhai [10]. The goal is to retrieve k experts that will act as reviewers for a given paper. The experts should not only be relevant to the paper but also should be able to work together to cover all the aspects of a paper. In their work, they present three methods that utilize the fact that a paper usually has multiple subtopics or aspects. Since the notion of multiple aspects is implicit, they face the problem of maximizing the aspect coverage. In [11], a solution to the problem is proposed by casting it as an integer programming problem. Also, the target is explicitly set as solving the Committee Review Assignment problem. Unlike our work, they consider as input directly the submitted papers and extract information using multi-aspect techniques. Moreover, our profile-based framework is more generic and can be employed for solving a variety of optimization problems related to selection of a group of experts.

3   Framework and Problem Statement

Consider a set of n candidate experts $\mathcal{E}$ each associated with a profile, i.e., $\mathcal{E} = \{(e_1, p_1), (e_2, p_2), \ldots, (e_n, p_n)\}$, where a pair (e_i, p_i) corresponds to an expert e_i and her profile p_i. At a high level, a profile is an abstract concept that describes the expertise of an expert. For instance, it can be unstructured (textual form) or structured (fields with values). Also, consider a query Q that consists of m terms $Q = \{q_1, \ldots, q_m\}$.

3.1   Framework for Expert Group Selection

We introduce a generic framework for expert group selection, which relies on the existence of a profile database that records data about experts $\mathcal{E}$, typically in the form of unstructured, textual information. The *Expert Group Selection* framework ($\mathcal{EGS}$) takes as input a query $Q = \{q_1, \ldots, q_m\}$ that consists of m terms, and a value k indicating the desired cardinality of the retrieved group of experts. The output is a set of k experts $\{e_1, \ldots, e_k\}$ that satisfy the input requirements. A graphical illustration of the $\mathcal{EGS}$ framework is shown in Fig. 1.

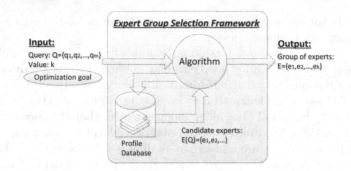

Fig. 1. The *Expert Group Selection* framework: input, output, and internal mechanism.

At the heart of our framework, an algorithm is employed that takes the input (Q, k), filters the profile database $\mathcal{E}$, and retrieves a set of *candidate experts* $\mathcal{E}(Q)$ that are potential results. The cardinality of $\mathcal{E}(Q)$ is query-dependent and at most equal to n, but in practice much smaller, since the query Q restricts the number of experts from the database that match Q. Then, the algorithm selects the subset of $\mathcal{E}(Q)$ of length k, which optimizes the goal of the query at hand.

Interestingly, the framework is parameterized by an *optimization goal* that is coupled with an appropriate algorithm. Obviously, by defining a specific optimization goal, we get different instantiations of the framework, which correspond to different problem statements. In the following, we focus on such an instantiation, namely we employ as optimization goal the *coverage* of all query terms by the retrieved expert group.

Other potential problem statements are produced by requesting a group of experts with maximum diversity, or even combinations of goals (multi-objective optimization): e.g., requiring a group of experts such that no other group has both higher coverage and diversity.

3.2 Problem Statement

We are now ready to define the specific problem instantiation, which is the focus of this paper. Our goal is to find a subset of k experts ($k < n$), whose combined expertise covers as many as possible of the query terms in Q (ideally all query terms).

Problem 1. ($\mathcal{EGS}^{cov}$ Problem) Given a profile database of experts $\mathcal{E} = \{(e_1, p_1), (e_2, p_2), \ldots, (e_n, p_n)\}$ and a query Q that consists of m terms $Q = \{q_1, \ldots, q_m\}$, find k experts ($k < n$) who cover as many as possible query terms Q.

Obviously, the $\mathcal{EGS}^{cov}$ problem is a combinatorial problem and it is NP-complete. As a result, its exact solution cannot be computed in polynomial time, unless $P = NP$. Therefore, we turn our attention to approximate algorithms that solve the problem.

4   Candidate Experts Retrieval

In this section, we describe how the candidate expert list $\mathcal{E}(Q)$ is generated, given a query $Q = \{q_1, \ldots, q_m\}$ as input. To this end, we also provide more implementation details on the framework, in particular on (a) profile generation from the profile database, and (b) how expert retrieval is performed.

Profile Generation. We consider the case that each expert e_i is associated with one pseudo-document d_i which describes her expertise. The way to generate the pseudo-document is orthogonal to our approach and not restrictive in any way. Several ways to generate d_i can be considered, for example providing a manually constructed list of the expert's skills or even her CV. In our evaluation, we focus on researchers as experts. Therefore, we build a profile database of researchers by adopting an IR-inspired approach, where we consider that d_i contains the titles of all papers co-authored by e_i. Intuitively, document d_i contains terms that reflect the researcher's interests and expertise. A by-product of this representation, is that the frequency of a term t in d_i relates to the number of papers authored by e_i on the topic t, thus it serves as an indicator of her expertise. As any researcher may have expertise of variable level on different topics, the term with the highest frequency in d_i indicates the most important research topic based on e_i's published record.

Expert Retrieval. Our profile database stores pseudo-documents and requires search functionality. Hence, each expert's pseudo-document is indexed by a search engine for efficient querying and access. We use Apache Lucene in order to store and index the pseudo-documents, and we use its powerful search mechanism for retrieval.

In more detail, for each query term $q_i \in Q$ ($1 \leq i \leq m$), we perform a search using Lucene and retrieve a ranked set of pseudo-documents (corresponding to experts), which is ranked based on the relevance to the corresponding query term q_i. As a result, after processing all queries $q_i \in Q$, we have retrieved a set $\mathcal{E}(Q)$ of experts, where each $e_j \in \mathcal{E}(Q)$ is associated with a subset $\{(q_1, w_{1j}), \ldots, (q_m, w_{mj})\}$ of the terms in Q, and each term q_i is accompanied by the relevance score w_{ij}.

Intuitively, the relevance score represents the quality of one's expertise and it is orthogonal to our approach. For the purposes of this paper we have used as relevance score the total number of occurrences of a certain query term appears in a pseudo-document, i.e., the term frequency, normalized by the length of the pseudo-document. For the coverage objective, good experts should satisfy the query, i.e., have many of the query terms in their profiles, and at the same time have high relevance score.

Example 1. Consider a query $Q = \{q_1, \ldots, q_9\}$ that consists of nine terms, then the following is a subset of the candidate experts set $\mathcal{E}(Q)$:

e_1 : $\{(q_4, 0.65), (q_7, 0.35), (q_1, 0.21), (q_9, 0.11)\}$
e_2 : $\{(q_2, 0.22), (q_4, 0.13), (q_3, 0.06)\}$
e_3 : $\{(q_5, 0.45), (q_1, 0.23), (q_2, 0.2), (q_8, 0.11), (q_6, 0.1)\}$
...

Algorithm 1. *Greedy Maximum Coverage (GMC)*

1: **INPUT:** k, $Q = \{q_1, \ldots, q_m\}$
2: **OUTPUT:** Group of experts: $\{e_1, \ldots, e_k\}$
3: $\mathcal{E}(Q) \leftarrow$ generate profiles based on Q
4: **for** $(i = 1 \ldots k)$ **do**
5:     select the expert e that covers the maximum number of uncovered query terms
6:     **if** (more than one experts are selected) **then**
7:         select the expert e with the highest total relevance score of the uncovered query terms
8: **return** the selected experts

Notice that a candidate expert e_j must have been returned as a result for at least one query term q_i. The above list is essentially a set of inverted lists constructed from the results of the queries $\{q_1, \ldots, q_9\}$.   □

The computation of relevance scores can be parameterized in order to follow any scoring scheme, such as term frequency (tf), term frequency inverted document frequency (tf-idf), BM25, or any other scheme. In our implementation, where the pseudo-documents are constructed by concatenating paper titles, using plain term frequency suffices since it practically reflects the number of papers authored by an expert that contain a specific query term on the title.

5   Mapping $\mathcal{EGS}^{cov}$ to the Maximum Coverage Problem

The $\mathcal{EGS}^{cov}$ problem is easily mapped to the well-known maximum coverage problem, which is defined as follows.

Definition 1. *(Maximum Coverage Problem) Given a number k and a collection of sets $S = \{S_1, \ldots, S_m\}$, find a subset $S' \subseteq S$ of sets, such that $|S'| \leq k$ and the number of covered elements $\bigcup_{S_i \in S'} S_i$ is maximized.*

It is straightforward to prove that $\mathcal{EGS}^{cov}$ can be mapped to the maximum coverage problem. Intuitively, the sets S_i correspond to the candidate experts $\mathcal{E}(Q)$ and each set (candidate expert) contains some of the query terms in Q. We need to return k experts whose profiles cover as many as possible query terms in Q.

As the maximum coverage problem is known to be NP-complete, a greedy algorithm is employed to provide an approximate solution. The greedy algorithm for maximum coverage iteratively selects the next set for inclusion in the answer set. Each time, the set containing the largest number of uncovered elements is chosen. After k iterations, the algorithm terminates.

There is a small modification in our approach because of the nature of our data. Since the candidate experts have been retrieved based on the input Q, it frequently occurs that two or more experts have the same set of expertise terms. In order to break the tie, we resort to the relevance score mentioned earlier, i.e., selecting the expert with the higher total value of relevance score. At the unlikely event that we will still have a tie between two experts, we choose the expert alphabetically.

Algorithm 1 lists the pseudocode of the greedy maximum coverage (GMC) algorithm. It has been shown that this algorithm achieves an approximation ratio of $1 - \frac{1}{e}$ [8].

6   Progressive Expert Selection

Although the greedy maximum coverage algorithm provides a solution to the $\mathcal{EGS}^{cov}$ problem, it often fails to produce a result of high quality, since it only exploits the relevance scores of each expert for the selection in the case of ties. Motivated by this shortcoming, in this section, we propose an alternative algorithm that takes advantage of the relevance scores in order to return k experts that cover the query terms Q. We call this algorithm *Progressive Expert Selection (PES)*, since it recalculates the scores of experts during its execution and refines the current list of k experts, in order to finally select the k experts with the highest scores.

Since *PES* is based on the selection of the k experts with the highest score values, we must first define an appropriate scoring function that computes each expert's score with respect to the query Q. Ultimately, our objective is to retrieve the k experts with highest score values.

Definition 2. *Given a set of candidate experts $\mathcal{E}(Q)$, the score $s(e_j)$ of expert $e_j \in \mathcal{E}(Q)$ is computed by the formula:*
$$s(e_j) = \sum_{i=1}^{m} log_{avgk_i+2}(w_{ij} + 1)$$
where m is the number of query terms, w_{ij} is the relevance score of expert e_j for term q_i, and $avgk_i$ is the average score among the k higher ranked experts for the i-th query term q_i.

The above definition assumes that at a given time during the algorithm execution, a ranked list of candidate results (experts) exists, i.e., those with highest score values. However, as the algorithm works iteratively and retrieves more data, the scores of the experts are recalculated, and this ranked list may change based on the new knowledge. This behavior is reflected in the $avgk_i$ which is the average score of query term q_i if the currently k highest scoring experts are taken into account. Notice that the value $avgk_i$ changes during the algorithm's execution.

Algorithm 2 describes the pseudocode of the *Progressive Expert Selection* algorithm. The algorithm maintains a priority queue that maintains experts sorted by their score. At any given time, the k first experts in the priority queue can be returned to the user. First, the set of candidate experts $\mathcal{E}(Q)$ is retrieved. Then, for each expert $e_j \in \mathcal{E}(Q)$, we examine each query term $q_i \in Q$ and compute its contribution to the score of expert e_j (lines 6–11). At the end, we insert e_j to the priority queue which is sorted according to the computed score.

Algorithm 2. *Progressive Expert Selection (PES)*

1: **INPUT:** k, $Q = \{q_1, \ldots, q_m\}$
2: **OUTPUT:** Group of experts: $\{e_1, \ldots, e_k\}$
3: $PQ \leftarrow \emptyset$ // Priority queue of experts sorted by score
4: **for** $(e_j \in \mathcal{E}(Q))$ **do**
5:      $score \leftarrow 0$
6:      **for** $(q_i \in Q)$ **do**
7:          $sum \leftarrow 0$
8:          **for** (the first k items $e_x \in PQ$ $(x \in [1, \ldots, k])$) **do**
9:              $sum \leftarrow sum + w_{ix}$
10:          $avgk_i \leftarrow sum/k$
11:          $score \leftarrow score + log_{avgk_i+2}(w_{ij} + 1)$
12:      $PQ.insert(score, e_j)$
13: **return** the first k items of PQ

The complicated part is how we compute the contribution of each query term to the total score of expert e_j. We use Definition 2 for this purpose, but before computing the score of an expert e_j, we compute the average relevance score $(avgk_i)$ for the query term at hand for the k (currently) best experts. This average score is used in order to smoothen the effect of expert's e_j relevance score (w_{ij}) for query term q_i, by taking the logarithm of w_{ij} with base that depends on $avgk_i$. In essence, an expert e_j will be assigned a high score in one of the following two cases:

1. The expert e_j covers a query term q_i that is not covered yet by another expert of the (current) best k experts. In this case, $avgk_i$ is equal to zero, the logarithm has a base equal to two, and the contribution of w_{ij} affects the score much.
2. The expert e_j has a high relevance score w_{ij}, when compared to the average score $avgk_i$ of the (current) k best experts.

Example 2. Consider a set of 4 candidates $\{e_1, \ldots, e_4\} \in \mathcal{E}(Q)$, from which we want to select a group of $k = 2$ experts for $m = 3$ query terms. Let $e_1 : \{(q_1, 0.7), (q_2, 0), (q_3, 0.6)\}$, $e_2 : \{(q_1, 0.3), (q_2, 0), (q_3, 0.2)\}$, $e_3 : \{(q_1, 0.2), (q_2, 0.5), (q_3, 0.4)\}$, and $e_4 : \{(q_1, 0.6), (q_2, 0.7), (q_3, 0.7)\}$. We show only the first 3 steps of the algorithm, due to lack of space.

The algorithm processes each expert one after the other, computes her score, and inserts her in the priority queue. First, the score of e_1 is computed as follows:
$$s(e_1) = log_{0+2}(0.7 + 1) + log_{0+2}(0 + 1) +$$
$$+ log_{0+2}(0.6 + 1) = 1.44$$
and $PQ = \{(e_1, 1.44)\}$. Notice that $avgk_i$ is equal to zero for all q_i, since the priority queue was empty when $s(e_1)$ was computed. Then, e_2 is processed:
$$s(e_2) = log_{0.7+2}(0.3 + 1) + log_{0+2}(0 + 1) +$$
$$+ log_{0.6+2}(0.2 + 1) = 0.45$$
where $avgk_1$ is equal to 0.7 and $avgk_3$ equals 0.6 because the priority queue contains e_1 and these are her relevance scores, w_{11} and w_{31} respectively. Now e_2 is

inserted in the priority queue: $PQ = \{(e_1, 1.44), (e_2, 0.45)\}$. Then e_3 is processed:
$$s(e_3) = log_{0.5+2}(0.2 + 1) + log_{0+2}(0.5 + 1) +$$
$$+ log_{0.4+2}(0.4 + 1) = 1.17$$
where the values $avgk_i$ are now calculated based on the relevance scores of experts e_1 and e_2 who are placed in the priority queue. In fact, $avgk_1 = \frac{0.7+0.3}{2} = 0.5$ and $avgk_3 = \frac{0.6+0.2}{2} = 0.4$. Notice that since no expert in the priority queue covers query term q_2, the contribution of q_2 for expert e_3 (i.e., w_{23}) uses the logarithm base of two, thereby increasing the weight of its contribution to the total score. At the end of the third step, the priority queue has three experts:
$$PQ = \{(e_1, 1.44), (e_3, 1.17), (e_2, 0.45)\}$$

$\square$

The design of *PES* aims to progressively improve the selection of experts during the algorithm's execution. In particular, *PES* always checks the list of k experts that would be returned, at any point during its execution, and inserts new experts with a score that is computed in an intentional manner, in order to reflect that experts with high scores in uncovered query terms are preferred. As will be demonstrated in the experimental evaluation, this is the main reason that explains its good result quality.

7 Experiments

In this section, we evaluate the effectiveness of the proposed framework for expert group selection. All algorithms have been implemented in Java.

7.1 Experimental Setup

Dataset. We parsed DBLP (the version dated October 2013) to extract a set of researchers and we created the pseudo-documents describing their expertise by concatenating the titles of their publications. The associated set of pseudo-documents was indexed using Apache Lucene (version 4.5.0) to allow efficient retrieval. At the end of this pre-processing step, we have stored 1,314,915 pseudo-documents, each one corresponding to one researcher, and the total number of paper titles is 6,498,625. This constitutes the profile database of our framework[1].

Queries. In order to evaluate our framework in a real setting, we created four different query sets $Q_1, \ldots, Q_4$ corresponding to real conferences that took place in 2013. We focused in the area of databases ($Q_1 = $ VLDB'13, $Q_2 = $ ICDE'13, $Q_3 = $ SIGMOD'13), but we also experimented with an information retrieval conference ($Q_4 = $ ECIR'13). These query sets consist of query terms that are collected from the conference web page, and correspond to the areas or topics of interest of each conference. The length (m) of each query set is: 34 for VLDB'13, 34 for ICDE'13, 55 for SIGMOD'13, and 62 for ECIR'13. As an example, part

---

[1] We intend to make data and queries publicly available after the publication of this paper.

of the query for ICDE is: {Data warehousing, Data analytics, MapReduce, Big data, Data Integration, Metadata management,...}. For each query, we set the value k equal to the length of the PC members list of the respective conference (149, 207, 125 and 37 for the aforementioned list respectively), in order to retrieve lists of experts of identical length with the PC members list.

Algorithms. We implemented both *GMC* and *PES*, and a baseline algorithm which creates a pseudo-document d_q that consists of all query terms, computes the cosine similarity $cos(d_q, d_i)$ of d_q with every pseudo-document d_i representing a researcher, and returns the top-k pseudo-documents (researchers) with highest cosine similarity to d_q. We refer to this baseline algorithm as *COS*. Notice that this is practically the approach proposed in [4]. Also, for comparison purposes, in all charts we show the values obtained when using the list of PC members of each conference as result. We refer to this alternative as *PC*.

Metrics. As the evaluation of the results is a hard task of subjective nature, we employ various metrics in order to have as much evidence as possible. All metrics take as input the set of experts produced by each algorithm, compute for each expert a value, and aggregate these values for all experts. Highest values are better. We emphasize that it is not clear how to define the ground truth, and the list of PC members cannot be used as golden standard, since we argue that our approach will produce better results. Thus, our metrics include:

- H-index: aggregates the H-index values of all experts (researchers) in the returned result. We obtained the H-index of researchers by scrapping the web site http://aminer.org/.
- Goodness: The goodness of a list of researchers $L = \{e_1, e_2, \dots\}$ is defined as: $Goodness(L) = \sum_{e_i \in L} score(e_i)$. The value $score(e_i)$ is an indicator of the quality of a researcher e_i and can be defined in various ways. In our evaluation, we define $score(e_i)$ as the number of publications of e_i in high quality venues, also called *reference set*. For example, when we evaluate a DB conference (such as VLDB, SIGMOD, or ICDE), we consider as high quality publications those appearing in {SIGMOD, VLDB, PVLDB, ICDE, EDBT, VLDBJ, TODS, TKDE}. In the case of IR conferences (such as ECIR), we consider {SIGIR, ECIR, CIKM, InfRetr, InfProcMan, SIGIRforum}.
- Lucene score: This score takes as input a list of experts $L = \{e_1, e_2, \dots\}$ produced by an algorithm, and examines each individual query term q_i. For each query term q_i, we aggregate the scores returned by Lucene of all experts in L for this specific query term. Then, we compare the scores for q_i that each algorithm (or result list) achieved, and the winner increases its Lucene score by one unit. Thus, the Lucene score of an algorithm takes integer values in the range: $[0, m]$, and the interpretation of its value is that it represents the number of query terms in which the list L of the algorithm achieved the highest value compared to the other algorithms.

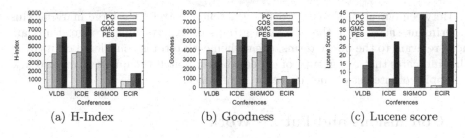

(a) H-Index   (b) Goodness   (c) Lucene score

Fig. 2. Experimental results for (a) H-index, (b) Goodness, and (c) Lucene score

7.2   Experimental Results

H-index Metric. Figure 2(a) shows the obtained results using the H-index metric. This metric is an indicator of the quality of the returned experts, as it measures the impact of the publications of a researcher by means of the gathered citations. In all cases, our algorithms *PES* and *GMC* outperform the baseline method and the list of PC members, often showing 100 % improvement in the value of H-index. The baseline (*COS*) cannot retrieve results of high quality. Perhaps surprisingly, the worst performing result is obtained when using the list of PC members. This result indicates that the H-index values of at least some of the PC members is lower than the experts returned by our algorithms.

Between our algorithms, *PES* achieves higher H-index values than *GMC*, due to the more elaborate scoring scheme which is computed adaptively. Instead, *GMC* often needs to make a greedy decision on which is the next expert to be included in the result, and this affects the selection of the next experts who must now compete for a subset of the query terms. In general, we have observed that *PES* performs better than *GMC* consistently, with respect to the H-index metric which is a widely accepted metric and is considered relatively objective.

Goodness Metric. Figure 2(b) depicts the results obtained using the Goodness metric. For SIGMOD and ICDE, our algorithms outperform both the baseline and the PC members list. This is consistent with the results of the H-index metric shown above. In the case of VLDB and ECIR our algorithms still outperforms the PC members list, however *COS* is marginally better. Also, we observe that only in one occasion (SIGMOD) *GMC* achieves a higher Goodness value than *PES*. In all other cases, *PES* is better, in consistence to Fig. 2(a).

Lucene Metric. Figure 2(c) shows the results acquired when the Lucene score is considered. This score reflects the number of query terms (out of m) that each algorithm achieved the highest score among the other algorithms, where the score is computed as the sum of scores of all returned experts for the specific query term q_i. Both *PES* and *GMC* are much better than the competitors, which demonstrate very low values, often equal to zero. The interpretation of this zero value is that the competitors did not manage to surpass our algorithms in any query term q_i. Put differently, always our algorithms returned experts with higher scores for every q_i.

Between our algorithms, *PES* is consistently better by a large margin. This experiment shows that the retrieved groups of experts by our algorithms are more relevant to the query terms, even when each term is considered on its own. This indicates that our groups of experts better match the query terms and also the topics of interest of the respective conference.

8   Conclusions and Future Work

In this paper, we introduced a framework for profile-based expert group selection, which relies on the existence of a profile database that characterized experts and is parameterized by an optimization goal. We focused on the problem of automatic construction of PCs for research conferences, where a set of k experts/researchers is required, and the optimization goal is coverage of the topics of the conference. The experimental evaluation on real-world data based on DBLP shows that the experts selected by our algorithms outperform both the result of a baseline algorithm and the list of PC members of the respective conference. In our future work, we intend to inject other optimization goals and integrate the necessary algorithms for solving the associated problems, such as maximizing diversity or dealing with combinations of optimization goals.

References

1. Balog, K., Azzopardi, L., de Rijke, M.: Formal models for expert finding in enterprise corpora. In: Proceedings of SIGIR, pp. 43–50 (2006)
2. Balog, K., de Rijke, M.: Non-local evidence for expert finding. In: Proceedings of CIKM, pp. 489–498 (2008)
3. Bast, H., Chitea, A., Suchanek, F.M., Weber, I.: ESTER: efficient search on text, entities, and relations. In: Proceedings of SIGIR, pp. 671–678 (2007)
4. Demartini, G., Gaugaz, J., Nejdl, W.: A vector space model for ranking entities and its application to expert search. In: Boughanem, M., Berrut, C., Mothe, J., Soule-Dupuy, C. (eds.) ECIR 2009. LNCS, vol. 5478, pp. 189–201. Springer, Heidelberg (2009)
5. Fang, H., Zhai, C.X.: Probabilistic models for expert finding. In: Amati, G., Carpineto, C., Romano, G. (eds.) ECIR 2007. LNCS, vol. 4425, pp. 418–430. Springer, Heidelberg (2007)
6. Gollapalli, S.D., Mitra, P., Giles, C.L.: Ranking experts using author-document-topic graphs. In: Proceedings of JCDL, pp. 87–96 (2013)
7. Merelo-Guervós, J.J., Castillo-Valdivieso, P.: Conference paper assignment using a combined greedy/evolutionary algorithm. In: Yao, X., Burke, E.K., Lozano, J.A., Smith, J., Merelo-Guervós, J.J., Bullinaria, J.A., Rowe, J.E., Tiňo, P., Kabán, A., Schwefel, H.-P. (eds.) PPSN 2004. LNCS, vol. 3242, pp. 602–611. Springer, Heidelberg (2004)
8. Hochbaum, D.S.: Approximating covering and packing problems: set cover, vertex cover, independent set, and related problems. In: Hochbaum, D.S. (ed.) Approximation Algorithms for NP-Hard Problems, pp. 94–143. PWS Publishing Co., Boston (1997)

9. Hofmann, K., Balog, K., Bogers, T., de Rijke, M.: Contextual factors for finding similar experts. JASIST **61**(5), 994–1014 (2010)
10. Karimzadehgan, M., Zhai, C.: Constrained multi-aspect expertise matching for committee review assignment. In: Proceedings of CIKM, pp. 1697–1700 (2009)
11. Karimzadehgan, M., Zhai, C., Belford, G.G.: Multi-aspect expertise matching for review assignment. In: Proceedings of CIKM, pp. 1113–1122 (2008)
12. Robertson, S.E.: The probability ranking principle in IR. In: Robertson, S.E. (ed.) Readings in Information Retrieval, pp. 281–286. Morgan Kaufmann Publishers Inc., San Francisco (1997)
13. Serdyukov, P., Rode, H., Hiemstra, D.: Modeling multi-step relevance propagation for expert finding. In: Proceedings of CIKM, pp. 1133–1142 (2008)

Tracking and Re-finding Printed Material Using a Personal Digital Library

Annika Hinze[(✉)] and Amay Dighe

Department of Computer Science, University of Waikato, Hamilton, New Zealand
hinze@waikato.ac.nz, aad11@students.waikato.ac.nz

Abstract. Most web searches aim to re-find previously known information or documents. Keeping track of one's digital and printed reading material is known to be a challenging and costly task. We describe the design, implementation and evaluation of our *Human-centred workplace* (HCW) – a system that supports the tracking of physical document printouts. HCW embeds QR codes in the document printout, stores the documents in a personal Digital Library, and uses cameras in the office to track changes in the document locations. We explored the HCW in three evaluations, using the system over several weeks in an office setting, a user study in a lab environment, and extensive functional tests.

Keywords: Search · Re-finding · Physical documents

1 Introduction

A study with knowledge workers based in the UK and the US found that 83 % of them felt that they wasted time each day on issues of document collaboration [16]. 73 % of knowledge workers reported wasting work time looking for files. Another study observed that knowledge workers spent 20 % of their time searching for hard copies of documents, and that 50 % of the time they did not find what they wanted [3]. It is estimated that the average organization makes 19 copies of each document (37 % being unnecessary, 45 % being duplicates) and loses one out of every 20 documents [15]. Our work addresses the superfluous printing and copying of duplicate documents, as well as the problem of re-finding previously printed copies.

Digital documents are typically managed electronically, while paper documents are mostly organised and managed manually. This leaves users to develop their own strategies for storage and retrieval of physical documents. Ironically, often the use of computers compounds this problem by making it easier to print a new version of a document that is not found immediately. Additionally, reading paper-based documents preferred by many as it offers the flexibility to read anywhere and is also easier to mark up [9]. So even though the majority of documents may now be digital, people still maintain physical copies, which then have to be kept track of and located.

This paper describes the design, implementation and evaluation of our Human-centred workplace (HCW) – a system that enables the tracking of physical printouts of documents using a personal digital library. The concept of

© Springer International Publishing Switzerland 2016
N. Fuhr et al. (Eds.): TPDL 2016, LNCS 9819, pp. 94–106, 2016.
DOI: 10.1007/978-3-319-43997-6_8

this system had been briefly introduced previously [8]. This paper contributes a description of the actual implementation, deployment and evaluation.

The remainder of this paper is organized as follows: Sect. 2 describes the design and implementation of the HCW, while Sect. 3 illustrates the interface and interaction design. Section 4 describes the system evaluation. Section 5 discuses related approaches, while Sect. 6 addresses differences to related approaches, insights of the evaluation for further research and the planned extensions and further steps in our research. The paper concludes with a brief summary.

2 Design and Implementation

The design concept of the HCW was briefly introduced in [8]; here we provide more details and implementation information. We identified five functional requirements, based on our discussion in the introduction (an extensive discussion of requirements and implementation can be found in [4]). These form the basis for our implementation as well as the exploration of related approaches (see Sect. 5). The first three requirements refer to the systems core functionality of tracking, search and recording printing: *(R1) Tracking Document Location:* Tracking physical document location is the core functionality we aim for to support the task of re-finding documents and avoiding having to re-print them. *(R2) Digital Search:* There needs to be a search interface to support re-searching and re-finding of physical documents. *(R3) Keeping record of printed documents*: We wish to track mostly printed documents, but also other physical documents. Keeping track of print-outs would avoid the need to reprint a document and thus avoid duplication of the document. The remaining two requirements refer to the manner in which the R1–R3 are to be achieved: *(R4) No Order to Follow:* Approaches that require users to follow a pre-defined archival methodology or to be generally orderly have been shown to fail; many people will not follow procedures, however sensible these may be. *(R5) No Special Hardware:* The system should not require any special hardware so that it can be installed in ordinary offices of knowledge workers.

Figure 1 shows the architecture of the HCW system, designed to fulfill these five requirements. It consists of three elements: Document Manager, Document Tracker, and Document Search. Not shown are the pre-existing elements of office document printer and web cams, which are used for monitoring documents. The dataflow sequence of HCW is as follows: as the user signals the intention to

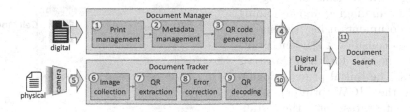

Fig. 1. HCW architecture and data flow

print a digital document (step 1), the document's metadata are obtained (step 2), and encoded into a unique QR code (step 3), and added automatically to the document's front page (step 4). The document metadata are added to the HCW database (see step 5), and the user may then read/move the document within the workplace. The cameras monitor the workplace (step 6) and continuously record images (step 7); the images are analysed for QR codes (step 8), and after error correction (step 9), they are decoded and the document's location is recoded in the digital library based on the areas covered by the cameras (step 10). The user searches for and re-finds the document via the HCW search interface (step 11).

The implementation uses Microsoft .net and C#. Two web cams are used: a simple wired web cam as found in typical office settings and a wireless high-resolution camera. The cameras' fields of vision are semantically encoded to refer to different office areas such as desk, floor, and table. Printer++ is used as a virtual printer to receive the user's print request (www.printerplusplus.com) and Stroke Scribe (http://strokescribe.com) is used for QR code generation. The QR code uses document header information as metadata. It is placed on a separate front page of the printed document. We experimented with different sizes for the QR code – the one seen in Fig. 2 is the minimal

Fig. 2. Front page

size for recognition in a typical office environment in which cameras are between two and five meters from the documents. QR decoding in the Document Tracker takes an image of the desk surface genrated by the camera and performs a simple five-step algorithm. First the image is converted to grey-scale to reduce the processing load. A Canny operation highlights the object edge (leaving the background black) and the barcode is extracted from the edge image. The barcode is read, and decoded (using the Aspose SDK, www.aspose.com), and sent to the Library to check for a matching document. The documents are included in a personal Digital Library (using Greenstone software, www.greenstone.org) with an extended metadata database to capture the location images and QR code information. More technical details are available in [4].

3   Interface and Interaction

In this section, we show the HCW interface and user interactions, and highlight the benefits of using the HCW for managing and re-finding paper documents using a scenario.

Let's consider a student printing documents for their Master's studies. The initial print dialogues (using the HCW printer) seamlessly integrates into the established workflow. The student is

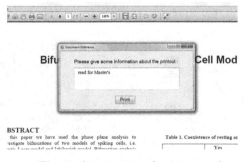

Fig. 3. Annotation of print record

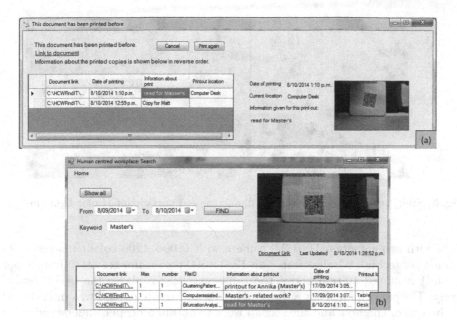

Fig. 4. (a) Duplicate printing warning, (b) search for printouts

prompted by the system to enter a short description about the print-out's purpose, see Fig. 3, indicating whether this copy is for their own reading or for someone else. If this document has already been printed, HCW warns about this potential duplication (see Fig. 4(a)), allowing our student to cancel printing and find the previous copy or to print again, e.g., to give a copy to someone else. Finally, when a previously printed document cannot easily be re-found, HCW can be used to trigger the student's memory with the purpose and last location of the print-out, see Fig. 4(b). Additionally, the use of HCW builds a personal digital library of reading material, which can be searched and browsed using the existing library interface.

4   Evaluation

We evaluated the HCW software prototype to explore to what extent it satisfies our goals of helping knowledge workers in ordinary office environments to re-find their documents. We carried out three studies: (1) an office-based single-user study over two weeks, (2) a lab-based study with 10 participants, (3) qualitative functionality tests.

4.1   Single-User Study (Office-Based)

The prototype was used by one academic knowledge worker regularly for two weeks for printing and tracking of student submissions, project work and publications. The software was set up in their office (see Fig. 5) on a Dell OptiPlex

Fig. 5. Study setup: cameras circled red, anonymised participant (Color figure online)

9020 with two cameras (USB 2.0 camera with 1600×1200 colour images at 25 frames per second; wireless web cam 1280×800 colour images at 30 frames per second) mounted 100 cm above the table and 125 cm above the desk, respectively. The participant kept a diary of events and incidents and was interviewed at the end of the first and the second week to obtain a deeper understanding of the participant's experiences and gain feedback about the system.

Feedback and Results. The study was performed during a very busy period in the participant's work. Even though they did not fill in the diary as diligently as was hoped by the researchers, detailed oral feedback was obtained. During the study period, more than 35 documents were printed (and thus entered into the HCW system). Four documents were purposely printed twice to be shared with colleagues. The participant expressed satisfaction with the front page of the document printout, stating it "provides sufficient information to identify the document" and "makes it easy to differentiate from other documents." They noticed that the print phase took a "little more time" than for ordinary printing, as the HCW processing delayed the printing start by a few seconds. The participant observed that they looked for a number of document print-outs several times "for referencing purposes" during the study period. As this was a very busy time the participant failed to note how many documents and printouts they tried to locate.

When a printout was not immediately visible on the desk, the participant confessed to the habit of reprinting the document. They found HCW's automatic warning about document duplication was a "useful feature" to reduce reprinting, and reported that the printout annotation and location information given by HCW helped trigger their memory as to the purpose of the document and also helped them find the printout if it was in the office. The participant explicitly praised the "simplicity of user interface" for finding physical documents, stating that "it was easy to understand" and a "simple to interact user interface." They noted that the availability of different searching parameters (such as keyword search and between-two-dates search) made the search "more accurate" and "targeted." They reported that "document search was generally successful" but that sometimes the recorded camera images would "show two documents at one place" (i.e., more than one document is shown in the image) in which case they "did not know which one is mine." They suggested highlighting the correct physical document in the image.

Overall, the participant found the HCW system "convenient" and "useful." They emphasized that "the software makes sense" and felt it helped them manage their documents in the workplace.

Functionality and Changes. During the two weeks of running the HCW prototype, occasional misfunctions were observed. Very long documents would sometimes not print – this was due to a malfunctioning print spooler service which was fixed during the study. Occasionally the QR code would appear to be shrunk, which led to difficulties in decoding. This was traced back to documents with more than 500 characters in the first few lines on the first page. This was addressed by lowering the error correction parameter in the QR encoding to allow for greater storage capacity of the QR code.

4.2   Lab-Based Study

The lab-based study used the improved software. Again, cameras were mounted 100 cm above a table and 125 cm above a main desk respectively, see Fig. 6. The study had 10 participants UP1–UP10 (6 female, 4 male) aged 18 to 50 years. We invited participants from a variety of backgrounds who were familiar with computer use (2 arts & social science, 4 management, 3 ICT and 1 earth science; 9 students and 1 professional). In an introductory interview, each of them reported often having to search for documents they had previously printed, spending up to three hours on document search in some cases. The study was designed around a set of tasks, and followed by a short interview. Each participant was given three tasks: (1) print the first copy of a document, (2) print a second copy of the document, and (3) find the location of the document.

Fig. 6. Study setup: cameras circled red (Color figure online)

Printing 1st Copy. All 10 participants found this process simple. Seven participants mentioned that while they appreciated the request for annotations on the print-out, they felt they needed greater familiarity with the system in order to better predict what sort of annotations would prove most helpful. Two participants felt the request for additional information held them back in their purpose of printing a document. They were not sure if the information they provided would help later. One participant wished to use language-specific characters, which were not supported. Four participants found the printing less convenient due to the delay in having to enter additional information and the short additional delay for QR code encoding. UP8 and UP9 suggested the use of a progress bar to indicate the impending commencement of printing. Eight participants found the front page sufficient to identify the document; the other two participants did not provide specifics about which information they would wish to include.

Printing 2nd Copy. All 10 participants noted that HCW's notification of an earlier printout together with its location caused them to reconsider whether a second copy was indeed needed. Nine participants re-found the previous printout and one participant reprinted the document. UP5 expressed that "avoiding unnecessary reprints of the document is a very useful feature, as it would help me to avoid having multiple copies of the same document around." UP8 commented that HCW "encourag[ed] using the existing copies [rather] than printing [a] new copy." Five participants felt that the process of re-finding a document was not time-consuming, the other five felt that re-printing would have been faster. UP4 observed that the front page of a reprinted document is identical to the original printout and suggested providing copy number and date of reprinting to distinguish physical copies of the same documents. UP8 suggested providing more information about the document on the front page.

Finding a Printed Copy. Seven participants were successful in finding a printout the researchers had placed in the lab based on the information provided in HCW. Eight found the process effective and was not time consuming. Three had difficulty using the search window efficiently and needed to ask the researcher for help; these users suggested that the user interface layout should be more informative. UP3 suggested an option to check the functionality of every connected camera placed in the workplace.

All ten participants stated they were excited about the idea of automatically keeping track of their desk papers. Five found the system convenient and described the system as "very useful." UP1 gave feedback that "the system is amazing; it will help to keep track of each and every document" and that the system made it "easier to find papers on the desk, simply by showing the picture of the desk the paper is on." UP9 expressed that they found the "system convenient and useful for a forgetful person like myself. Not only does it help to find printed document or where my file is, it also helps the environment by avoiding re-printing." UP8 found the system "very useful as I could see which documents have been printed earlier."

4.3 Functional Quality Evaluation

Reading QR codes at an angle was found to have a higher reading error rate. We tested a 10×10 cm QR code at a distance of 110 cm from the camera. A document presented to the camera at an angle of $0°$ deviation was read successfully in all tested cases. An angle of $10°$ read 4 of 5 documents and $20°$ was successful in 3 of 5 attempts. At $30°$ or more, successful reading cannot be guaranteed (only 1 in 5 for $30°$, none for $45°$). When the document is positioned at 125 cm from the camera, the success rate at $20°$ dropped to 2 of 5. These can be improved by enlarging the QR code, but 16×16 cm is a natural limitation for QR code on A4 paper. The system still takes about 2 to 3 s to recognize the QR code. Best results are therefore achieved when the users pause briefly between adding each document to a pile of papers. Additional tests are described in [8].

5 Related Approaches

We here present an analysis of related work based on our five requirements (see Sect. 2). The subsequent Sect. 6 then provides a comparison to our HCW system and discusses implications and open research issues.

SOPHYA is a physical document collection system which utilises a wired technology for managing and retrieval of physical documents and artefacts within the collection [13]. SOPHYA thus provides a means of linking the management of real world document artefacts (e.g. folders) with their electronic counterparts, so that document management activities such as filing, locating, retrieving document can be supported. The system uses specially designed hardware shelves and physical document containers for holding documents. SOPHYA supports unordered (piling) [12] and ordered (filing) [13] document collections in two different system implementations. Our notion of *filing* and *piling* of documents follow Henderson and Srinivasan's concepts [7]. The connection between the container and the location of the container is established with electronic circuitry. Each folder has an allocated physical location within a container. An LED on the surface of the container acts as a user interface to indicate that the required document is in the container. Firmware embedded in the physical storage location communicates with the container (e.g., by reading IDs of the containers and controls the user interface). The firmware also communicates with the middleware, which maintains a simple database to keep track of information in the container and the physical location of the container. For our scenario of non-disciplined knowledge workers, SOPHYA has a number of limitations. First the documents still have to be placed in a particular container to be located so it does not provide flexibility and a particular procedure needs to be used. Secondly, metadata need to be entered and maintained manually and this is time consuming. HCW aims to cater for real-life situations in which people deposit their physical documents anywhere in the office and need to recover them easily.

PaperSpace is a document management system that maintains a link between the printed document and its digital counterpart [17,18]. PaperSpace works with operation codes (in the shape of small graphic icons) printed in the margins of each page of the document. PaperSpace uses a medium resolution webcam to recognise the papers. The system features other functionalities such as capturing and parsing gestured operation performed on the (paper) command bar. The bar image provides linking functions between the paper document and its digital counterpart, and users can directly manipulate the digital document using their printouts. The PaperSpace system provides an innovative interface for linking physical and digital copy. Its approach to enhance the print copy with annotations is closest related to HCW's use of QR codes. However, PaperSpace does not provide any assistance to *re-find* the paper version of a document once printed.

Video-based document tracking identifies paper documents on a desk and automatically links them to the corresponding electronic documents [14]. A camera is mounted above the desk to capture and track the document movements. The video is analysed using a computer vision technique for document recognition that enables every paper document on the desk to be linked to its electronic copy. In the system, the document representations can be searched using keywords or by manipulating the image of the desk. The system's advantage is its technical simplicity: it does not involve tags or special readers. However, only one document can be placed or removed from the stack at a time. It is also assumed that every document placed on the desk is unique.

DocuDesk uses interactive desk technology to establish relationships between the digital and physical documents [5]. The DocuDesk uses an interactive desk and overhead video Infra-Red camera. In DocuDesk there are two ways of linking the document with its digital counterpart, by 2D barcode or 1D barcode. A camera above the desk records an image of the document and, using image recognition, a link with the digital counterpart is created. On placing the document on the DocuDesk, the user is given various options such as email and link. The email option sends the digital copy of the physical book, while the link option attaches additional digital media to the book. DocuDesk does not provide tracking and search functions for physical documents.

Limpid Desk is a visualization tool that allows its users to "see" the contents of a stack of documents; in particular, it allows a user to "see" contents of documents further down in the stack without the top layer needing to be removed [10,11]. The upper layer is *transparentized* and users can find desired documents even if they are hidden in the document stack. The hardware used in Limpid Desk includes Projector, Camera and thermo-camera. When the user touches a document on the desk the system detects the touch (via the thermo-camera) and then the upper layer document is virtually transparentized by projection. The Limpid Desk supports physical search interaction techniques, such as 'stack browsing' in which the upper layer documents are transparentized one by one through to the bottom of the stack. The Limpid Desk system meets our requirement of giving simple access to physical documents. As the user can visually access a lower layer document without removing the document on the top, the limpid desk is a possible solution to the problem of finding a document in a pile.

The Fused Library uses RFID tags to link physical items with content in a digital library [2]. RFID tags are placed underneath a desk, allowing identification of the user's location (using laptop-based RFID readers). Depending upon the user's current location, the library catalogue will present the user with a tailored home page including a quick link to related useful sections in that location. The library catalogue will highlight the books near the user's location. The fused library uses concepts of *physical hypermedia*, for which a user's context (e.g., their location) triggers links to digital material [6]. The Fused Library is a library-based system that meets our requirements of tracking location of user and documents. However it does not keep track of printed documents as such. As offices are typically much less structural than say a traditional library, locating physical and digital object across the workplace would be challenging using the fused library approach.

6 Discussion

This section brings together the discussion of related work in light of the requirements and the HCW system, further comments on the user studies, and aspects of future work.

Related Work. Table 1 provides an overview of the main results of our related work discussion with respect to the system requirements. For comparison, the table also contains information about our HCW system (last row). As can be seen from the table, most related systems provide document tracking and digital search. However, only PaperSpace (in addition to HCW) keeps records of printed documents. Additionally, most systems require the user to employ special hardware and/or to follow some pre-defined methodology. Some hardware is required in all cases, however, PaperSpace, Video tracking and HCW use simple hardware already existing or easily installed in ordinary offices instead of custom-built gear. Tracking document locations using these low-key hardware options is harder to implement and remains quite challenging. Overall, none of the existing systems were suited to address the problems described and the requirements as identified previously. HCW addresses all five requirements and, similar to issues discussed for PaperSpace, its tracking of document locations could be improved through further research.

Implications of User Studies. Although 10 participants are not sufficient for statistical evaluation, they provide indicative observations. The participants with backgrounds other than computer science focused more on the overall outcome

Table 1. Systems for re-finding physical documents

	Tracking Document Locations	Digital Search	Keeping Record of Printed Document	No Order to Follow	No Special Hardware
SOPHYA	✓	✓			
Fused Library	✓	✓			
Limpid Desk				✓	
Paper Space	(✓)	✓	✓		(✓)
Document Video Tracking		✓		✓	(✓)
DocuDesk	✓	✓			
HCW	(✓)	✓	✓	✓	(✓)

and benefit of the system (e.g., "It is very cool, [it] will be of great help to organize and search the physical documents"), while participants with IT background were more critical of the operational aspects. They seemed to find it harder to accommodate even a small system delay and were more analytical of the systems performance. Participants from other backgrounds on average took 20 min to complete the user study; the CS participants took about 30 min. The researchers had the impression that both studies were somewhat hampered by the use of the system in a one-off limited-time manner. The true benefit will only become apparent after sufficient time has elapsed so that the location of paper copies and the purpose of printouts had been forgotten. This would change the motivation for the participants, especially if they could be sure that the system functionality, the digital copies in the Digital Library and the provided information about printouts would be available in future. In this respect the system is akin to augmented memory systems that encounter similar challenges for effective evaluation. Furthermore, the aspect of building a personal library is not yet studied in any detail as similarly the benefits would be of a more long-term nature.

QR Code Quality. Similar to Sallam's observations [17], we noted that even small delays, as caused by our QR code reading and their tag reading, are irritating to users and will not be easily accommodated through changed user behaviour. We are therefore exploring a number of ideas for improving the readability of QR codes from a distance beyond the simple (and limiting) increase in QR size. Alternative methods for marking paper print-outs for tracking to be explored are marginal markings, similar to the tags used in PaperSpace [17], in combination with QR codes.

Integration into Personal Digital Library. The HCW system would be best used not as a stand-alone digital library merely for printed documents but for tracking reading material. In [1], we introduced such a system for tracking academic reading, which currently only covers digital documents. Merging these two approaches to personal digital libraries is one of our future research goals. Similarly, a closer integration into scholarly workflows (finding, reading, annotating, writing) is desirable. We wish to improve the current user interface and explore whether a closer integration into the Digital Library interface would be beneficial. The current annotation of locations is only very rudimentary – greater flexibility seems desirable but its impact on non-technical end-users needs to be explored.

7   Summary and Conclusions

We live in a digital age though many still use paper copies of documents every day for convenience. Our research is motivated by a number of factors: lost documents with valuable annotations, time wasted searching for print copies of documents, and the wish to save trees by reducing the number of duplicate paper printouts. We aimed to find a solution that does not require knowledge workers to follow yet another well-intentioned new methodology or structure in ordering their material, nor does it necessitate the acquisition of expensive hardware. We are further interested in automatically building a personal digital library not through explicit ingest of documents but through the use of previously available information from the users' workflow.

This paper described our HCW prototype that supports the management and re-finding of physical documents. We implemented a software prototype and explored its effectiveness in two user studies and together with an exploration of its functional qualities. Our current studies focused on testing convenience and feasibility of HCW system itself and the explicit interactions. Studies of longer term use of the system would allow an exploration of annotation types used to describe the print-outs (possibly allowing for predefined categories to speed up this step), and to test the impact of workflow patterns on the personal digital library and its use. However, already from these three studies it becomes clear that the concept of the *Human-centred workplace* may successfully address the issues of re-finding printed documents and help avoiding repeated re-printing.

Its better integration with a personal digital library for managing reading material opens up further applications beyond tracking documents, and would make this system a useful element in the established workflow of academics and other knowledge workers. We also identified areas for software improvement such as more effective frame rate for QR recognition, and support for reading documents at greater distance and at an angle. Future work plans are manifold, such as the exploration of methods to track the document piles, and the plans outlined in the discussion.

References

1. Al-Anazi, M., Hinze, A., Vanderschantz, N., Timpany, C., Cunningham, S.J.: Personal digital libraries: keeping track of academic reading material. In: Tuamsuk, K., Jatowt, A., Rasmussen, E. (eds.) ICADL 2014. LNCS, vol. 8839, pp. 39–47. Springer, Heidelberg (2014)
2. Buchanan, G., Pearson, J.: An architecture for supporting RFID-enhanced interactions in digital libraries. In: Lalmas, M., Jose, J., Rauber, A., Sebastiani, F., Frommholz, I. (eds.) ECDL 2010. LNCS, vol. 6273, pp. 92–103. Springer, Heidelberg (2010)
3. Center, C.M.: Document management overview: document imaging in the new millenium. Technical report, Laserfiche (2007)
4. Dighe, A.: Human-centred workplace: re-finding physical documents in an office workplace. Master's thesis, Computer Science, University of Waikato (2014)
5. Everitt, K.M., Morris, M.R., Brush, A.B., Wilson, A.D.: Docudesk: an interactive surface for creating and rehydrating many-to-many linkages among paper and digital documents. In: 3rd IEEE International Workshop on Horizontal Interactive Human Computer Systems, 2008, TABLETOP 2008, pp. 25–28. IEEE (2008)
6. Grønbæk, K., Kristensen, J.F., Ørbæk, P., Eriksen, M.A.: Physical hypermedia: organising collections of mixed physical and digital material. In: Proceedings of the Fourteenth ACM Conference on Hypertext and Hypermedia, pp. 10–19. ACM (2003)

7. Henderson, S., Srinivasan, A.: Filing, piling & structuring: strategies for personal document management. In: 2011 44th Hawaii International Conference on System Sciences (HICSS), pp. 1–10. IEEE (2011)
8. Hinze, A., Dighe, A.: Re-finding physical documents: extending a digital library into a human-centred workplace. In: Zaphiris, P., Buchanan, G., Rasmussen, E., Loizides, F. (eds.) TPDL 2012. LNCS, vol. 7489, pp. 57–63. Springer, Heidelberg (2012)
9. Hinze, A., McKay, D., Vanderschantz, N., Timpany, C., Cunningham, S.J.: Book selection behavior in the physical library: implications for ebook collections. In: 12th ACM/IEEE-CS Joint Conference on Digital Libraries, JCDL 2012, pp. 305–314. ACM (2012)
10. Iwai, D., Sato, K.: Limpid desk: see-through access to disorderly desktop in projection-based mixed reality. In: Proceedings of the ACM Symposium on Virtual Reality Software and Technology, pp. 112–115. ACM (2006)
11. Iwai, D., Sato, K.: Document search support by making physical documents transparent in projection-based mixed reality. Virtual Real. 15(2–3), 147–160 (2011)
12. Jervis, M., Masoodian, M.: Digital management and retrieval of physical documents. In: Proceedings of the 3rd International Conference on Tangible and Embedded Interaction, pp. 47–54. ACM (2009)
13. Jervis, M., Masoodian, M.: SOPHYA: a system for digital management of ordered physical document collections. In: Proceedings of the Fourth International Conference on Tangible, Embedded, and Embodied Interaction, pp. 33–40. ACM (2010)
14. Kim, J., Seitz, S.M., Agrawala, M.: Video-based document tracking: unifying your physical and electronic desktops. In: Proceedings of the 17th Annual ACM Symposium on User Interface Software and Technology, pp. 99–107. ACM (2004)
15. Nollkamper, P.E.: Fundamentals of Law Office Management: Systems, Procedures, and Ethics. Cengage Learning, Boston (2004)
16. Perforce: the case for better document collaboration. Technical report (2013). perforce.com
17. Sallam, S.: Paperspace: a novel approach to document management by combining paper and digital documents. Master's thesis, Computer Science, University of Saskatchewan (2006)
18. Smith, J., Long, J., Lung, T., Anwar, M.M., Subramanian, S.: Paperspace: a system for managing digital and paper documents. In: CHI 2006 Extended Abstracts on Human Factors in Computing Systems, pp. 1343–1348. ACM (2006)

ArchiveWeb: Collaboratively Extending and Exploring Web Archive Collections

Zeon Trevor Fernando[(✉)], Ivana Marenzi, Wolfgang Nejdl, and Rishita Kalyani

L3S Research Center, Hannover, Germany
{fernando,marenzi,nejdl,kalyani}@L3S.de

Abstract. Curated web archive collections contain focused digital contents which are collected by archiving organizations to provide a representative sample covering specific topics and events to preserve them for future exploration and analysis. In this paper, we discuss how to best support collaborative construction and exploration of these collections through the ArchiveWeb system. ArchiveWeb has been developed using an iterative evaluation-driven design-based research approach, with considerable user feedback at all stages. This paper describes the functionalities of our current prototype for searching, constructing, exploring and discussing web archive collections, as well as feedback on this prototype from seven archiving organizations, and our plans for improving the next release of the system.

Keywords: Working with web archives · Collaborative search and exploration

1 Introduction

The web reflects a considerable part of our society and is becoming an important corpus for studying human society by researchers in the humanities, social sciences, and computer sciences alike. Web archives collect, preserve, and provide ongoing access to ephemeral web pages and hence encode important traces of human thought, activity, and history. Curated web archive collections contain focused digital content from specific organizations, related to specific topics or covering specific events, which are collected to provide representative samples and preserve them for future exploration and analysis. However, there have been only a few concerted efforts to provide tools and platforms for exploring and working with such web archives.

Within the ALEXANDRIA project[1] we aim to develop models, tools and techniques not only to archive and index relevant parts of the web, but also to retrieve and explore this information in a meaningful way. This paper focuses on the ArchiveWeb platform which supports searching, collecting, exploring and discussing web archive collections such as the ones provided through the web archiving service of the Internet Archive, Archive-It[2].

---

[1] http://alexandria-project.eu/.
[2] http://archive-it.org/.

© Springer International Publishing Switzerland 2016
N. Fuhr et al. (Eds.): TPDL 2016, LNCS 9819, pp. 107–118, 2016.
DOI: 10.1007/978-3-319-43997-6_9

Archive-It provides a subscription web archiving service that helps organizations to harvest, build, and preserve collections of digital content. Currently the Archive-It system is mainly used by librarians and curators in order to build their collections. Less support is given to users and domain experts who actually want to work with the collections, or to the general public. ArchiveWeb[3] aims to provide facilities to collaboratively explore and work with web collections in an interactive and user friendly way, both for research and for learning.

In summer 2015 we carried out a preliminary analysis of user requirements during which we interviewed representatives of the Internet Archive as well as experts working in seven archiving organizations and libraries active in web archiving. In most cases the motivation for starting the archiving of digital collections had been to preserve institutional or government websites, or collecting and preserving research outputs (for example faculty members hosting their funded projects on external websites including assets related to the project such as documents and research datasets). Several institutions also curate special collections received as donations of materials from individuals or organizations. Traditionally these materials were paper manuscripts, but they increasingly include digital and multimedia materials as well as social media and web presence.

All respondents confirmed that the current efforts at their institutions are mainly carried out by individual curators or subject specialists who are responsible for individual collections and archiving requests. In some cases there is a collection development executive (CDE) group, as part of the library organization, that coordinates collection building and development. Collaboration is an important aspect in order to avoid duplicating resources and efforts, one of the challenges being "figuring out how to not duplicate efforts that are going on elsewhere, but also how to still build things that are useful". Some curators are trying to find ways of collaborating both in terms of collecting and displaying information, sharing metadata and schema creation, user folksonomies and tagging, to make the collections come together and be more alive. Easy annotation of resources in collections is a feature that many experts would be happy to explore.

On the basis of this preliminary study, and building on previous work we did for collaborative learning environments [1,10–12,19] we designed and built the ArchiveWeb system to support collaborative creation and enrichment of web archive collections, with a focus on user interface and searching/sharing functionalities. We ingested 200 web archive collections from Archive-It, and asked our archiving partners for their input. Evaluation was done based on a task-based evaluation design, and analyzing both quantitative interaction data (query log), as well as qualitative feedback in the form of interviews.

In the following section, we provide a short overview of related work, and in Sect. 3 we describe some web archive collections we are working with. Section 4 includes both a description of the main functionalities of our system and of our evaluation design, a detailed discussion of the evaluation results, as well as directions for future work.

---

[3] http://archiveweb.l3s.uni-hannover.de/aw/index.jsf

2 Related Work

Tools for supporting search and exploration in web archives are still limited [4]. Most desired search functionalities in web archives are fulltext search with good ranking, followed by URL search [17]. A recent survey showed that 89 % of web archives provide URL search access and 79 % give metadata search functionalities [7]. Some existing projects that provide limited support for web archive research are discussed below.

The Wayback Machine[4] is a web archive access tool supported by the Internet Archive. It provides the ability to retrieve and access web pages stored in a web archive through URL search. The results for each URL are displayed in a calendar view which displays the number of times the URL was crawled by the Wayback Machine. Archive-It and ArchiveTheNet[5] are web archive services provided by the Internet Archive and the Internet Memory Foundation. These services enable focused archiving of web contents by organizations, such as universities or libraries, that otherwise could not manage their own archives. The Memento Project[6] enables the discovery of archived content from across multiple web archives via URL search.

A few researchers have worked on providing new interfaces and visualizations for searching, exploring and discovering insights from web archives. Odijk et al. [15] present an exploratory search interface to improve accessibility of digital archived collections for humanities scholars, in order to highlight different perspectives across heterogeneous historical collections. The motivation for this work derives from the huge amount of digital material that has become available to study our recent history, including books, newspapers and web pages, all of which provide different perspectives on people, places and events over time. In their paper, the authors connect heterogeneous digital collections through the temporal references found in the documents as well as their textual content, in order to support scholars to detect, visualize and explore materials from different perspectives. Padia et al. [16] provide an overview of a web archive collection by highlighting the collection's underlying characteristics using different visualizations of image plots, wordle, bubble charts and timelines. Lin et al. [9] present an interactive visualization based on topic models for exploring archived content, so that users can get an overview of the collection content. The visualization displays a person-by-topic matrix that shows the association between U.S. senators websites and the derived topics. It also provides drill-down capabilities for users to examine the pages in which a topic is prevalent.

All of the above tools and interfaces help support the exploration and search of web archives for individual users and researchers. In addition, ArchiveWeb aims at supporting the collaborative exploration of web archives. Previous research on helping users keep track of their resources include tools that provide better search and organizational facilities based on metadata/time [5] or tagging [3]. Our system provides similar organizational functionalities refined

---

[4] http://archive.org/web/.
[5] http://archivethe.net/.
[6] https://tools.ietf.org/html/rfc7089.

through several learning communities and previous work, LearnWeb [12], thus gaining advantage from several years of development and user feedback in that context. ArchiveWeb builds on the LearnWeb platform which already supports collaborative sensemaking [6,18] by allowing users to share and collaboratively work on resources retrieved from various web sources [2,8,13,14].

3   Web Archive Collections

Archive-It is a subscription web archiving service from the Internet Archive that helps organizations to harvest, build, and preserve collections of digital content. It was first deployed in 2006 and is widely used as a service to collect, catalog, and manage collections of archived web content. Fulltext search is also available, even though effective ranking is still an open issue. All content is hosted and stored at the Internet Archive data centers.

Currently about 200 collections from Archive-It have been integrated into the ArchiveWeb system with full metadata indexing, and can be explored through a visual interface. Main topics covered by the Archive-It collections available through ArchiveWeb are: Human Rights, Contemporary art, Global Events as well as various Web resources related to society, history, culture, science, statistics, and governments of various countries.

3.1   Human Rights

Columbia University Human Rights Web Archive
By Columbia University Libraries

This collection is dedicated to preserving web-based information of importance to the global community of human rights researchers, students, policymakers and advocates.

The collection *Human Rights Web Archive* (HRWA) by *Columbia University Libraries* is made up of searchable archived copies of websites related to human rights created by various non-governmental organizations, national human rights institutions, tribunals and individuals. The collection was started in 2008 and is still being continued, adding new websites on a regular basis.

Identification of websites for archiving is done by subject specialists with expertise in human rights and different regions of the world at Columbia University Libraries. Public nominations are provided by human rights researchers, advocates, organizations and individuals who are involved in the creation of human rights related websites. Priority is given to websites hosted in countries that do not have any systematic web archiving initiatives in place. Websites of intergovernmental organizations such as the United Nations are not included in the collection.

Archive-It services are used to maintain the collection, and the Internet Archive and Columbia University Libraries store copies of the resulting data. The collection includes over *711 websites* with more than 50 million searchable documents and over *115 million archived documents* with an archived data size of more than *5TB*.

The HRWA collection provides a good balance between websites of large and well known human organizations based in North America and Europe, and websites of smaller organizations from other regions. It includes websites from organizations such as Human Rights Watch, Amnesty International, Transparency International, International Crisis Group, and websites from regions with less web archiving activity and more political/social unrest. It also covers websites and organizations which are at greater risk of disappearing. One example of a website that no longer has a live version but can still be accessed by researchers via the Web archive collection is TibetInfoNet[7], which monitors the situation in Tibet. This page has been captured between May 2008 and July 2015, the ArchiveWeb screenshot shows the human rights group restricted to "Tibet" related resources, with the TibetInfoNet resource on the right side (Fig. 1).

The URL provided in ArchiveWeb below the screenshot on the right side points to the original web site, which is no longer available on the live web, but can be retrieved in its previous version through the Wayback Machine (as described in Sect. 4.3).

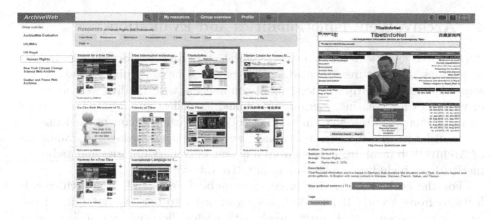

Fig. 1. Human rights example resource

3.2   Occupy Movement 2011/12

Occupy Movement 2011/12
By Internet Archive Global Events

This collection documents and aggregates information related to the Occupy Movement starting in the Autumn of 2011 and continuing in 2012. This demonstration inspired similar protests and demonstrations around the world

The collection *Occupy Movement 2011/12* by *Internet Archive Global Events* started in November 2011 with members from the Archive-It team and a large web archiving community collaboratively identifying and capturing websites related to the *"Occupy Movement"*. The movement started in Autumn 2011 and continued in 2012. The collection includes movement-wide or city specific occupy websites, blogs, social media sites and news articles from alternating or traditional media outlets.

---

[7] http://www.tibetinfonet.net/.

The selection of seeds involved public participation by content curators and individuals, as well as crawling of websites from community generated feeds (commencing on December 2011) with *933 seed URLs* selected (e.g. "Occupy Feeds"). In the subsequent years the collection was crawled frequently, capturing about *26 million documents* and archiving over *900* GB of data.

The Archive-It team is also collecting web content on global events at risk of disappearing such as *"Earthquake in Haiti"*, *"North Africa & Middle East 2011–2013"*, *"Ukraine Conflict"*, *"2013 Boston Marathon Bombing"* and others.

The collection consists of seeds organized into groups such as blogs, international news sites and articles, other sites and social media. An interesting analysis[8] of the collection was carried out in 2014 to see how much of the seed contents still live on the web. The seeds were categorized into news articles, social media URLs, and "movement sites". Only 41 % of the 582 "movement sites", 85 % of 203 social media URLs and 90 % of 163 news articles are still alive on the web. All "movement sites" that were no longer available on the web, were either showing 404 errors or taken down by cyber squatters.

4   ArchiveWeb: Functionalities and Evaluation

ArchiveWeb facilitates collaborative exploration of multiple focused web archive collections. The features provided by the system to support this exploration are: *searching* across multiple collections in conjunction with the live web, *grouping* of resources for creating, merging or expanding collections from existing ones and *enrichment* of resources in existing collections using comments and tags.

In order to collect relevant feedback, we carried out a task-based evaluation of ArchiveWeb involving experts from different university libraries and archiving institutions, including a follow-up qualitative interview.

For the evaluation, we imported some publicly accessible web archive collections from Archive-It into ArchiveWeb in order to test the potential of the system to support collaborative work with such collections. We invited eight experts from university libraries and archiving organizations to evaluate the system. Evaluators were from the following institutions: Columbia University, University of Toronto, Cornell University, Stanford Libraries, National Museum of Women Arts, New York Arts Resources Consortium and the Internet Archive. We asked them to carry out two sets of tasks both *individual* and *collaborative*. The *individual tasks* included creating a sub-collection from an existing collection of their institution, enriching an existing collection with additional relevant resources from the live web and annotating resources with information about why they should be included and how often they should be archived. The *collaborative tasks* involved selecting ten featured resources from their collections, including them into a joint group shared with all other evaluators, and discussing the reasons why such resources were included. The final task involved creating a new collection about a shared topic among the evaluators, searching for relevant

---

[8] https://archive-it.org/blog/post/only-41-of-occupy-movement-urls-accessible-on-liv e-web/.

seed URLs for this collection using ArchiveWeb, and agreeing with the rest of the evaluators about which seeds should be included, what should be the archiving frequency and why.

The evaluation period spanned about three weeks, after which we invited the evaluators to give us feedback through a questionnaire and through follow-up interviews to fully understand their experience.

The following sections give an overview over the main insights based on both quantitative analysis of query logs and qualitative feedback from the interviews, along with the description of the system functionalities.

4.1   Searching for Resources in Collections and on the Web

ArchiveWeb provides a keyword based search system that returns results from Archive-It collections as well as from the live web (using the Bing Search API[9]). If users search for a keyword (e.g. "human rights"), ArchiveWeb returns a list of results from these sources, indicating whether the resource comes from a specific Archive-It collection or from the live web (Fig. 2b). The results being returned from Archive-It collections are based on a full metadata index for these collections (a fulltext index of all content is planned for the next version). Besides web-pages, also images (from Bing, Flickr, Ipernity) and videos (from YouTube, TED, Yovisto, Vimeo) can be searched, but not yet from Archive-It collections. For Archive-It collections each search result provides a pointer to the resource item saved in ArchiveWeb (Fig. 2c) and displays information about when it was captured in Archive-It (Fig. 2d). Live web results are not yet saved in ArchiveWeb and can be copied by the user into a group (Fig. 2e); in this case, the Option window provides details about when/if this page has been indexed by the Internet Archive, using the Wayback CDX server API[10].

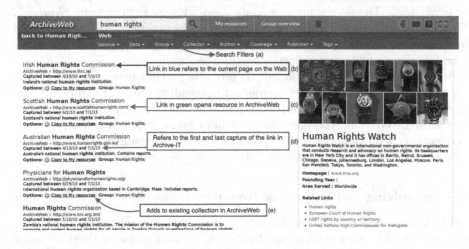

Fig. 2. Search page

[9] http://datamarket.azure.com/dataset/bing/search.
[10] https://github.com/internetarchive/wayback/tree/master/wayback-cdx-server.

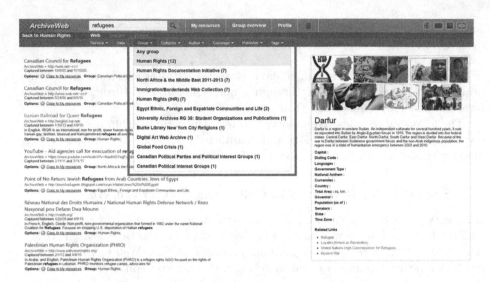

Fig. 3. Refinement through faceted search filters

Search results can be refined using faceted search filters visible below the search box (Fig. 3). For example, for our evaluation it was important to provide filters to show only the resources in a specific collection (Filter: *Group*) or that were archived by a specific institution (Filter: *Collector*).

Feedback. The integration of searching across web archive collections as well as the live web provides the ability to suggest additional material that could be archived. This is a new feature not present in existing systems used for curation/creation of web archive collections, and it was much appreciated by the evaluators. The search filters mainly used while searching were "Author", "Tags", "Group" and "Service". These filters were mainly used to understand how the results would change with different important filter options while either searching across Archive-It collections or the live web.

Suggestions we received included: (i) providing a transparent explanation of how the relevance ranking is determined, (ii) providing the total number of captures of a page along with the capturing period in order to determine the popularity of an archived result, (iii) providing as default just search results from Archive-It collections with the option to expand the results to the live web in order to facilitate curators who expect to see only archived resources in the first step.

4.2 Organizing and Extending Resource Collections

ArchiveWeb provides the functionality to organize resources into collections (groups of resources) according to clearly defined and coherent themes/topics. This functionality allows working with existing groups, creating new collections/groups and sub-groups, adding new resources to groups, and moving resources between groups.

A group overview interface allows browsing through existing collections available within ArchiveWeb and the collections that a user has created or joined. Descriptions for every collection provide information about the topic/theme and what kind of resources it contains. For Archive-It collections, the description reports the details available on the public Archive-It interface along with a reference link to the specific collection in the ArchiveWeb platform.

After joining specific groups/collections users have the opportunity to edit metadata of existing resources and to contribute new resources from other existing collections or from the live web. The ArchiveWeb collections, derived directly from their Archive-It counterparts, are "read-only". Users can browse through the resources of each collection using the advanced visualization and exploration functionalities of ArchiveWeb, but they cannot change the original collection. They can create a copy of the entire "read-only" group, or add individual resources to their own collections by selecting resources individually, as well as merge multiple existing collections into one new collection.

Users can also organize their resources into sub-groups within collections in order to group resources that are related to a similar subtopic. Resources can be uploaded to a collection either from the desktop or by suggesting a URL. Each group has a specific interface that visualizes all thumbnails of the most recent snapshot of each resource when it was added to the group (Fig. 4). Resources that no longer exist on the web, or do not have a redirect available, display a thumbnail with the message "The page is no longer available on the web" (we are currently implementing the functionality to upload an earlier thumbnail of the page taken from the Internet Archive Wayback Machine). The overview interface within a group provides a summary of the activities of various members of the group, including the actions of new added resources, resources which were edited or deleted, and users who have joined/left the group.

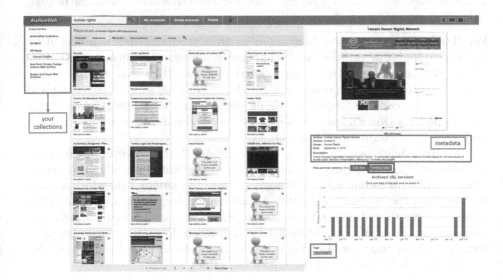

Fig. 4. Group resources interface

Feedback. One of the evaluators stated, that *"ArchiveWeb increases curatorial functionality over that of Archive-It, e.g.: on-the-fly creation of groups, moving resources between groups and easily annotating resources"*. Several of the new functionalities of ArchiveWeb were positively highlighted by the evaluators including (i) the ability to curate new arbitrary collections of seed records from across multiple web archive collections, (ii) the possibility of having resources exist at multiple levels (e.g. personal, group, subfolder), and (iii) the ability to create collaborative collections with colleagues from various institutions.

4.3   Enriching Resource Metadata and Discussing Resources

After a resource is added to an ArchiveWeb collection the resource can be enriched with additional comments and tags (in addition to the metadata already provided). The comments on a resource can be used to discuss as to why this resource has been chosen as a seed for a collection, and to decide upon the crawl frequency and crawl depth. By exchanging comments, collaborators can discuss the relevance of a suggested seed resource for a particular collection. The use of tags helps categorize or label a resource with subjects or topics covered by the resource, making it possible to browse collections by filtering based on certain tags. Users can also edit metadata such as title, description and author fields.

The system allows users to archive a single page or a website (resource) by clicking on the *"Archive Now"* button which sends a request to the Wayback Machine to archive it. This functionality is similar to the *"Save Page Now"* feature of the Wayback machine, but it facilitates users to gather captures easily as they work within the system. All the captures from Archive-It as well as the wayback captures added using the *"Archive Now"* functionality are visualized both as a *list* and in a *timeline* view to help users navigate through the different archived versions that are available. Figure 4 shows a timeline view on the bottom right.

Feedback. One of the new functionalities that was positively highlighted by the evaluators is the possibility to add schema-agnostic classification and other information using tags. Other appreciated functionalities were (i) the ability to use notes/comments in order to highlight the motivations for collecting and to discuss the capture frequency, highly useful when developing collaborative collections across institutions, (ii) *list* and *timeline* visualizations were considered useful to explore individual seed records.

Metadata fields which were generally modified while editing a resource included *title* and *description* (as not all of them are automatically filled appropriately while uploading a URL to a group), and the *author* field which is often missing from the original metadata. An interesting suggestion from one of the evaluators was to add open/customizable metadata fields besides *title*, *description* and *author* as they would be needed if curators from different institutions build collections together. Tagging is helpful but not considered a substitute for using custom field names and corresponding values.

4.4   Future Improvements

The evaluators also provided additional suggestions for functionalities that are missing in the ArchiveWeb system which we plan to incorporate in the next release, for example:

Exporting Resources - the ability to export resources and metadata from collections for research or usage outside of ArchiveWeb.

Bulk Operations - the ability to select multiple resources by clicking, searching or filtering, and then adding them to a collection. The system should also support bulk editing of resources within a collection such as collective tagging (e.g. to assign a tag to multiple seed records), as well as tagging similar resources across multiple collections and then allowing the option to group them together into a new collection.

Advanced Search - the ability to limit the search within certain collections, specific domains or paths, and to search within the title only. We also plan to incorporate fulltext search of Archive-It websites instead of just the metadata, even though not all evaluators realized that this functionality was not provided, as rich metadata descriptions are available for many collections.

5   Conclusion

In this paper we discussed the ArchiveWeb system which supports collaborative exploration of web archive collections. We provided a description of the main features of the system such as (i) searching across multiple collections as well as the live web, (ii) grouping of resources for creating new collections or merging existing ones, as well as (iii) collaborative enrichment of resources using comments and tags.

The system has been developed based on an iterative evaluation-driven design based research approach. Starting from a platform which already supported collaborative search and sharing of web resources, ArchiveWeb was designed to address expert users' requirements (e.g. librarians and curators in archiving institutions). The resulting ArchiveWeb prototype, fully functional, was evaluated through a task-based evaluation study carried out with the same experts who participated in the preliminary investigation. After the quantitative analysis of the logs and the qualitative feedback from the evaluations, we are now incorporating new features such as exporting resources, bulk editing operations and advanced search which will be available in the next release of the system.

Acknowledgments. We thank Jefferson Bailey from the Internet Archive who provided us with the contacts to his colleagues at university libraries and archiving institutions. We are also grateful to all experts, who participated with enthusiasm in our evaluation, providing valuable feedback and useful suggestions to improve the system. This work was partially funded by the European commission in the context of the ALEXANDRIA project (ERC advanced grant no 339233).

References

1. Abel, F., Herder, E., Marenzi, I., Nejdl, W., Zerr, S.: Evaluating the benefits of social annotation for collaborative search. In: 2nd Annual Workshop on Search in Social Media (SSM 2009) (2009)
2. Amershi, S., Morris, M.R.: Cosearch: a system for co-located collaborative web search. In: Proceedings of CHI 2008, pp. 1647–1656 (2008)
3. Cutrell, E., Robbins, D., Dumais, S., Sarin, R.: Fast, flexible filtering with phlat. In: Proceedings of CHI 2006, pp. 261–270 (2006)
4. Dougherty, M., van den Heuvel, C.: Historical infrastructures for web archiving: annotation of ephemeral collections for researchers and cultural heritage institutions. In: MIT6 Conference, Boston, MA (2009)
5. Dumais, S., Cutrell, E., Cadiz, J., Jancke, G., Sarin, R., Robbins, D.C.: Stuff I've seen: a system for personal information retrieval and re-use. In: Proceedings of SIGIR 2003, pp. 72–79 (2003)
6. Evans, B.M., Chi, E.H.: Towards a model of understanding social search. In: Proceedings of CSWC 2008, pp. 485–494 (2008)
7. Gomes, D., Miranda, J., Costa, M.: A survey on web archiving initiatives. In: Gradmann, S., Borri, F., Meghini, C., Schuldt, H. (eds.) TPDL 2011. LNCS, vol. 6966, pp. 408–420. Springer, Heidelberg (2011)
8. Held, C., Cress, U.: Learning by foraging: the impact of social tags on knowledge acquisition. In: Cress, U., Dimitrova, V., Specht, M. (eds.) EC-TEL 2009. LNCS, vol. 5794, pp. 254–266. Springer, Heidelberg (2009)
9. Lin, J., Gholami, M., Rao, J.: Infrastructure for supporting exploration and discovery in web archives. In: Proceedings of WWW 2014, pp. 851–856 (2014)
10. Marenzi, I., Kupetz, R., Nejdl, W., Zerr, S.: Supporting active learning in CLIL through collaborative search. In: Luo, X., Spaniol, M., Wang, L., Li, Q., Nejdl, W., Zhang, W. (eds.) ICWL 2010. LNCS, vol. 6483, pp. 200–209. Springer, Heidelberg (2010)
11. Marenzi, I., Nejdl, W.: I search therefore I learn - active and collaborative learning in language teaching: two case studies, pp. 103–125. Collaborative Learning 2.0: Open Educational Resources (2012)
12. Marenzi, I., Zerr, S.: Multiliteracies and active learning in CLIL - the development of LearnWeb2.0. IEEE Trans. Learn. Technol. (TLT) 5(4), 336–348 (2012)
13. Morris, M.R.: A survey of collaborative web search practices. In: Proceedings of CHI 2008, pp. 1657–1660 (2008)
14. Morris, M.R., Horvitz, E.: Searchtogether: an interface for collaborative web search. In: Proceedings of UIST 2007, pp. 3–12 (2007)
15. Odijk, D., Gârbacea, C., Schoegje, T., Hollink, L., de Boer, V., Ribbens, K., van Ossenbruggen, J.: Supporting exploration of historical perspectives across collections. In: Kapidakis, S., et al. (eds.) TPDL 2015. LNCS, vol. 9316, pp. 238–251. Springer, Heidelberg (2015). doi:10.1007/978-3-319-24592-8_18
16. Padia, K., AlNoamany, Y., Weigle, M.C.: Visualizing digital collections at archive-it. In: Proceedings of JCDL 2012, pp. 15–18 (2012)
17. Ras, M., van Bussel, S.: Web archiving user survey. Technical report, National Library of the Netherlands (Koninklijke Bibliotheek) (2007)
18. Russell, D.M., Stefik, M.J., Pirolli, P., Card, S.K.: The cost structure of sensemaking. In: Proceedings of the INTERACT 1993 and CHI 1993, pp. 269–276 (1993)
19. Zerr, S., d'Aquin, M., Marenzi, I., Taibi, D., Adamou, A., Dietze, S.: Towards analytics and collaborative exploration of social and linked media for technology-enchanced learning scenarios. In: Proceedings of the 1st International Workshop on Dataset PROFIling & fEderated Search for Linked Data (2014)

Web Archives

Web Archive Profiling Through Fulltext Search

Sawood Alam[1]([⊠]), Michael L. Nelson[1], Herbert Van de Sompel[2],
and David S.H. Rosenthal[3]

[1] Computer Science Department, Old Dominion University, Norfolk, VA, USA
{salam,mln}@cs.odu.edu
[2] Los Alamos National Laboratory, Los Alamos, NM, USA
herbertv@lanl.gov
[3] Stanford University Libraries, Stanford, CA, USA
dshr@stanford.edu

Abstract. An archive profile is a high-level summary of a web archive's holdings that can be used for routing Memento queries to the appropriate archives. It can be created by generating summaries from the CDX files (index of web archives) which we explored in an earlier work. However, requiring archives to update their profiles periodically is difficult. Alternative means to discover the holdings of an archive involve sampling based approaches such as fulltext keyword searching to learn the URIs present in the response or looking up for a sample set of URIs and see which of those are present in the archive. It is the fulltext search based discovery and profiling that is the scope of this paper. We developed the *Random Searcher Model (RSM)* to discover the holdings of an archive by a random search walk. We measured the search cost of discovering certain percentages of the archive holdings for various profiling policies under different RSM configurations. We can make routing decisions of 80 % of the requests correctly while maintaining about 0.9 recall by discovering only 10 % of the archive holdings and generating a profile that costs less than 1 % of the complete knowledge profile.

Keywords: Web archive · Memento · Archive profiling · Random searcher

1 Introduction

The number of public web archives supporting the Memento protocol [17] natively or through proxies continues to grow. Currently, there are 28[1], with more scheduled to support Memento in the near future. The Memento Aggregator [13], the Time Travel Service[2], and other services, need to know which archives to poll when a request for an archived version of a file is received. Efficient Memento routing in aggregators is desired from both the aggregators' and archives' perspective. Aggregators can reduce the average response time, improve overall throughput, and save network bandwidth. Archives benefit by the reduced number of requests

---

[1] http://labs.mementoweb.org/aggregator_config/archivelist.xml.
[2] http://timetravel.mementoweb.org/.

© Springer International Publishing Switzerland 2016
N. Fuhr et al. (Eds.): TPDL 2016, LNCS 9819, pp. 121–132, 2016.
DOI: 10.1007/978-3-319-43997-6_10

for which they have no holdings, hence saving computing resources and band-width. In December 2015, soon after the surge of OldWeb.today[3] many archives struggled with the increased traffic. We found that fewer than 5 % of the queried URIs are present in any individual archive other than the Internet Archive. In this case, being able to identify a subset of archives that might return good results for a particular request becomes very important.

We envision following four approaches to discover the holdings of an archive:

CDX Profiling – If an archive's CDX (Capture/Crawl inDeX) files[4] are avail-able, then we can generate profiles with complete knowledge of their holdings. In our recent work [2] we established a generic archive profiling framework and explored the *CDX Profiling*. We examined an extended set of 23 different policies to build archive profiles and measured their routing efficiency.

Fulltext Search Profiling – Random query terms are sent to the fulltext search interface of the archive (if present) and from the search response we learn the URIs that it holds. These URIs are then utilized to build archive profiles. In this paper we are exploring the *Fulltext Search Profiling* within the framework established by our recent work. It is important to note that some web archives (including the Internet Archive) do not provide fulltext search, hence this approach is not applicable for them.

Sample URI Profiling – A sample set of URIs are used to query the archive and build the profile from the successful responses. This is quite wasteful, as <5 % of the sample URIs are found in any archive. AlSum et al. explored the *Sample URI Profiling*, but only on Top-Level Domains (TLDs) [3]. We want to further explore this approach with many profiling policies.

Response Cache Profiling – This approach depends on the response data col-lected by an aggregator over a period of time as queries are made to the archive. Cached responses are analyzed to learn about their holdings. As a result the *Response Cache Profiling* is based on what people were looking for as opposed to what is in the archives. This approach is different from the other three in a way that it is a usage-based profiling approach while the other three are content-based.

In this paper we introduce a *Random Searcher Model (RSM)* to randomly explore a collection and discover K distinct documents. Unlike the *Random Surfer Model* [4] which constructs the graph of web pages based on the hyper-links, RSM assumes that the documents in the collection are connected with co-occurring terms. Randomly picking a term from a document and searching for that term in the collection yields many other documents that share the term. Selecting a document from the returned document set to choose the next ran-dom term and repeat this process allows us to know about the holdings of the collection. To generalize the RSM and to accommodate several possibilities we experimented with some variations in how to select the next search term.

---

[3] http://oldweb.today/.

[4] http://archive.org/web/researcher/cdx_file_format.php.

2   Related Work

Query routing is common practice in various fields including meta-searching and search aggregation [7–10,15,16,19]. Sanderson et al. first explored Memento query routing by exhaustive CDX profiling [14]. They collected CDX files from various IIPC member archives and generated profiles based on the complete URI-Rs (original Resource URIs) from them (we denote it as *URIR Profile* in this paper). This approach gives complete knowledge of the holdings in each participating archive, hence queries can be routed precisely to archives that have any mementos (URI-M) for the given URI-R. It is a resource and time intensive task to generate such profiles and some archives may be unwilling or unable to provide their CDX files. Such profiles are large enough in size (typically, a few billion URI-R keys) that they require special infrastructure to support fast lookup. Acquiring fresh CDX files from various archives and updating these profiles regularly is difficult.

Many web archives tend to limit their crawling and holdings to some specific TLDs, for example, the British Library Web Archive prefers sites with .uk TLD. AlSum et al. explored *Sample URI Profiling* based on TLD [3] in which they recorded *URI-R Count* and *URI-M Count* under each TLD for twelve public web archives. Their results show that they were able to retrieve the complete TimeMap [17] in 84 % of the cases using only the top 3 archives and in 91 % of the cases when using the top 6 archives. This simple approach can significantly reduce the number of queries generated by a Memento aggregator with some loss in coverage.

The two efforts described above have explored extreme cases of profiling. We believe that an intermediate approach that gives flexibility with regards to balancing accuracy and effort can result in better and more effective routing. Our earlier work [2] establishes the general framework of flexible archive profiling. It describes the general model of partial URI, date, language, and composite keys and associated statistical measures. We also introduced a profile serialization format called CDXJ (CDX-JSON) [1] to make it easy to store, disseminate, merge, and perform efficient lookups. In our earlier work we generated profiles from CDX files where we knew the complete holdings of the archive. We then examined costs and routing efficiencies of 23 different profiling polices.

Bornand et al. explored *Response Cache Profiling* and implemented Memento routing by building binary classifiers from the aggregator cache data (in this case the classifier is functionally equivalent to an archive profile) [5]. They report a 77 % reduction in the number of requests and a 42 % reduction in response time while maintaining a 0.847 recall value.

There have been many efforts on crawling the hidden web that have no hyperlinks and are accessible only by filling out HTML forms [11,12,18]. Our keyword search based archived content discovery is related to these efforts because archived contents span over a long period of time, which causes a disconnect between old and contemporary pages. As a result, hyperlink based shallow crawling might only discover a temporal sub-graph of the holdings.

3   Methodology

If a web archive provides fulltext searching in its collection, it can be leveraged to discover the holdings of the archive. In this approach, we send some random query terms to the archive and collect the resulting URIs returned from the archive. Although with each successful response we learn some new URIs, the learning rate slows down as we go forward because of the fact that some of the URIs returned in a response may have been already seen in earlier attempts. This follows Heaps' Law [6].

With carefully chosen values of various advanced search attributes and selection of suitable list of keywords we can affect the parameters of the Heaps' Law and maximize the learning rate. In this section we analyze different fulltext approaches to discover holdings of an archive. We also explore different ways to perform the search when the language of the archive is unknown or a list of suitable keywords is not available for that language.

For our experiments we chose the Archive-It hosted North Carolina State Government Web Site Archive collection[5] which is the largest collection in Archive-It and provides fulltext search. Since we have Archive-It data (up to 2013) hosted in our dark archive at Old Dominion University (ODU), it was easier for us to perform the coverage analysis on this dataset.

We started our experiment by searching for stop-words. For example when we searched for the term "a" it returned 26M+ results, which is very close to the number of the HTML resources in the collection up till March 2013. However, each page only contains 20 results, hence, in order to learn all the 26M+ URIs we will have to make 1.3M+ HTTP GET requests. Our goal was to make as few requests as possible to learn enough diverse set of URIs in the collection. Additionally, this approach cannot be generalized, as not all archives will behave the same on stop-words.

In the next step we built static and dynamic word lists and searched for those words as query terms. For each term we only record the URIs in the first resulting page and move on to the next word in the list. Having a configurable pagination would have benefited us by choosing a larger number of results per page, but Archive-It has it fixed to 20 results and does not allow any changes.

Top Words – We collected top the 2,000 English nouns[6] and used them as the query terms. We accumulated the first page results to plot the learning curve, but also extracted other information such as the result count for each search term. As expected, these top terms yielded high values for the result count for each terms. Additionally, each term yielded more than one page of results, hence no effort was unsuccessful. We also observed that the response time is correlated with the result count, so the same number of top terms would take longer to fetch than a random word list.

Random Linux Dictionary – We ran the same procedure on a randomly chosen set of 2,000 unique words from the built-in Linux dictionary. This time

---

[5] https://www.archive-it.org/collections/194.
[6] http://worddetail.org/most_common/nouns.

the average number of results per terms was lower, but there were many terms for which the collection returned no results or fewer than 20 (page size) results.

Dynamically Discovered Word List – The above two experiments were based on the static lists of words. This static word list approach requires the knowledge of the language (and field) of the collection to choose a static list of words for that language, which may not be easily available. Additionally, the list would be finite and may not be enough to discover a sufficient number of the collection holdings. To overcome this issue, we build a model (discussed in Sect. 4) that can dynamically discover new words from the searched pages and utilize those words to perform further searching. We introduced different policies in the dynamic discovery model based on how the new words are learned and how the next word for searching is chosen. We then analyzed the learning rate and the number of HTTP requests for each policy.

4   Random Searcher Model

To perform searches on a static or dynamic word list, we developed a general *Random Searcher Model (RSM)* that can be configured to operate in one of the four modes with the help of configurable *Vocabulary Seeding* and *Word Selection* policies. To understand the model we will discuss different data structures, policies, operation modes, and procedures separately then put them together to describe the overall working.

4.1   Data Structure

The RSM has the following data structures to hold various intermediate statistics and states:

Vocabulary – A data structure to hold the list of words to be searched. Depending on some policies it may or may not allow duplicates. This data structure should provide the number of total or unique words in the list. The data structure should support functions to overwrite the list, add more words to the list, pop a random word from the list, and remove all occurrences of the popped word in the list.

SearchLog – A data structure to hold the record of each search attempt. It contains the searched word, attempt result (success or failure), and any additional meta information such as the response result count and newly discovered URIs. An implementation may choose to offload some results to a separate data structure or include more attributes for analysis. This data structure should provide the number of total or successful searches. The data structure should support functions to add new records, querying if a given word is present, and randomly selecting a successful searched word.

ResultBank – A dictionary like data structure to hold all the discovered URI-Rs and their respective memento counts. This data structure should provide the number of total URI-Rs (or dictionary keys). The data structure should support functions to add new records and iterate over all the records.

4.2 Policies

Policies control the behavior of the RSM. We have defined two different policies with various valid configuration values as described below:

Vocabulary Seeding Policy – This policy controls how the *Vocabulary* is seeded and how often. Valid values are:

- *Static* – The *Vocabulary* is manually seeded in the beginning with a static list of search keywords. It allows seeding only once and the procedure terminates when all the seeded keywords are consumed.
- *Progressive* – The *Vocabulary* is initialized with a few words, but it is aggressively overwritten after each successful search with the newly discovered words, except when there are no new words discovered.
- *Conservative* – The *Vocabulary* is initialized with a few words and new words discovery is only performed when all the words from the *Vocabulary* are consumed. The *Vocabulary* is only reseeded when it is empty.

Word Selection Policy – This policy controls how search words are selected from the *Vocabulary*. Valid values are:

- *Popularity Biased* – Randomly select one word from the *Vocabulary* where the probability of a word being selected is proportional to the term frequency in the *Vocabulary* normalized by the total number of terms in the *Vocabulary*. A simple implementation of this policy would consider the *Vocabulary* as a bag of random words (with duplicates allowed) to select a random word from it. There are other memory efficient approaches possible, but this simple approach illustrates the concept naturally.
- *Equal Opportunity* – Randomly select one word from the list of unique words in the *Vocabulary*.

4.3 Operation Modes

An operation mode is a valid combination of *Vocabulary Seeding* and *Word Selection* policies (as mapped in Table 1). Not all combinations of the two policies are valid. We recognize the following four as valid modes of the *RSM*:

Static – This mode operates on a static word list that is known in advance so it does not have to make additional fetches to discover more words. However, finding a suitable word list for a collection could be difficult.

PopularityBiased – In this mode the searcher randomly picks a URI from the returned result and fetches that page to discover new words. Once the words are fetched, it randomly picks a word from that and repeats the search operation. In this way the searcher has to fetch a page after every search attempt to search for the next word. Selection of the words is random, but the duplicates are not removed so the words with higher frequency in the page have higher chance of being selected.

Table 1. RSM operation mode mapping with policies

Operation mode	Vocabulary seeding	Word selection
Static	Static	Equal opportunity
PopularityBiased	Progressive	Popularity biased
EqualOpportunity	Progressive	Equal opportunity
Conservative	Conservative	Equal opportunity

EqualOpportunity – This mode works the same way as the *PopularityBiased* mode does, but it picks the next word from the unique list of words so a rare word on the page has the same probability of being selected as a high frequency one.

Conservative – The *PopularityBiased* and the *EqualOpportunity* modes have the drawback of being twice as costly in terms of number of HTTP requests as compared to a static word list of the same size. The reason for this is that after every successful search, there is a page fetch to learn new words. However, this *Conservative* mode does not throw away previously learned words and consumes each of them. It makes a page fetch to learn new words only when all the previously learned words are consumed. This reduces the number of HTTP requests significantly.

4.4 Procedures

The RSM has the following public procedures:

Initialize() – This method initializes a Random Searcher instance with supplied configuration options and *Vocabulary* seed.

TerminationCondition() – This method returns True if the conditions are met to terminate the *NextWord* iterator. It returns False otherwise.

NextWord() – This method pops the next word from the *Vocabulary* based on the Word Selection Policy (and removes all the occurrences of the word in the *Vocabulary*, if any). If the *Vocabulary* is empty and the terminating condition is not met then it randomly picks a successful searched word from the *SearchLog*. This is a generator function that can act as an iterator until the configured termination condition is met.

Search() – This method searches the collection for a given search keyword and populates the *SearchLog* and the *ResultBank* with appropriate values from the search result. Depending on the configured policies it may also select a URI from the result to extract words from it to seed the *Vocabulary*.

ExtractWords() – This method fetches the page at the given URI, sanitizes it (by stripping off the markup, scripts, and styles), tokenizes the text to split in non-empty words, and returns the bag of words (duplicates included, if any) in the order they appeared in the document.

GenerateProfile() – This method iterates over all the items in the *ResultBank* and generates an *Archive Profile* based on the supplied profiling policy.

To run the *Random Searcher* an instance of the RSM is initialized with one of the four possible modes, some initial seed words, termination condition configuration, tokenizer pattern, and other configurations. Then the *NextWord* generator is iterated over to discover next word for searching until it hits the termination condition. For every word, the *Search* method is called which internally performs the collection lookup and updates various data structures of the RSM instance. Once the iterator terminates, *GenerateProfile* method is called to serialize collected statistics in an *Archive Profile*.

5   Implementation

We implemented the RSM in Python and made the code available[7]. However, due to the lack of a uniform search API across archives, the implementation is not generic enough to run against any archive. The page scraping part of the code that extracts useful pieces of information from the search result page and assembles them in a data structure needs custom implementation for each archive. The Archive-It search interface has an undocumented JSON response that we used to avoid HTML parsing. Our implementation has some additional intermediate data structures and logging in place for the sake of analysis which is not needed for production purposes.

6   Evaluation

To evaluate the RSM we estimated the search cost of different operation modes for discovering certain portion of the archive holdings in terms of a given profiling policy. To evaluate generated archive profiles we measured how much of the archive holdings we must discover in order to gain a satisfactory Recall and corresponding routing efficiency for a given profiling policy. In this analysis we have used four different profiling policies. Examples are derived from the URI https://www.news.BBC.co.uk/Images/Logo.png?width=80&height=40.

H1P0 – Only TLDs are used as keys (212 unique keys in the collection). E.g., uk)/.

DDom – Only registered domain name is used as keys (91,629 unique keys in the collection). E.g., uk,co,bbc)/.

HxP1 – All host segments and one path segment are used as keys (1,724,284 unique keys in the collection). E.g., uk,co,bbc,news)/Images.

URIR – SURTed URI-Rs are used as keys (30,800,406 unique keys in the collection). E.g., uk,co,bbc,news)/Images/Logo.png?height=80&width=200.

---

[7] https://github.com/oduwsdl/archive_profiler.

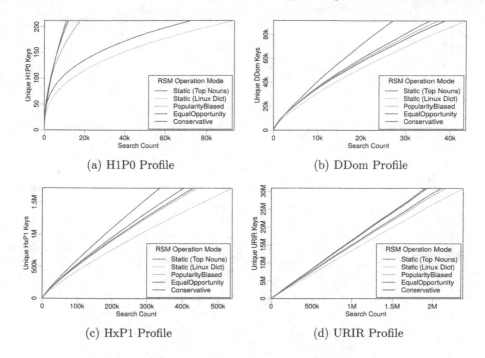

(a) H1P0 Profile

(b) DDom Profile

(c) HxP1 Profile

(d) URIR Profile

Fig. 1. Searches needed vs. required coverage

6.1   Estimating Searches Needed

Figure 1 illustrates the projected estimate of the search cost for each of the four profiling policies based on the initial 2,000 searches. Each graph is extended up to the 100 % limit of each profiling policy on the Y-axis. For example, there are total 212 unique TLDs in the collection, hence the upper Y-axis limit in Fig. 1(a) is set to 212. The RSM operation modes in which additional HTTP request is made to fetch a Memento to extract words cause additional HTTP GET overhead. Table 2 shows the HTTP cost of each RSM operation mode with respect to the corresponding query cost. The value of δ is quite small as compared to C. In our experiments we found that for $C = 2,000$ the value of δ was 8. Which means for making 2,000 queries we only needed eight additional HTTP GETs to extract new words in *Conservative* mode. The *Conservative* mode is an overall winner. It shows a higher learning rate and does not require a static list of words to be supplied. It is also good because it does not have much overhead HTTP request cost for the term discovery.

Table 2. Cost comparison of RSM operating modes

Operation mode	Query cost	HTTP cost
Static	C	C
PopularityBiased	C	$2 * C$
EqualOpportunity	C	$2 * C$
Conservative	C	$C + \delta$ (where $\delta \ll C$)

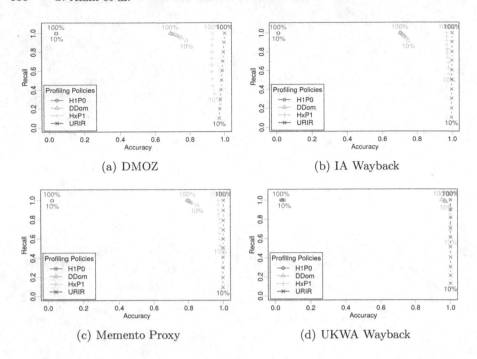

Fig. 2. Incremental accuracy vs. recall as a function of archive knowledge

6.2   Profile Routing Efficiency

Unlike CDX analysis, a fulltext search based archive profile will often be created based on partial knowledge of the archive holdings. This may cause incomplete recall in Memento routing, which means that an archive might have some Mementos unknown to the profile and as a result it may not have a matching key in it to route the query there. Hence, it is important to analyze the incremental changes in the confusion matrix as we know more and more of the archive's holdings. Table 3 illustrates the confusion matrix in the Memento routing context. For this analysis we selected the early ten years of UKWA dataset as the gold dataset. Then we extracted the unique URI-Rs from it and randomized it. The randomized URI-R list was then split into ten equal chunks. We then profiled these chunks incrementally using different policies to see how these policies perform against a sample set of query URIs when we know only 10 %, 20 %, 30 %... to 100 % of the archive. We repeated this process for four different query URI sample sets of one million each. In each case we calculated the confusion matrix and plotted Accuracy vs. Recall graph to estimate the routing efficiency. Accuracy here means how often an archive profile is correct in routing or not routing a query to the archive.

Figure 2 shows that if the complete archive is known, we should choose higher order profiles such as HxP1. The HxP1 profile has the Accuracy almost as good as the URIR profile. From our earlier work we know that the cost of the HxP1 profile is less than one sixth of the cost of the URIR profile. Additionally, the

Table 3. Confusion matrix of memento routing

Predicted\actual	Present in the archive	Not in the archive
Routed to the archive	True positives (TP)	False positives (FP)
Not routed to the archive	False negatives (FN)	True negatives (TN)

HxP1 profile has higher Recall than the URIR profile when the complete archive is not known. However, since we cannot afford to lose much of the recall, we favor smaller profiles such as DDom when we have partial access to the archive (such as when using fulltext search). The DDom profile (that has less than 1 % cost as compared to the URIR profile) can have about 0.9 Recall while correctly routing (or not routing) more than 80 % of the URIs by only knowing 10 % of the archive. Cases where only a tiny fraction of the archive is known by sampling (such as when neither CDX is available nor fulltext search), we favor the smallest profile H1P0/TLD-only to maintain an acceptable Recall value. Even though the URIR profile always yields 100 % Accuracy, it suffers from poor Recall. With the complete CDX accessible, the URIR policy costs a lot compared to other policies. Additionally, the URIR policy does not have any predictive powers for unseen URIs, hence maintaining the freshness of the profile is challenging as the archives acquire more URI-Rs.

7 Future Work and Conclusions

So far we only worked on profiles based on URIs. Going forward we want to examine other dimensions such as the Memento date and the content language and a combination of more than one of these. We want to examine how these profiles generated for various archives can be digested into a system that ranks the archives based on the probability of the availability of each queried URI in various archives.

In this paper we examined the *Fulltext Search Profiling*. We developed a *Random Searcher Model* to discover the holdings of archives that support fulltext search. We evaluated the query and HTTP costs to learn certain percentage of the holdings of an archive using RSM under different profiling policies. We also evaluated the routing efficiency in terms of *Accuracy* and *Recall*. We concluded that the DDom profile (that has less than 1 % cost as compared to the URIR profile) can have about 0.9 Recall while correctly routing (or not routing) more than 80 % of the URIs by only knowing 10 % of the archive. Finally, we open-sourced our RSM implementation code.

Acknowledgements. This work is supported in part by the International Internet Preservation Consortium (IIPC).

References

1. Alam, S., Kreymer, I., Nelson, M.L.: Object Resource Stream (ORS) and CDX-JSON (CDXJ) Draft (2015). https://github.com/oduwsdl/ORS
2. Alam, S., Nelson, M.L., Van de Sompel, H., Balakireva, L.L., Shankar, H., Rosenthal, D.S.H.: Web archive profiling through CDX summarization. In: Kapidakis, S., et al. (eds.) TPDL 2015. LNCS, vol. 9316, pp. 3–14. Springer, Heidelberg (2015). doi:10.1007/978-3-319-24592-8_1
3. AlSum, A., Weigle, M.C., Nelson, M.L., Van de Sompel, H.: Profiling web archive coverage for top-level domain and content language. Int. J. Digit. Libr. **14**(3–4), 149–166 (2014)
4. Blum, A., Chan, T.H., Rwebangira, M.R.: A random-surfer web-graph model. In: Proceedings of the Meeting on Analytic Algorithmics and Combinatorics, ANALCO 2006, pp. 238–246. Society for Industrial and Applied Mathematics (2006)
5. Bornand, N., Balakireva, L., Van de Sompel, H.: Routing memento requests using binary classifiers. In: Proceedings of the 16th ACM/IEEE-CS Joint Conference on Digital Libraries, JCDL 2016 (2016)
6. Egghe, L.: Untangling Herdan's law and Heaps' law: mathematical and informetric arguments. J. Am. Soc. Inf. Sci. Technol. **58**(5), 702–709 (2007)
7. Gravano, L., Chang, C.C.K., García-Molina, H., Paepcke, A.: STARTS: stanford proposal for internet meta-searching. SIGMOD Rec. **26**(2), 207–218 (1997)
8. Levy, A.Y., Rajaraman, A., Ordille, J.J.: Querying heterogeneous information sources using source descriptions (1996)
9. Liu, L.: Query routing in large-scale digital library systems. In: Proceedings of 15th International Conference on Data Engineering, pp. 154–163 (1999)
10. Meng, W., Yu, C., Liu, K.L.: Building efficient and effective metasearch engines. ACM Comput. Surv. (CSUR) **34**(1), 48–89 (2002)
11. Ntoulas, A., Zerfos, P., Cho, J.: Downloading textual hidden web content through keyword queries. In: Proceedings of the 5th ACM/IEEE-CS Joint Conference on Digital Libraries, JCDL 2005, pp. 100–109 (2005)
12. Raghavan, S., Garcia-Molina, H.: Crawling the hidden web. Technical report 2000-36, Stanford InfoLab (2000). http://ilpubs.stanford.edu:8090/456/
13. Sanderson, R.: Global web archive integration with memento. In: Proceedings of the 12th ACM/IEEE-CS Joint Conference on Digital Libraries, pp. 379–380. ACM (2012)
14. Sanderson, R., Van de Sompel, H., Nelson, M.L.: IIPC Memento Aggregator Experiment (2012). http://www.netpreserve.org/sites/default/files/resources/Sanderson.pdf
15. Sugiura, A., Etzioni, O.: Query routing for web search engines: architecture and experiments. Comput. Netw. **33**(1), 417–429 (2000)
16. Tran, T., Zhang, L.: Keyword query routing. IEEE Trans. Knowl. Data Eng. **26**(2), 363–375 (2014)
17. Van de Sompel, H., Nelson, M.L., Sanderson, R.: HTTP Framework for Time-Based Access to Resource States - Memento. RFC 7089 (2013)
18. Wu, P., Wen, J.R., Liu, H., Ma, W.Y.: Query selection techniques for efficient crawling of structured web sources. In: Proceedings of the 22nd International Conference on Data Engineering, ICDE 2006, pp. 47–47 (2006)
19. Xu, J., Callan, J.: Effective retrieval with distributed collections. In: Proceedings of the 21st Annual International ACM SIGIR Conference on Research and Development in Information Retrieval, pp. 112–120. ACM (1998)

Comparing Topic Coverage in Breadth-First and Depth-First Crawls Using Anchor Texts

Thaer Samar[1]([✉]), Myriam C. Traub[1], Jacco van Ossenbruggen[1],
and Arjen P. de Vries[2]

[1] Centrum Wiskunde & Informatica, Amsterdam, The Netherlands
samar@cwi.nl
[2] Radboud University, Nijmegen, The Netherlands

Abstract. Web archives preserve the fast changing Web by repeatedly crawling its content. The crawling strategy has an influence on the data that is archived. We use link anchor text of two Web crawls created with different crawling strategies in order to compare their coverage of past popular topics. One of our crawls was collected by the National Library of the Netherlands (*KB*) using a *depth-first* strategy on manually selected websites from the *.nl* domain, with the goal to crawl websites as completes as possible. The second crawl was collected by the *Common Crawl* foundation using a *breadth-first* strategy on the entire Web, this strategy focuses on discovering as many links as possible. The two crawls differ in their scope of coverage, while the *KB* dataset covers mainly the Dutch domain, the *Common Crawl* dataset covers websites from the entire Web. Therefore, we used three different sources to identify topics that were popular on the Web; both at the global level (entire Web) and at the national level (*.nl* domain): Google Trends, *WikiStats*, and queries collected from users of the Dutch historic newspaper archive. The two crawls are different in terms of their size, number of included websites and domains. To allow fair comparison between the two crawls, we created sub-collections from the *Common Crawl* dataset based on the *.nl* domain and the *KB* seeds. Using simple exact string matching between anchor texts and popular topics from the three different sources, we found that the *breadth-first* crawl covered more topics than the *depth-first* crawl. Surprisingly, this is not limited to popular topics from the entire Web but also applies to topics that were popular in the *.nl* domain.

1 Introduction

The World Wide Web offers rich means for its users to publish, share, create, discuss, collaborate and even earn a living. Web data, however, is surprisingly volatile. Ntoulas et al. found that 80 % of the Web pages disappear within one year [21]. In order to preserve (at least a fraction of) this data, many national libraries and archives have set up Web archiving initiatives. However, it is impossible to archive the entire Web due its increasing size, and the dynamic and ephemeral nature of its content. Therefore, institutes have to make decisions on the websites to be included in the archive, the crawling frequency, and the crawling strategy. One strategy is to crawl a manually selected set of websites (called

© Springer International Publishing Switzerland 2016
N. Fuhr et al. (Eds.): TPDL 2016, LNCS 9819, pp. 133–146, 2016.
DOI: 10.1007/978-3-319-43997-6_11

the crawler's *seeds*) and to harvest these websites in depth (*depth-first* crawl). Another strategy automatically crawls as many websites as possible (usually the national domains), but not in depth (*breadth-first* crawl). Both crawling strategies result in incomplete crawls, as both strategies exclude websites. *Depth-first* ignores websites outside the seeds list, and *breadth-first* archives websites incompletely as it does not follow the links to sub-pages. On top of the content of websites, Web archives also preserve information registered by crawlers such as the date of the crawl, the timestamp of the last modification of the page, the MIME-type, and information that can be derived from the archived pages, for example hyperlinks and anchor texts.

Web archives preserve content which may no longer be available on the Web. We explore how well the collections resulting from different crawling strategies cover content related to topics that were in the focus of Web users in a particular time period. We perform our analysis on two Web archive collections harvested in 2014 using different crawling strategies. The first collection is a crawl from the entire Web harvested by the *Common Crawl* foundation using the *breadth-first* crawling strategy. The second collection is the Dutch Web archive collection preserved by the National Library of The Netherlands[1] (*KB*). Here, the *depth-first* strategy was applied to manually selected websites (KB *seeds*) related to the Dutch history, social, and culture heritage. We propose to use anchor text specified in hyperlinks extracted from the two collections to investigate their coverage of the topics that were of interest to users in the same year (2014). Users of Web search engines express their information needs by issuing queries. User queries collected from major search engines would be the best record of popular topics. However, these queries were not available for us. Therefore, we used different sources as indicators of the trending topics on the Web at the time when the crawls we used were collected (2014). Since our crawls originate from the entire Web (*Common Crawl* crawl) and from the Dutch domain (*KB* crawl), we looked for popular topics both worldwide and on the national level. Our first source is Google Trends. Google provides a list of the top searched terms on the entire Web, and in the given country domain. The second source is the *WikiStats* which aggregates page views of Wikipedia pages. Again we focus on all Wikipedia pages (in all languages), and the pages written in Dutch. Finally, we use queries collected from users searching the Dutch digital newspaper archive via the *KB*'s Delpher[2] interface. These are three heterogeneous sources, the first and the third are real user queries, the second consists of Wikipedia titles associated with their frequency of views over time. We use these sources to represent users interests, which we refer to as topics.

2   Related Work

The structure of the Web graph is defined by its links which consist of a source URL, a destination URL and an anchor text describing the link. Several studies

---

[1] www.kb.nl.
[2] www.delpher.nl.

explored the structure of the Web graph based on crawls from the entire Web [2,19,23]. The link structure was used to study the evolution and the structure of Web crawls of national domains [1,5,25]. Properties of the Web graph, such as the PageRank and out-degree, were used to propose algorithms for seed selection of Web crawlers [24], and to improve the effectiveness of Web search. An empirical study showed that anchor texts exhibit characteristics similar to real user queries [7]. They also showed that anchor texts are similar to titles of webpages. This is based on the observation that titles can be used as an approximation of queries [10]. Anchor texts enrich the representation of a Web page's content to improve Web search effectiveness [4,6,8,11,14–16,18]. Kanhabua and Nejdl studied the evolution of anchor texts extracted from the edit history of Wikipedia [12]. They found that anchor texts with temporal information can be candidates for capturing and tracing the entity evolution.

The link structure and anchor texts constructed from the archived pages play an important role in assessing the completeness of Web archives. It is impossible to archive the entire Web due its increasing size and evolving content. Therefore, the archived parts of the Web are incomplete. Web archiving theorists acknowledge that the archived parts of the Web is both incomplete and over complete [3,17]. It is impossible to crawl the Web in a way that all websites and pages are included, for example the *depth-first* crawling strategy excludes websites not in the seeds list, and the *breadth-first* strategy does not crawl discovered websites in depth. Thus both strategies result in an incomplete crawl. On the other hand, Web archives are over complete, as they do not only contain the raw content but also metadata, such as the MIME-type and the date of the crawling time. More over, information that can be constructed from the archived pages, for example, the link structure and anchor texts. The wealth of information available in the Web archives has been discussed in [22]. Links and anchor texts can be used to locate missing webpages, of which the original URL is not accessible anymore. Klein and Nelson [13] computed lexical signatures of lost webpages, using the top n words of link anchors, and used these and other methods to retrieve alternative URLs for lost webpages. The use of the link structure and anchor texts to uncover and reconstruct target pages that were not archived was studied in [9], based on a *depth-first* crawl of manually selected websites. They used the link structure extracted from archived Web pages to uncover target URLs that were not archived. Links extracted from the archived pages contain evidence of the existence of unarchived target URLs. Based on the link evidence, Huurdeman et al. found that the number of unarchived Web pages is roughly as high as the number of the archived Web pages. Then, they used link evidence to reconstruct basic representations of target URLs. This evidence includes the aggregated anchor text, crawl date, and source URLs.

3   Setup

In this section we describe the two crawls on which we base our analysis. Then, we introduce the pipeline of extracting hyperlinks and anchor texts from the crawls. After that, we discus how we zoom in the link structure of *Common Crawl* dataset to generate subsets based on filters synthesized from the *KB* dataset in order to allow a fair comparison. Finally, we introduce the sources that we used to identify popular topics.

3.1   Data

KB **Dataset:** The *KB* archives a pre-selected set of more than 10,000 websites (*seeds*) with the aim to crawl these websites as complete as possible. The selection is based on categories related to Dutch historical, social and cultural heritage. The websites are categorized by curators of the *KB* using the *UNESCO* classification code. The crawling frequency varies between yearly, biannually, quarterly, and daily, for example news agency websites (such as nu.nl). Our snapshot of the Dutch Web archive between February 2009 and May 2015 consists of 150,557 files in ARC^3 format, which contain aggregated web content. Each ARC file contains multiple Web objects, in total, 251,591,618 objects exist in the ARC files. We focus on data crawled in 2014, as we have only access to *Common Crawl* pages crawled in that year.

Common Crawl **Dataset:** Common Crawl[4] is a non-profit organization aiming to build and maintain an openly accessible repository of archived Web crawls. We use the crawl collected in March 2014, which consists of 2.8 billion Web pages.

3.2   Anchor Links Extraction

From the two datasets, we extracted hyperlinks from the archived objects with *text/html* as MIME-type. For that we used MapReduce to process all archived web objects contained in the archive's ARC files. During the processing of the archived objects, we used JSoup[5] to extract anchor links (a) in order to be able to focus on links between textual content. For each anchor link, we kept the URL of the page that contains the link *source*, the URL of the *target*, and the anchor text specified in the link. Based on the crawl-date, we keep pages crawled in 2014. The anchor texts pointing to the target pages were used in that year. Depending on the source URL and target URL, the link can be an internal link or external link. An internal link has the same domain-name for both source and target (intra-domain), while for an external link the domain-name of the source URL is different from that of the target URL (an inter-domain link). We limit our analysis to the external links as it is of more interest to look into links between different hosts (sites). By discarding internal links we exclude links from menus and other non-content information. The exact URLs may change frequently, while we are really interested in anchor text used by one site to link to another site. Therefore, we replace both the source URL and the target URL by their hosts (site name) before we analyze the data. This pre-processing can be

---

³ http://archive.org/web/researcher/ArcFileFormat.php.
⁴ http://commoncrawl.org/.
⁵ http://jsoup.org/.

viewed as a process to smooth the graph structure to maintain the most salient information. We deduplicate the links based on their values for source, target, and anchor text for *KB* dataset (*Common Crawl* dataset consists of one crawl). This prevents the differences in crawling frequency to influence our analysis. At the end of this pipeline, we keep (*sourceHost, targetHost, anchorText*). We refer to the links extracted from the *KB* dataset as KB_{links}, and links extracted from the *Common Crawl* dataset as CC_{links}.

3.3   Link Subsets from *Common Crawl*

The two crawls differ in terms of size, number of crawled websites and web pages, and the domains of the crawled websites. These differences are reflected in the extracted links structure. The number of links extracted from the *Common Crawl* dataset is $559x$ times larger than the number of *KB* links, (see Table 1). Therefore, in addition to performing one-to-one comparison between the two crawls, we generate subsets from the CC_{links} by mapping it to the Dutch domain in two different ways: First, we focused on pages that originate from the *.nl* domain. This was done by keeping only links from the CC_{links} whose *source hosts* are from the *.nl* domain. We refer to the set as $CC_{links} \cap .nl$. Second, the *KB* crawl is based on a list of manually selected websites (*KB* seeds). We used the *KB* seeds to generate another subset of links from the CC_{links}, based on links with *source hosts* from the *KB* seeds. We refer to this subset as ($CC_{links} \cap KB_{seeds}$). Finally, we investigate the impact of anchor texts associated with targets of links in the *KB* dataset on the topic coverage of the CC_{links}. In order to do that, we dropped links from CC_{links} in which the *target hosts* are targets of links in the KB_{links}. We refer to this set of filtered links as ($CC_{links} \setminus KB_{targets}$). These subsets allow us to investigate whether the *KB* seeds list comprises the part of the Dutch Web that is essential from the perspective of topic coverage, or whether a broader and less deep crawl would still contain sufficient information.

Table 1. Number of unique links in each dataset.

Links dataset	Num. of links
KB_{links}	3,033,855
CC_{links}	1,696,102,933
$CC_{links} \cap .nl$	5,128,501
$CC_{links} \cap KB_{seeds}$	2,629,765
$CC_{links} \setminus KB_{targets}$	1,174,261,413

3.4   Sources of Topics

Our assumption is that the *Common Crawl* (a *breadth-first* crawl) covers more global topics, and that the *KB* (a *depth-first* crawl) covers more topics from the *.nl* domain. In order to validate our assumption, we use different sources to identify which topics were popular on the Web, topics that attracted attention in the entire Web (global) and topics that were only picked up in the .nl domain.

Google Trends. Google Trends[6] is a public resource, which lists the most searched queries in the global Web or per country in a given year. For our analysis, we use global trends and the trends searched in the Netherlands in 2014 (the year of our crawls).

Wikipedia Page Views Statistics. The *WikiStats* dataset [20] consists of the number of views for Wikipedia pages. The goal is to show how the interest in Wikipedia pages changes over time, and allows comparison between chosen Wikipedia pages. The views are aggregated from the *Page view statistics for Wikimedia projects*[7], which aggregates the request history of articles from Wikimedia projects[8]. For each page, this project provides the page title, the number of requests (on hourly basis), the language in which the page is written, and the name of the project. The *WikiStats* data set consists of the weekly views of Wikipedia pages in the period from January 2008 to January 2015. We select Wikipedia pages viewed in 2014, then aggregate their page view counts, and those pages viewed more than 1,000 times. Finally, we created two datasets: the first contains all Wikipedia pages from all domains (*WikiStats* global), and the second contains only pages written in Dutch language (*WikiStats* .nl).

User Queries. Under conditions of strict confidentiality, the *KB* made anonymized user logs available, collected between March 2015 and December 2015 from users visiting the public digital newspaper archive on a webservice called Delpher. The collection consists of newspapers articles published in the Netherlands since 1618. The data set made available consists of 10 million OCRed newspaper pages in DIDL XML format[9].

Sources Summary. We processed all topics from the sources mentioned with the same pre-processing pipeline, which includes lower casing, stopwords (English and Dutch) removal, and the removal of short terms with a length of less than three characters. The resulting dataset statistics are summarized in Table 2.

Table 2. Number of unique topics per source.

Topics source	Count
Google global trends	84
Google .nl trends	68
WikiStats global	3, 293, 749
WikiStats .nl	99, 396
Real queries	1, 580, 386

---

[6] http://www.google.com/trends/topcharts?hl=en#date=2014&geo=.

[7] http://dumps.wikimedia.org/other/pagecounts-raw/.

[8] These projects are: wikibooks, wiktionary, wikinews, wikivoyage, wikiquote, wikisource, wikiversity, and wikipedia.

[9] http://www.xml.com/pub/a/2001/05/30/didl.html.

4   Analysis

Using anchor texts we investigate the coverage of topics in *Common Crawl* (a *breadth-first* crawl), and *KB* (a *depth-first* crawl). Since anchor texts usually describe target pages, we first provide a deep analysis of them with regard to their hosts and top-level domains (*TLDs*). Then, we present a detailed analysis of anchor texts associated with hyperlinks. Finally, we investigate the anchor texts coverage of topics from the three sources described in Sect. 3.4.

4.1   Target Pages

For all link datasets, the number of unique hosts in the target pages is higher than the number of unique hosts of source pages (see Table 3). In KB_{links}, the number of unique target hosts is 442,296, which is 14 times higher than the number of source hosts (31,829). In CC_{links}, the ratio between the target hosts (30,416,854) and the source hosts (9,715,414) is lower, here, the number of target hosts is only 3 times higher than the number of source hosts. These numbers of source hosts and target host shows the big difference between the two dataset. However, subsets from *Common Crawl* dataset have comparable numbers. The crawling strategy clearly affects the percentage of target hosts that have been crawled. The percentage of the crawled target hosts differ between the link datasets, (see Table 3). For example, only 6.5 % of KB_{links} target hosts were crawled, whereas 23.9 % of target hosts in CC_{links} were crawled. However, both crawling strategies showed that large fractions of target hosts were not crawled, and we cannot find their raw content. This suggests that the use of target hosts, and anchor texts as a means to describe them is a valuable resource. We also looked into the overlap of target hosts between the datasets. A high percentage (71.4 %) of target hosts in KB_{links} were also targets of links in CC_{links}. The percentage of overlap decreases to 38.5 % after subsetting the *Common Crawl* dataset based on source pages from the .*nl* domain ($CC_{links} \cap .nl$), and decreases to 24.2 % after projecting the *KB* seeds on CC_{links} ($CC_{links} \cap KB_{seeds}$). Recall, that there is no overlap between KB_{links} and $CC_{links} \setminus KB_{targets}$, because all links whose target hosts are the same as the target hosts in KB_{links} were dropped from CC_{links}. In terms of the source hosts not only the number of hosts is lower compared to the number of target hosts, but also the overlap between KB_{links} and the other datasets is smaller (see Table 3).

Top-level Domains. Another way of looking at the difference between the link datasets is based on the TLDs of the target pages. The TLDs represent the target domains of the crawled pages. In CC_{links}, a high percentage of links points to the pages from the .*nl* domain, and the majority (60.5 %) of the target pages are from the .*com* TLD, see Table 4. The majority of target pages (45.6 %) in KB_{links} are from the .*nl* domain, which is expected because the *KB* crawl was harvested based on websites mainly from the Dutch Web. The target pages in $CC_{links} \cap .nl$ has the same distribution of top-ranked TLDs of target pages in KB_{links}. In the distribution of TLDs for $CC_{links} \cap KB_{seeds}$, the .*com* is the

Table 3. Analysis of hosts: For both the target and source pages, we present the absolute count of unique hosts (*first row*), the fraction of hosts from KB_{links} that were found in the corresponding dataset in column header (*second row*), and the percentage of target hosts that has been crawled in each link dataset (*third row*).

	KB_{links}	CC_{links}	$CC_{links} \cap .nl$	$CC_{links} \cap KB_{seeds}$	$CC_{links} \backslash KB_{targets}$
Target hosts	442,296	30,416,854	800,957	529,962	30,100,936
	100.0 %	71.4 %	38.5 %	24.2 %	0 %
Source hosts	31,829	9,715,414	120,498	2,942	8,237,940
	100.0 %	57.8 %	28.5 %	8.5 %	42.9 %
Crawled target hosts	6.5 %	23.9 %	8.5 %	0.4 %	20.5 %

Table 4. TLDs of target pages: The count of unique TLDs, and the top-10 TLDs.

	KB_{links}	CC_{links}	$CC_{links} \cap .nl$	$CC_{links} \cap KB_{seeds}$	$CC_{links} \backslash KB_{targets}$
Count (unique)	293	456	267	268	451
	nl	com	nl	com	com
	com	org	com	org	org
	org	net	org	nl	net
	net	de	net	net	de
	de	info	de	de	info
	be	nl	be	be	nl
	eu	ru	eu	it	it
	info	it	info	ro	ru
	fr	fr	it	fr	fr
	it	pl	fr	info	pl

most prevalent TLD; 49 % of target pages belong to this domain, not all websites in the *KB* seeds were found in *Common Crawl* dataset, only 43.6 % (unique) were found. The *KB* seeds are not all from the *.nl* domain, only 88 % of the seeds belong to the *.nl* domain. The remaining seeds (12 %) belong to different TLDs: 5 % from the *.org* domain, 3.4 % from the *.com* domain, 1.2 % from the *.net* domain, 0.6 from the *.eu* domain, and 0.5 % from the *.info* domain. The distribution of the top TLDs is similar in CC_{links} and $CC_{links} \backslash KB_{targets}$. The only difference is the number of target pages per TLD, which decreases for some TLDs in $CC_{links} \backslash KB_{targets}$ compared to CC_{links}. This is caused by dropping links whose target hosts are the same as the target hosts in KB_{links}. Thus the highest relative decrease was for the *.nl* domain.

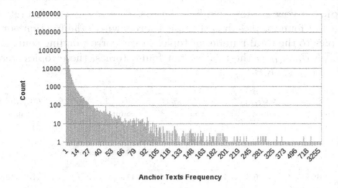

Anchor Texts Frequency

Fig. 1. Anchor texts frequency distribution of KB_{links} in log scale representation.

Table 5. Anchor texts summary: For each link dataset, we present the number of unique anchor texts, and the overlap of anchor texts between KB_{links} and the corresponding dataset. Considering all anchor texts in KB_{links} (%overlap_all), and by considering anchor texts used at least twice in KB_{links} (%overlap_GT1).

Links dataset	Count	%overlap_all	%overlap_GT1
KB_{links}	1,581,013	100.0	13.0
CC_{links}	83,920,299	23.6	49.9
$CC_{links} \cap .nl$	2,613,774	13.7	40.5
$CC_{links} \cap KB_{seeds}$	1,289,803	9.2	26.7
$CC_{links} \backslash KB_{targets}$	61,153,447	15.3	34.4

4.2   Anchor Texts

Some anchor texts are used by multiple links and the frequency of the anchor texts represents their popularity in the archive. We processed the anchor texts with the same pre-processing pipeline we used for the topics (Sect. 3.4) and computed the frequencies of all unique anchor texts for each link dataset. The number of unique anchor texts varies strongly among the datasets (see Table 5). When we compared the percentage of overlap between anchor texts in KB_{links} and all other link datasets based on exact string matching, we found that 23.6% of the unique anchor texts in KB_{links} exist in the unique anchor texts of CC_{links}. The frequency of anchor texts in KB_{links} shows a long tail distribution (Fig. 1). A high percentage (87%) of the anchor texts in KB_{links} occurs only once. We investigated the overlap considering only anchor texts with a frequency larger than one. This results in an increase of the percentage of overlap between KB_{links} with all datasets. We can use the frequency as threshold to focus on most popular anchor texts.

Table 6. Topic Coverage: for each link dataset, we present the absolute count and the fraction (%) of found topics in each topic source, where the fraction is the number of matched topics to the total number of topics in the corresponding source. The %*lost* under $CC_{links} \backslash KB_{targets}$ is the relative not found topics, these topics were found in CC_{links} but in $CC_{links} \backslash KB_{targets}$.

Topics source	KB_{links}		CC_{links}		$CC_{links} \cap .nl$		$CC_{links} \cap KB_{seeds}$		$CC_{links} \backslash KB_{targets}$		
	count	%	count	%	count	%	count	%	count	%	%lost
Google global trends	24	28.6	51	60.7	25	29.8	23	27.4	51	60.7	0.0
Google .nl trends	22	32.4	27	39.7	25	36.8	18	26.5	24	35.3	−11.1
WikiStats global	80,043	2.4	1,376,222	41.8	122,659	3.7	116,259	3.5	1,122,767	34.1	−18.4
WikiStats .nl	24,726	24.9	48,825	49.1	31,742	31.9	19,098	19.2	43,304	43.6	−11.3
Real queries	26,099	1.7	77,152	4.9	38,033	2.4	15,839	1.0	66,874	4.2	−13.3

4.3   Topic Coverage

An anchor text describes the target page with a brief text which is known to resemble user queries. Therefore analyzing the anchor texts' overlap with queries is a good proxy for assessing whether the crawls are likely to contain answers to user queries and popular topics. Not all target pages that are linked to from the crawled pages are harvested by the crawler. As mentioned earlier, Web archives are incomplete, and the advantage of anchor texts is their availability for both crawled and not crawled target pages. In order to investigate the topic coverage, we used exact string matching between pre-processed anchor texts from the five link datasets with topics from the sources (described in Sect. 3.4). Topic coverage varies among the datasets for the different sources of topics, (see Table 6). For some cases we found high coverage, for example anchor texts from CC_{links} matched 60.7 % of Google global trends and 49.1 % of the Dutch Wikipedia pages in the *WikiStats*. After sorting the anchor texts in descending order based on their frequencies, we investigated the relation between the percentage of topics covered and the frequency (popularity) of anchor texts. We report on exact string matches between the top anchor texts and both, Wikipedia titles and real user queries, considering different rank cutoffs c; $c = 1k$, $10k$, and $100k$. The percentage of matched anchor texts with *WikiStats* (global and .nl domain) decreases as c increases for all link datasets (see Table 8). The lowest overlap corresponds to the case when all anchor texts are used to match Wikipedia titles.

In general, the percentage of overlap between anchor texts from the different datasets and the user queries is low. For example, we found that only 1.7 % of the user queries had a match in KB_{links} when we applied exact string matching with all anchor texts. We found the highest percentage of overlap with user queries (4.9 %) for anchor texts in CC_{links} (see Table 6). When we compared the top-c anchor texts instead of the complete set of anchor texts, we found a relation between the top ranked anchor texts and the percentage of the topic coverage. A high percentage of the most frequently used anchor texts matched user queries, and the percentage of overlap decreases while the cutoff c increases, see Table 9.

Table 7. Unique Topic Coverage in KB_{links}: in comparison with topics found in other datasets. Under every link dataset x, we present the percentage of topics found in the KB but not found in x, and the percentage of topics found in x but not found in KB_{links}.

Topics source	CC_{links}		$CC_{links} \cap .nl$		$CC_{links} \cap KB_{seeds}$		$CC_{links} \backslash KB_{targets}$	
Google global trends	0.0%	54.9%	29.2%	32.0%	33.3%	30.4%	0.0%	54.9%
Google .nl trends	0.0%	18.5%	9.1%	20.0%	31.8%	16.7%	13.6%	20.8%
WikiStats global	6.1%	94.5%	46.1%	64.8%	56.3%	69.9%	12.2%	93.7%
WikiStats .nl	16.5%	57.7%	28.2%	44.0%	53.6%	39.9%	26.4%	58.0%
Real queries	22.3%	73.7%	31.4%	53.0%	60.4%	34.7%	31.8%	73.4%

Table 8. The fraction of top-c ranked anchor texts matching document titles in *WikiStats*, fraction in percentage (*num. matches/c*). In addition to the percentage of matching when all anchor texts are used (*num. matches/num. all anchor texts*).

Links dataset	WikiStats global				WikiStats .nl domain			
	top-1k	top-10k	top-100k	all	top-1k	top-10k	top-100k	all
KB_{links}	44.6	39.1	22.1	5.1	32.4	24.3	10.2	1.6
CC_{links}	51.3	56.9	38.2	1.6	11.5	14.4	7.4	0.1
$CC_{links} \cap .nl$	45.2	37.3	24.7	4.7	32.0	23.9	11.8	1.2
$CC_{links} \cap KB_{seeds}$	70.6	65.6	33.9	9.0	32.0	23.3	8.3	1.5
$CC_{links} \backslash KB_{targets}$	71.2	63.5	28.1	1.8	19.5	14.7	4.2	0.1

Table 9. The fraction of top-c ranked anchor texts matching user queries, same notation as in Table 8.

Links dataset	Real queries			
	top-1k	top-10k	top-100k	All
KB_{links}	26.4	23.2	9.7	1.7
CC_{links}	17.2	18.2	7.0	0.9
$CC_{links} \cap .nl$	33.5	26.0	12.8	1.5
$CC_{links} \cap KB_{seeds}$	30.0	20.5	6.9	1.2
$CC_{links} \backslash KB_{targets}$	28.5	23.0	6.9	0.1

We found the highest overlap of topics and anchor texts in CC_{links}, suggesting that the *breadth-first* crawl covers more topics than the *depth-first* crawl. This result holds for both, the global and the national (*.nl*) topics. Focusing on the Dutch part of the *Common Crawl* dataset ($CC_{links} \cap .nl$) showed that this part covers more topics than topics covered in KB_{links}. However, the comparison is based on the absolute count of found topics in each links dataset. That does not necessarily mean that all topics covered by KB_{links}, are identical with those found, for instance, in CC_{links}. For all topic sources, we analyzed the topics that were found in KB_{links} but not in the other datasets (see Table 7). For example, we found that all Google trends (both the global and the *.nl* domain) that were found in KB_{links}, were also found in CC_{links}. On the other hand, 54.9 % of Google's global trends and 18.4 % of Google's *.nl* trends found in CC_{links} were not found in KB_{links}. Regarding the *WikiStats* dataset, not all topics found in KB_{links} were found in CC_{links}. The percentage of topics that were found in KB_{links} is higher for the Wikipedia pages from the *.nl* domain (16.5 %), while 6.1 % of Wikipedia pages (global) found in KB_{links} were also found in CC_{links}.

These results suggest that anchor texts can be used as a resource for finding topics that were popular with users from the past. The coverage of topics was higher for the most frequently used anchor texts in the crawls. Anchor texts from the *breadth-first* crawl cover more topics than the anchor texts from the *depth-first* crawl. However, some topics were only covered by the *depth-first* crawl.

5 Conclusions

We studied the influence of the crawling strategy on the coverage of topics that were of interest to users on the Web. We performed our analysis on two Web crawls created by following different crawling strategies; the *Common Crawl* dataset, (a *breadth-first* crawl) collected from the entire Web, and the *KB* dataset (a *depth-first* crawl) harvested by the *KB* based on manually selected websites). We made use of anchor texts to investigate the topic coverage in the two crawls. We extracted anchor texts from the raw content of documents in crawls, and compared them with other sources that identify popular topics on Web at the time of the crawls (2014). The two crawls differ in terms of scope. While *Common Crawl* covers domains from the entire Web, *KB* covers mainly the Dutch domain. Therefore, we used different sources as a proxy of topics that were popular in 2014, both worldwide (entire Web) and national (*.nl* domain).

Using exact string matching between anchor texts and topics from different sources, we found that the percentages of matches vary between the topic sources and the two crawls. For example, CC_{links} covers 61 % of Google global trends, and 5 % of real queries (submitted by users to the search system of the Dutch digital newspaper archive). KB_{links} covers 32 % of Google *.nl* trends, and 2 % of the real queries. This suggests that anchor texts are a useful resource for investigating popular topics from the past. We found a correlation between the frequency of anchor texts in the archive and the percentage of topic matches.

When we compared the topic coverage between the *Common Crawl* and the *KB* datasets, we found that the percentage of overlapping topics is higher in the *Common Crawl* dataset, for both global and *.nl* topics. This result holds for the $CC_{links} \cap .nl$ (only focusing on links in *Common Crawl* originating from the *.nl* domain). More over, using the $CC_{links} \cap KB_{seeds}$ (was created using *KB* seeds to subset CC_{links}) has comparable result to KB_{links}. However, not all topics found by the *depth-first* crawl were found by the *breadth-first* crawl. We conclude that the coverage in the *breadth-first* crawl is higher even for topics of national interest, but there are topics that are covered only by *depth-first* crawl.

In future work, we can investigate the topic coverage in the crawls taking the importance of topics into account, in this analysis all topics were weighted equally.

Acknowledgments. We would like to thank the National Library of the Netherlands for their support. This research was funded by the Netherlands Organization for Scientific Research (NWO CATCH program, WebART project), and Dutch COMMIT/program (SEALINCMedia project). Part of the analysis work was carried out on the Dutch e-infrastructure with the support of the SURF Foundation.

References

1. Baeza-Yates, R.A., Poblete, B.: Evolution of the Chilean web structure composition. In: LA-WEB, pp. 11–13 (2003)
2. Broder, A.Z., Kumar, R., Maghoul, F., Raghavan, P., Rajagopalan, S., Stata, R., Tomkins, A., Janet, L.: Wiener.: graph structure in the web. Comput. Netw. **33** (1–6), 309–320 (2000)
3. Brügger, N.: Historical network analysis of the web. Soc. Sci. Comput. Rev. **31**(3), 306–321 (2013)
4. Craswell, N., Hawking, D., Robertson, S.: Effective site finding using link anchor information. In: SIGIR, pp. 250–257. ACM (2001)
5. Donato, D., Leonardi, S., Millozzi, S., Tsaparas, P.: Mining the inner structure of the web graph. In: WebDB, pp. 145–150 (2005)
6. Dou, Z., Song, R., Nie, J.-Y., Wen, J.-R.: Using anchor texts with their hyperlink structure for web search. In: SIGIR, pp. 227–234 (2009)
7. Eiron, N., McCurley, K.S.: Analysis of anchor text for web search. In: SIGIR, pp. 459–460 (2003)
8. Fujii, A.: Modeling anchor text and classifying queries to enhance web document retrieval. In: WWW, pp. 337–346 (2008)
9. Huurdeman, H.C., Kamps, J., Samar, T., de Vries, A.P., Ben-David, A., Rogers, R.A.: Lost but not forgotten: finding pages on the unarchived web. Int. J. Digit. Libr. **16**, 247–265 (2015)
10. Jin, R., Hauptmann, A.G., Zhai, C.: Title language model for information retrieval. In: SIGIR, 11–15 August 2002, Tampere, Finland, pp. 42–48 (2002)
11. Kamps, J.: Web-centric language models. In: CIKM, pp. 307–308 (2005)
12. Kanhabua, N., Nejdl, W.: On the value of temporal anchor texts in wikipedia. In: SIGIR Workshop on Temporal, Social and Spatially-Aware Information Access (2014)
13. Klein, M., Nelson, M.L.: Moved but not gone: an evaluation of real-time methods for discovering replacement web pages. Int. J. Digit. Libr. **14**, 17–38 (2014)
14. Kleinberg, J.M.: Authoritative sources in a hyperlinked environment. J. ACM **46**, 604–632 (1999)
15. Koolen, M., Kamps, J.: The importance of anchor text for ad hoc search revisited. In: SIGIR, pp. 122–129 (2010)
16. Kraft, R., Zien, J.: Mining anchor text for query refinement. In: Proceedings of the 13th International Conference on World Wide Web, WWW 2004, pp. 666–674. ACM, New York (2004). doi:10.1145/988672.988763. ISBN: 1-58113-844-X
17. Masanès, J.: Web Archiving. Springer, Berlin (2006)

18. Metzler, D., Novak, J., Cui, H., Reddy, S.: Building enriched document represen-
    tations using aggregated anchor text. In: SIGIR (2009)
19. Meusel, R., Vigna, S., Lehmberg, O., Bizer, C.: Graph structure in the web -
    revisited: a trick of the heavy tail. In: WWW, pp. 427–432 (2014)
20. Mühleisen, H.: Wikistats – wikipedia pageviews (2013). http://wikistats.ins.cwi.nl
21. Ntoulas, A., Cho, J., Olston, C.: What's new on the web? The evolution of the
    web from a search engine perspective. In: WWW, pp. 1–12 (2004)
22. Rauber, A., Bruckner, R.M., Aschenbrenner, A., Witvoet, O., Kaiser, M.: Uncov-
    ering information hidden in web archives: a glimpse at web analysis build-
    ing on data warehouses. D-Lib Mag. 8(12), 1082–9873 (2002). doi:10.1045/
    december2002-rauber
23. Ángeles Serrano, M., Maguitman, A.G., Boguñá, M., Fortunato, S., Vespignani, A.:
    Decoding the structure of the www: a comparative analysis of web crawls. ACM
    Trans. Web (TWEB) 1(2), 10 (2007). doi:10.1145/1255438.1255442
24. Zheng, S., Pavel Dmitriev, C., Giles, L.: Graph-based seed selection for web-scale
    crawlers. In: Proceedings of the 18th ACM Conference on Information and Knowl-
    edge Management, CIKM, Hong Kong, China, 2–6 November 2009, pp. 1967–1970
    (2009)
25. Zhu, J.J.H., Meng, T., Xie, Z., Li, G., Li, X.: A teapot graph and its hierarchical
    structure of the Chinese web. In: WWW, pp. 1133–1134 (2008)

How to Search the Internet Archive
Without Indexing It

Nattiya Kanhabua[1]([✉]), Philipp Kemkes[2], Wolfgang Nejdl[2],
Tu Ngoc Nguyen[2], Felipe Reis[2], and Nam Khanh Tran[2]

[1] Department of Computer Science, Aalborg University, Aalborg, Denmark
nattiya@cs.aau.dk
[2] L3S Research Center/Leibniz Universität Hannover, Hannover, Germany
{kemkes,nejdl,tunguyen,reis,ntran}@L3S.de

Abstract. Significant parts of cultural heritage are produced on the web
during the last decades. While easy accessibility to the current web is a
good baseline, optimal access to the past web faces several challenges.
This includes dealing with large-scale web archive collections and lacking
of usage logs that contain implicit human feedback most relevant for
today's web search. In this paper, we propose an entity-oriented search
system to support retrieval and analytics on the Internet Archive. We
use Bing to retrieve a ranked list of results from the current web. In
addition, we link retrieved results to the WayBack Machine; thus allowing
keyword search on the Internet Archive without processing and indexing
its raw archived content. Our search system complements existing web
archive search tools through a user-friendly interface, which comes close
to the functionalities of modern web search engines (e.g., keyword search,
query auto-completion and related query suggestion), and provides a
great benefit of taking user feedback on the current web into account
also for web archive search. Through extensive experiments, we conduct
quantitative and qualitative analyses in order to provide insights that
enable further research on and practical applications of web archives.

1   Introduction

Traditional institutions, e.g., national libraries, keep our cultural heritage and
need to be complemented with facilities for preservation and public access to
online cultural assets. This is critical given that even for the presumably interesting resources shared through social media like Twitter were estimated that
27 % of those are lost and not archived after $2\frac{1}{2}$ years [8]. National and international initiatives have recognized this need and started to collect and preserve parts of the web. The Internet Archive has by far the largest web archive
collection among the institutions active in web preservation, where it has collected more than 2.5 Petabyte of web content since 1996. Another important
European initiative is the Internet Memory Foundation, active in several EU-
funded research projects on web archiving, with a set of smaller crawls for specific
topics, domains and projects. Two important national libraries engaged in web
preservation are the British Library and the German National Library, with the
aim to preserve their national web content.

© Springer International Publishing Switzerland 2016
N. Fuhr et al. (Eds.): TPDL 2016, LNCS 9819, pp. 147–160, 2016.
DOI: 10.1007/978-3-319-43997-6_12

Easy access to historical web information becomes more and more important as the means for accessing and exploring these archives, but the current facilities are severely underdeveloped [1,3]. None of the archive initiatives is able to provide their collections through an interface, which comes close to the functionalities we see on today's web search engines. The Wayback Machine[1] provides the ability to retrieve and access web pages stored in the Internet Archive. However, it requires users to represent their information needs by specifying the URLs of web pages to be retrieved; thus demanding a lot of manual effort compared to the current web search engines, such as, Google, Yahoo! and Bing. Clearly, a simple, yet effective search interface is needed to retrieve information, which is stored in the Internet Archive and other web archive collections.

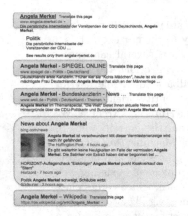

Fig. 1. URLs in blue areas are *long-term relevant*; URLs in the red area are *short-term relevant*. (Color figure online)

One major problem with web archive search is an absence of query logs. Without search logs in web archives, it is difficult to understand users' information needs and thus not able to provide a good ranking of search results without bias. Consider the top search results of the entity Angela Merkel in Fig. 1, which was retrieved from Bing on August 20, 2015. Top search results for a popular entity like the German politician consist of both long-term relevant pages, e.g., the Wikipedia page or bibliography pages in news websites, and short-term relevant pages, i.e., news articles about the entity. It can be seen that no news articles aged over one day appear in the top results, the rest of the search results are long-term relevant web pages associated with static URLs (those that are unchanged over a long time period).

To compensate this shortcoming, we built a prototype archive search system on top of Bing, which already provides a good mix of long-term and short-term relevant results. Our assumption is that, on the current web search, there are certain types of query intents that are similar to information needs on web archive search. In particular, we are interested in supporting web archive searches for *named entity queries*, which represent a significant fraction of current web search queries [5,11].

The goal of this work is to provide a scalable and responsive search system that supports entity-oriented search on web archives. We propose a novel web archive search system that leverages a current web search engine and the Internet Archive. Relying on commercial web search engine technologies for accessing web archives help us to achieve good quality ranking results (with high precision) based on search sessions and implicit human feedback. While providing entity-based indexing of web archives is crucial, we do not address the indexing issue in this work, but instead extend the WayBack Machine API in order to retrieve archived content.

---

[1] https://archive.org/web/.

For the best of our knowledge, we are the first to provide entity-oriented search on the Internet Archive, as the basis for a new kind of access to web archives, with the following contributions: (1) We propose a novel web archive search system that supports entity-based queries and multilingual search. (2) We make our search system publicly accessible for enabling further research on and practical applications for web archives. (3) Through extensive experiments, we conduct qualitative and quantitative studies and provide detailed analysis on the results returned through our web archive search system. (4) Finally, we outline the next steps towards more advanced retrieval and exploration of web archive content.

The organization of the rest of the paper is as follows. In Sect. 2, provides a discussion of related work. In Sect. 3, we present our problem statement. In Sect. 4, we describe our proposed web archive search and the underlying methodology. In Sect. 5, we present the evaluation of our proposed approaches and discuss the experimental results. Finally, we conclude our work in Sect. 6.

2   Related Work

The Internet Archive is a non-profit organization with the goal of preserving digital document collections as cultural heritage and making them freely accessible online. Another important European initiative is the Internet Memory Foundation, active in several EU funded research projects on web archiving. Two important national libraries engaged in web preservation are the British Library and the German National Library, with the aim to preserve national web content. Despite an enormous amount of information is stored in web archives, there are a few search prototypes to provide access to these archives, but all come with a number of limitations [2]. In 2009, the Internet Archive ran a pilot in providing full-text searchability for parts of their archive, making the first five years of their web archive (1996–2000) available for searching. However, the search ranking mechanisms available at that time were not adequate, and the search results were full of spam; thus limiting users from advanced search and exploration of archived content.

The Wayback Machine, a web archive access tool developed by the Internet Archive, provides the ability to retrieve and access web pages stored in a web archive, but it requires a user to access data by specifying the URL of a web page to be retrieved. For example, given the query URL http://www.usa.gov, search results are displayed in a calendar view showing the number of times the URL was crawled (not how many times it was actually updated). To date, it is not possible to search by keywords. Similarly, the Memento project[2] provides access to previous versions of a web page existed at some dates in the past, by entering the web page's URL, and by specifying the desired date in a browser plug-in. In this manner, Memento makes archived content discoverable via the original URL that the searcher already knew about, and redirecting the user to the archive, which hosts the page at the time indicated by the user. Archive-IT[3]

---

[2] http://timetravel.mementoweb.org.
[3] https://archive-it.org.

is a web archiving service for collecting and accessing cultural heritage sites on the web, built by the Internet Archive. The service supports organizations to harvest, build, and preserve collections of digital content, as well as full-text search on the archived collections. Nevertheless, this search functionality is only limited to a set of smaller crawls for specific topics, domains and projects.

In the context of searching web archives, Nguyen et al. [7] proposed an approach to discovering important documents along the time-span of the web archives by combining relevance, temporal authority, diversity and time in a unified ranking framework. Singh et al. [9] proposed a novel algorithm, HistDiv, that explicitly models the aspects and important time windows for supporting historical search intent. For an application of web archives, Tran et al. [10] studied a timeline summarization of an entity served as important memory cues in a retrospective event exploration.

3   Problem Statement

Information Needs. Entity queries, e.g., person, organization and location, comprise a significant fraction in web search logs [5,11]. Due to a lack of web archive search logs, we assume similar search behavior, i.e., information needs are either (1) exploring entity-related information, or (2) seeking a specific event in which entities involve. We allow users to search archives using entity queries that exist in Wikipedia for both English and German. To enhance query formulation, we provide two techniques, namely, auto-completion and related-entity suggestion. Finally, we also determine the categories for a given entity, by using a heuristic approach [4] for a further analysis.

Search Results. In our context, we define two main types of relevant results for entity queries: (1) general or static pages about the entity that do not change much over time (which are relevant over a long period of time), so-called *long-term relevant* and (2) dynamic pages such as news articles or blogs (which are only relevant for a short-time period), denoted *short-term relevant*.

Ranking Method. In this work, we aim at providing high precision search results, rather than optimizing recall. In order to achieve our goal, we build on the ranking of search results provided by Bing, given that the current web search engines already try to provide a suitable mix of long-term and short-term relevant pages, while taking a lot of user feedback into account. Although we assume Bing returns relevant results, we still need to investigate search results for different kinds of entities in a principled way.

Result Coverage. Coverage of results returned by Bing is concerned about how many of these results are archived on the Internet Archive. We hypothesize that many of these general pages about the entity are archived already, while news or recent pages are not be indexed yet. In fact, important news sites or domains themselves should all be archived. Another aspect to be addressed is the *temporal dynamics* of search results. We assume that the top-ranked results returned by Bing at different time points and rate of their changes vary by entity categories. Finally, we will analyze result variations over time in order to gain more insight.

4   Our Approach

The huge size of web archives has not only created great challenges for indexing them, but also increased the difficulty for users to express their information needs. More precisely, it is not easy for users to compose a succinct and precise query because of the temporal dimension. Our archive search system provides keyword-based search functionalities, similar to existing web search engines. Users can issue an entity-based query for any entity described in Wikipedia in English and German. The system returns a ranked list of search results, which provides links to both the current page on the web, as well as the archived versions in the Internet Archive, using blue and green links. Blue link refers to the current page on the web. Green link refers to the archived versions in the Internet Archive. Our system is publicly accessible[4]. Figure 2 shows the overview of our system framework and a search user interface. In the following, we will describe the proposed approach underlying different components, which comprise query auto-completion, Bing search API, archive linking, caching, and related entity suggestion.

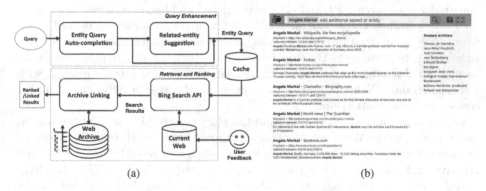

(a)   (b)

Fig. 2. Entity-oriented web archive search: (a) system framework and (b) user interface.

4.1   Query Auto-Completion

We support the query formulation process using query auto-completion. When a user types a query, we suggest a short list of relevant entities in order to help the user complete his/her information needs. We use a Wikipedia entity index comprised of all Wikipedia entities and store it in a trie data structure to allow fast prefix lookup. Additionally, we split all entities at white and special characters. All strings starting at each token are added to the trie as an additional reference to the original entity. Furthermore, we also take into account simplified versions of all

---

[4] http://alexandria-project.eu/archivesearch/.

tokens which contained letters with accents in our index. This allows our application to suggest entities even if the user does not know the exact name or cannot type the name in a foreign alphabet. As an example, for the query "schroder", we would suggest the former German chancellor "Gerhard Schröder". We further rank the suggested query completions by their popularity using the cumulative page views (see the detailed description below). To penalize the time-sensitive popularity of the entities, the daily page views are accumulated over a long period. Finally, the entity selected by the user is sent as input to the search API.

Wikipedia page views[5] are statistics consisting of the number of times a particular Wikipedia page has been requested over time. In Wikipedia, the view counts for pages that redirect to a given page are not combined with page views of the page being redirected to. In this work, we aggregate all these related views to present the popularity (reflected by the page views) of an entity query for all its query variants. We computed the aggregated statistics of page views approximately a period of 4 years (2011 to 2015).

4.2 Bing Search API, Archive Linking and Result Caching

Bing is Microsoft's search engine providing access to their current web index through a RESTful API available at the Azure Marketplace. Bing returns results in XML or JSON data formats and offers two different API endpoints: (1) the full featured *Bing Search API*, and (2) the restricted and less expensive *Bing Search API - web Results Only*. The later lacks of few meta data like the overall result count. Nevertheless, it provides all basic search result information, such as URL, title and a text snippet. By specifying parameters, we can request optimized results for different languages/countries. We therefore use the endpoint with web results only. Yahoo! and Google provide similar search APIs but at higher costs with more restriction.

After obtaining search results from the Bing search API, we link the ranked list of results to the WayBack Machine to support browsing through the archived versions of web pages. The WayBack Machine is a tool provided by the Internet Archive that allows access to its web archives by specifying a URL. The URL-based access can be programmatically used through an API provided by the Internet Archive. For a given URL it returns a list of all dates when the URL has been archived. When a URL has been archived many times in the past, retrieval can take very long time. Therefore, we use two requests to retrieve only the first and last capture dates to display the time span at which the web pages has been archived. When the temporal intent of the user is provided, we narrow down to return only the revisions around the interested time point.

To avoid recurring requests to the Bing and WayBack Machine APIs, we store the search results locally in our cache, using a simple relational database. In order to take into account the fact that search results change over time, we update search results monthly to keep our cache up-to-date and to track changes at both sources. Besides the queries entered by our users, we use also the 10,000 most viewed English Wikipedia entities as queries. As a side effect, this procedure

---

[5] https://en.wikipedia.org/wiki/Wikipedia:Pageview_statistics.

results in building a corpus of past search results, which will support promising, longitudinal studies of web archive search, investigating how results change over time, triggered by events or changing user behavior. In Sect. 5, we will analyze the cache in order to reveal several important aspects, e.g., how long a web page stays relevant and when it fades away from top-ranked results. This insight will help improving the next version of our web archive search prototype.

4.3 Related Entity Suggestion

A traditional web search engine supports exploration by suggesting related queries, which is based on analyzing search sessions and identifying the co-occurrences of the issued query. For web archive search, we do not have query logs for obtaining search sessions, thus we leverage a dump of Wikipedia articles and build an entity graph in order to find related queries for our entity-oriented search. We follow the approach to determining the link-based entity relatedness originally proposed in [6]. Relatedness between two entities e_1 and e_2 is measured based on the overlap between the set of Wikipedia articles that link to e_1 and the set of articles that link to e_2.

$$relatedness(e_1, e_2) = \frac{log(max(|S_1|, |S_2|)) - log(|S_1 \cap S_2|)}{log(|W|) - log(min(|S_1|, |S_2|))} \quad (1)$$

where S_i is the set of articles that links to entity e_i, and W is the set of all articles in Wikipedia. Our entity graph is constructed from a recent Wikipedia dump (downloaded from September 2015 for both English and German Wikipedia), with the assumption that the link-based relationships between a pair of two entities can be accumulated; thus relaxing time sensitivity (in this case, the relatedness does not bias to any time point). An interesting extension will be to take time dependent relationships into account, which is planned for the next version of our system.

4.4 Multilingualisms

The described components are designed to support different languages. Dependent on the language selected by the user, we use the corresponding Wikipedia version to generate the query auto-completion and the related entities suggestions. More importantly, we request search results from Bing, which are optimized for the selected language and region. Furthermore, we leverage Wikipedia inter-language links when a user changes the front-end language. For example: When an English user searched the term "climate change" and switches to German, he will be redirected to "Klimawandel". Currently, we support English and German, but will add additional languages in the future.

5 Experiments

We conducted extensive experiments in order to gain insight into our assumptions presented in Sect. 3. We seek to answer two main research questions as follows.

RQ1. What is the coverage of archived content retrieved by the current search engine?

RQ2. To what extent the search results change over time, and why they change?

In the following, we will divide our experimental results into two main parts, where we describe our quantitative and qualitative analyses for each of the afore-mentioned research questions.

Part I: Analysis Results for RQ1. Our system relies on the assumption that many pages returned as search results for entity queries are archived by the Internet Archive. To check this assumption, we took all English Wikipedia enti-ties sorted by their view count and selected buckets of 1,000 entities at different positions in this list to represent entities from different popularity categories. We started with the 1,000 most viewed entities and continued with the entities from position 50,001 to 51,000 and so on. For each entity in an individual bucket, a search query was conducted and we checked how many results on the first five pages (10 results per page) were archived by the Internet Archive. Table 1 and Fig. 3 show the average results per page and bucket.

The results of popular entities (rank: 1–1000) show a very high coverage with the Internet Archive. On page one, 94% of the results are archived. On pages two to five, still 87% to 89% are available at the Internet Archive. Overall the coverage declines for less popular entities. Interestingly, it drops faster for the first pages of the search results than for the posterior ones. Upon inspection, this seems to be caused by the fact that Bing ranks recent results higher (for example on the first page of its search results), while the Internet Archive needs much more time to archive less popular pages.

To gain more insight, we conducted a coverage study by entity categories. More precisely, we analyzed top-100 search results for 300 popular entities from 14 different categories, for example, actor, journalist, painter, and politician, where there are approximately 20 entities per category. In this study, we only considered search results with .DE domains (German web pages), and checked the coverage with our local German web archive, instead of web archives of the

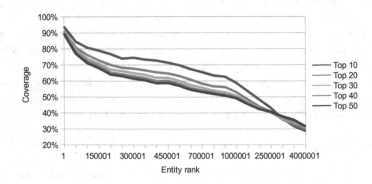

Fig. 3. Coverage (percentage) of archived content at different top-k results over entities ranked by their popularity.

Table 1. Coverage (percentage) of archived content at different top-k results over entities ranked by their popularity.

Entity rank	Top 10	Top 20	Top 30	Top 40	Top 50
1–1000	94 %	91 %	91 %	90 %	89 %
50001–51000	85 %	81 %	79 %	78 %	77 %
100001–101000	81 %	76 %	74 %	72 %	71 %
150001–151000	79 %	73 %	71 %	69 %	68 %
200001–201000	77 %	70 %	67 %	65 %	64 %
250001–251000	74 %	68 %	66 %	64 %	63 %
300001–301000	74 %	68 %	65 %	63 %	61 %
350001–351000	73 %	67 %	64 %	62 %	60 %
400001–401000	73 %	65 %	62 %	60 %	59 %
450001–451000	71 %	65 %	62 %	60 %	59 %
500001–501000	70 %	63 %	60 %	58 %	57 %
600001–601000	67 %	61 %	58 %	56 %	55 %
700001–701000	66 %	58 %	56 %	54 %	53 %
800001–801000	63 %	57 %	54 %	53 %	52 %
900001–901000	63 %	56 %	53 %	51 %	51 %
1000001–1001000	59 %	53 %	51 %	50 %	49 %
1500001–1501000	54 %	48 %	46 %	46 %	45 %
2000001–2001000	48 %	44 %	43 %	43 %	42 %
2500001–2501000	43 %	41 %	40 %	40 %	41 %
3000001–3001000	36 %	36 %	36 %	37 %	38 %
3500001–3501000	32 %	33 %	34 %	35 %	36 %
4000001–4001000	29 %	30 %	31 %	31 %	31 %

Internet Archive. As shown in Fig. 4, the coverage statistics on the German web archive shows significantly lower results than the one based on comparison with the Internet Archive due to search result bias towards English web pages, in general, even for the German version of Bing. It can be observed that categories with lower coverage tend to associate with recent and dynamic web content, whereas the results of the categories with higher coverage are rather static and less changed. Note that, result URLs are not always archived (e.g., for newspaper articles), but nearly all domains (i.e., news sites) are archived, regardless of entity categories.

We also performed search result annotation of 9 entities that were manually selected. We employed 5 human annotators to label top-100 search .DE results (by filtering out non .DE domains). For each (query, URL) pair, we asked at least 4 assessors to give a label based on relevance assessment criteria consisting of three scales: *long-term relevant, short-term relevant* and unknown. The results are shown in Table 2, where we can notice that non-active entities, such as,

Table 2. Percentage of long-term relevant and short-term relevant pages in top-100 results (filtered out non .DE domains).

Entity	Category	Long-term	Short-term
Leonard_Nimoy	Actor	52.00 %	48.00 %
Elon_Musk	Business people	37.50 %	62.50 %
Costa_Concordia_disaster	Incidents	52.40 %	47.60 %
Ernest_Hemingway	Journalist	81.48 %	18.52 %
Giuliana_Rancic	Journalist	40.00 %	60.00 %
Pablo_Picasso	Painter	97.60 %	2.40 %
Banksy	Painter	48.10 %	51.20 %
Vietnam	Politics	100.00 %	0 %
Ku_Klux_Klan	Politics	12.50 %	87.50 %

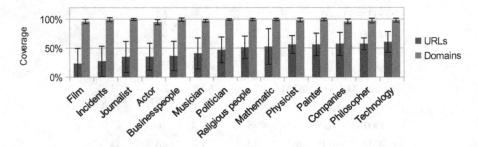

Fig. 4. Coverage (percentage) of archived URLs and domains at top-100 results for different entity categories.

Pablo Picasso and Ernest Hemingway, have more long-term relevant results than active entities like Elon Musk and Leonard Nimoy (to be expected).

In the following, we provide analyses for two selected entities in order to better understand the coverage aspect of search results and web archives.

Lady Gaga (view count rank: 108; views: 9,453,966; querying date: 19.01.2016). Lady Gaga is a famous American artist. Her Bing top-50 results have almost complete coverage on the Internet Archive. More specifically, 98 % of the results are archived. Among the results, only one URL[6] is not archived. This URL points to a gossip news article inside the entertainment section of the New York Daily News web site that only mentions the entity in question. As the entity is not the core mention, the article has relatively low relevance. We checked again on January 25, 2016, and the URL was still not archived. The URL was published online on January 18, 2016. Thus, we interpret that low relevance news are unlikely to be archived.

---

[6] http://www.nydailynews.com/entertainment/gossip/linda-perry-slams-lady-gaga-article-1.2500319.

Battle of Rathmines (view count rank: 1,000,798; views: 8,723; querying date: 18.01.2016). The Battle of Rathmines was fought in the area what is now the Dublin suburb of Rathmines in August 1649, during the Irish Confederate Wars, the Irish theatre of the Wars of the Three Kingdoms. The number of page views for this entity is 8,723, which is not very popular. Nevertheless, about 56 % of the top-50 Bing results are archived. Among the other 44 % of results, which are not archived, is the (4th ranked) page[7] from a website that shows information from Wikipedia, and the (48th ranked) page[8] that is a blog post. This entity is related to a real event in Ireland history that took place in almost 500 years ago, but is very local and rather unimportant outside of Ireland.

Part II: Analysis Results for RQ2. In this section, we present the analysis of Bing results for entity-based queries executed in three time periods: June 2015, August 2015 and January 2016. For a given entity, we computed the overlap between the top-ranked results at different time periods. Through this, we gain insight into our question RQ2, on how Bing query results change over time. We conducted a study on 300 popular queries of 14 different categories as explained in RQ1. We discuss a few samples ranging from low to high overlapping rates.

Figure 5 illustrates the change of search results (as measured by the overlap statistics) over time for different entity categories over the period of 2, 5 and 7 months, respectively. In general, result change after 2 months results in approximately 47 % of the top-100 URLs not returned any more. After 7 months, result change increases to 60 %. Across different categories, there is not much difference in temporal dynamics for the search results. The *Actor* category varies most, whereas the result variation is least for *Philosopher*. The main explanation here is that the entities in our *Actor* category are more active than entities in the *Philosopher* category, thus having more short-term relevant web pages in the top-100 result list, which change over time. We observe in Fig. 6 that the overlap in the top-50 results is slightly higher than in the top-100 results. This indicates there is less result change in the first 50 results than in the next 50 results. However, the difference is not significant.

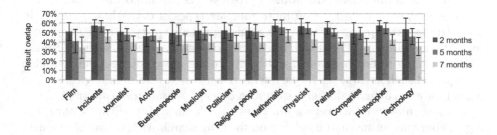

Fig. 5. Result overlap for different entity categories.

---

[7] http://www.thefullwiki.org/Battle_of_Rathmines.

[8] http://irelandinhistory.blogspot.de/2014/08/blog-post_11.html.

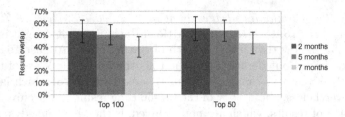

Fig. 6. Result overlap for top-50 and top-100 results.

Entity: Barack Obama. For the President of the United States 47.5 % of the results from August overlap with the results retrieved in June. These shared results are URLs pointing to biography or permanent pages, and most of them appear in high ranks. The remaining 52.5 % are URLs mostly to news or categories inside news web portals, which change over time. The overlapping of search results in this context decreases proportional to the number of recent events related to the entity: the higher the proportion of news results, the lower the overall overlapping with future result sets.

Entity: Donald Trump. For Donald Trump, more than 80 % of the URLs are news (National Journal, newsobserver.com, NBC). Most of these news articles published in June and do not overlap in August and January. Our data analysis in this experiment shows that the URLs from the news category, which is listed in results from June will probably not be shown in August or January. Total overlap average is 24 %, which is very low. Even for the shorter periods, June to August, and August to January, a result overlap is low with 30 %. Search results for Trump are a very clear sample for fast changing results, caused by frequent news articles.

6   Discussion and Conclusion

Although the system described in this paper already provides interesting functionalities, it is obviously still work in progress. As one important extension of functionality, we are working on more complex types of entity-based queries in order to support exploratory search, e.g., giving a main entity *Donald Trump* and related search intents. The search intent can consist of an entity, such as, *Hillary Clinton* aiming to find all events, which involve these two entities, and a specific time period such as *2015–2016* narrowing down search results to a specific time period, or any contextual query, such as a concept *presidential campaign*.

Another important aspect for our future development is to advance our ranking. As our current method is relying on the Bing search API, we sacrifice recall. Learning from Bing over time as well as from our user logs, we will be able to provide more sophisticated ranking taking different features into account. Bing results can act as 'soft' ground truth for learning the model. Our ranking model will then be able to return relevant documents, which are not longer available on the current Web.

Finally, we also work on improving our suggestion components, i.e., related entity suggestion to deal with queries having time as another aspect (e.g., Obama 2008). In the current system, we exploited a state-of-the-art method to suggest related entities to the query entity, with the assumption that their relationship strengths are accumulated over time. This relationship measure is reasonable to serve for queries with arbitrary relevant time. However, in reality relationships between entities do change over time, typically triggered by events. We can therefore return different related entities for different time periods to the input entity. For instance, the entities mostly related to *Hillary Clinton* in 2008 should differ from those in 2012, because of her different political positions. Moreover, with an exploratory query for example *Donald Trump* and search intent *Hillary Clinton*, it is more helpful to recommend the entities which are related to both *Donald Trump* and *Hillary Clinton*.

To summarize, we discussed a web archive search prototype that for the first time supports entity-oriented queries on the Internet Archive. Our system leverages Bing and the WayBack Machine to allow users to search the past Web. We provided search functionalities including keyword search, query auto-completion, query suggestion, and a ranked list of results, which are close to the current search engine systems. We conducted extensive analyses that shed light on web archive search, and included a discussion of future work as well as ideas/challenges for the next steps.

Acknowledgments. This work was partially funded by the European Commission for the ERC Advanced Grant ALEXANDRIA under the grant number 339233.

References

1. Costa, M., Gomes, D., Couto, F., Silva, M.: A survey of web archive search architectures. In: Proceedings of the 22nd International Conference on World Wide Web (Companion), WWW 2013, pp. 1045–1050 (2013)
2. Dougherty, M., van den Heuvel, C.: Historical infrastructures for web archiving: annotation of ephemeral collections for researchers and cultural heritage institutions. In: Proceedings of Media in Transition MIT6 Conference 2009 (2009)
3. Gomes, D., Miranda, J.A., Costa, M.: A survey on web archiving initiatives. In: Proceedings of the 15th International Conference on Theory, Practice of Digital Libraries: Research and Advanced Technology for Digital Libraries, TPDL 2011, pp. 408–420 (2011)
4. Kanhabua, N., Nørvåg, K.: Exploiting time-based synonyms in searching document archives. In: Proceedings of the 10th Annual Joint Conference on Digital Libraries, JCDL 2010, pp. 79–88 (2010)
5. Miliaraki, I., Blanco, R., Lalmas, M.: From "selena gomez" to "marlon brando": understanding explorative entity search. In: Proceedings of the 24th International Conference on World Wide Web, WWW 2015, pp. 765–775 (2015)
6. Milne, D., Witten, I.H.: Learning to link with Wikipedia. In: Proceedings of the 17th ACM Conference on Information and Knowledge Management, pp. 509–518 (2008)

7. Nguyen, T.N., Kanhabua, N., Niederée, C., Zhu, X.: A time-aware random walk model for finding important documents in web archives. In: Proceedings of the 38th International ACM SIGIR Conference on Research and Development in Information Retrieval, SIGIR 2015, pp. 915–918 (2015)
8. SalahEldeen, H.M., Nelson, M.L.: Losing my revolution: how many resources shared on social media have been lost? In: Zaphiris, P., Buchanan, G., Rasmussen, E., Loizides, F. (eds.) TPDL 2012. LNCS, vol. 7489, pp. 125–137. Springer, Heidelberg (2012)
9. Singh, J., Nejdl, W., Anand, A.: History by diversity: helping historians search news archives. In: Proceedings of the 2016 ACM on Conference on Human Information Interaction and Retrieval, CHIIR 2016, pp. 183–192 (2016)
10. Tran, T.A., Niederée, C., Kanhabua, N., Gadiraju, U., Anand, A.: Balancing novelty, salience: adaptive learning to rank entities for timeline summarization of high-impact events. In: Proceedings of the 24th ACM International on Conference on Information and Knowledge Management, CIKM 2015, pp. 1201–1210 (2015)
11. Yin, X., Shah, S.: Building taxonomy of web search intents for name entity queries. In: Proceedings of the 19th International Conference on World Wide Web, WWW 2010, pp. 1001–1010 (2010)

Semantics

BIB-R: A Benchmark for the Interpretation
of Bibliographic Records

Joffrey Decourselle[1], Fabien Duchateau[1(✉)], Trond Aalberg[2],
Naimdjon Takhirov[3], and Nicolas Lumineau[1]

[1] LIRIS, UMR5205, Université Claude Bernard Lyon 1, Lyon, France
{joffrey.decourselle,fabien.duchateau,nicolas.lumineau}@liris.cnrs.fr
[2] NTNU, Trondheim, Norway
trondaal@idi.ntnu.no
[3] Westerdals - Oslo School of Arts,
Communication and Technology - Faculty of Technology, Oslo, Norway
taknai@westerdals.no

Abstract. In a global context which promotes the use of explicit
semantics for sharing information and developing new services, the
MAchine Readable Cataloguing (MARC) format that is commonly used
by libraries worldwide has demonstrated its limitations. The semantic model for representing cultural items presented in the Functional
Requirements for Bibliographic Records (FRBR) is expected to be a
successor of MARC, and the complex transformation of MARC catalogs
to FRBR catalogs (FRBRization) led to the proposition of various tools
and approaches. However, these projects and the results they achieve are
difficult to compare on a fair basis due to a lack of common datasets and
appropriate metrics. Our contributions fill this gap by proposing the first
public benchmark for the FRBRization process.

Keywords: Benchmark · Migration · Record interpretation · FRBRization · LRM · FRBR · MARC · Dataset · Evaluation metric

1 Introduction

Cultural institutions are responsible for cataloging and offering access to a
large number of cultural items. The most popular format for libraries, MAchine
Readable Cataloguing (MARC), available in different implementations such as
MARC21 or UNIMARC, has shown some limitations in terms of interoperability, reuse, or information disambiguation [10]. The Functional Requirements
for Bibliographic Records (FRBR) and its updated version Library Reference
Model (LRM) [17] have been designed to provide a sound and more explicit
semantics which will enable new enhancements such as improved navigation and
enrichment features [3,5,11]. However, more than twenty years after the original specifications of FRBR, the model is still not widely used in libraries [6].
A major obstacle to the adoption of FRBR is the interpretation of records (e.g.,
FRBRization), which consists of migrating cultural heritage data from legacy
formats (e.g., MARC) to models based on the FRBR semantics[1].

---

[1] For instance, RDA, the LD4L project or BIBFRAME.

© Springer International Publishing Switzerland 2016
N. Fuhr et al. (Eds.): TPDL 2016, LNCS 9819, pp. 163–174, 2016.
DOI: 10.1007/978-3-319-43997-6_13

In the last decades, the proliferation of FRBRization tools has demonstrated the need for specific enhancements (e.g., clustered deduplication, exploitation of added entries) to improve the process [9]. Despite this progress, it is still very complicated to evaluate and compare FRBRization tools for several reasons. First, the experiments described in papers are rarely reproducible, mainly because the tools and the datasets are not available. A few catalog excerpts are provided, but they do not reflect the reality and the challenges of library catalogs because they are mainly used for illustrating specific cases [2]. Last but not least, the current metrics are not sufficient to evaluate all possible cases that might occur during FRBRization. In addition, these metrics imply that the user has to wait until the end of the FRBRization process before obtaining any insight about the resulting quality. To summarize, we advocate that the lack of a common FRBRization benchmark is an obstacle to the adoption of FRBR.

In this paper, we propose BIB-R^2, the **first benchmark for evaluating FRBRization**. It is composed of two datasets and a set of evaluation metrics. The goal of the **first dataset T42** is to identify the weak and strong points of a tool by testing all possible issues that libraries may face during FRBRization. The **second dataset BIB-RCAT** is extracted from catalogs of three different cultural institutions and can be used for comparing or experimenting with the data quality that is typically found in real world catalogs. The assessment of the process relies on a **set of metrics** to predict the quality of the output and to evaluate the quality of a FRBRized catalog. An **experimental study** with three recent FRBRization solutions shows the benefits of our benchmark.

In the rest of the paper, we present related work in Sect. 2 and an overview of our benchmark BIB-R in Sect. 3. Evaluation metrics for pre-FRBRization and post-FRBRization are presented respectively in Sects. 4 and 5 and the datasets are described in Sect. 6. The experimental study is detailed in Sect. 7. Section 8 concludes and outlines future work.

2   Related Work

Issues related to FRBRization and identification of challenging bibliographic patterns are described in recent surveys [2,16,19]. In this section, we present rule-based FRBRization tools, and we focus on the evaluation of this process.

Tools. Due to page limitation, we refer to a recent survey for an exhaustive list of FRBRization tools [9]. The last decade has seen the emergence of rule-based FRBRization tools, since grouping-records tools are not able to process complex structures [2,12]. We focus on three rule-based tools that are publicly available for experiments. The tool VFRBR, developed in the context of the Variations project, aims at FRBRizing catalogs with a focus on the music domain [14,15]. Since music items are often described using added entries, VFRBR's strategy is to interpret these added entries as separate entities. The online catalog Sherzo is

---

^{2} http://bib-r.github.io/.

the proof of concept that lets users explore musical works, composers and related entities issued from 185,000 MARC records. Extensible Catalog (XC) is an open-source project for a complete Integrated Library System, which includes the FRBRization tool Metadata Service Toolkit [4]. XC exploits added entries and is therefore able to detect complementary works for instance. The third tool FRBR-ML [18] is based on Aalberg's approach [1]. The authors discuss the possible structures of the FRBR output catalog, and the tool provides enhancements to disambiguate some complex cases by exploiting other catalogs or Linked Open Data knowledge bases.

Evaluation. Only the output of the FRBRization process is evaluated, under various forms: the most frequent option requires a ground truth or gold standard, i.e., an expert FRBRized catalog [13]. The comparison between the expert FRBRized catalog and the FRBRized catalog produced by a tool indicates whether the tool is able to perform an acceptable transformation. One of the main issues is the manual construction of the expert FRBRized catalog. The FRBR-ML approach avoids the tiresome construction of a gold standard by converting the FRBRized catalog back to a MARC catalog [18]. The evaluation is performed between the initial MARC catalog and the converted one. With this type of evaluation, the drawback is that the last transformation (into MARC) may have a negative impact on the quality of the catalog. Besides, the rules that enable this last transformation have to be written too. To evaluate a process, metrics are required. In TelPlus, an aggregation metric is proposed to measure the percentage of aggregated content (e.g., Works, Persons, Places). FRBR-ML is evaluated with three metrics: redundancy, completeness and extension respectively measures duplicate data, loss of data and amount of enrichment.

Discussion. The digital library community has successfully identified the bibliographic patterns, as well as a few FRBRization issues. But there is no collection publicly available for testing each of these challenges. Thus, most FRBRization tools have been tested against private datasets, whose characteristics are not clearly defined. Available metrics either assess the deduplication (aggregation) or compare two MARC collections (redundancy, completeness and extension). Thus, there is no metric which compares the quality of a generated FRBR collection, especially in terms of bibliographic patterns. And the whole FRBRization process is currently not evaluated (e.g., the tuning task). Contrary to other research domains, there is no benchmark for one of the most crucial task in the digital library community. Yet, we advocate that understanding the weak and strong points of the FRBRization process tends to promote novelty and enhancements in the future implementations. In addition, common datasets and evaluation metrics enable a fair comparison between the tools. In the next section, we describe our benchmark for FRBRization.

3   Overview of the Benchmark

The FRBrization process has been described and enhanced in the last decade [1,9]. Figure 1 depicts an overview of the FRBRization process. It is composed of three main steps (pre-FRBRization, FRBRization and post-FRBRization). During pre-FRBRization, librarians are in charge of preparing the input catalog (traditionally in MARC) and tuning the tool. The optional preparation allows to clean fields, to delete empty records, etc. For the tuning task, librarians can configure some parameters (e.g., setting a decision threshold for deduplication), but the main challenge is to add, modify or delete rules. Next, the FRBRization starts using a clean catalog and a customized set of rules. The transformation of each record produces a set of entities and relationships according to the rules applied. A deduplication task is necessary to detect and merge entities that represent the same concept. Finally, the last step is post-FRBRization, during which optional tasks are performed on the raw FRBR collection [9]. We only mention validation and enrichment. The former enables expert to verify and correct the generated FRBR catalog while the latter refers to the task of adding information from external sources. Most FRBRization approaches only evaluate the last step (post-FRBRization, using the FRBRized catalog), but the initial step has a strong impact on the final result in terms of quality and performance.

Our benchmark BIB-R provides metrics and datasets to evaluate pre-FRBRization and post-FRBRization. It focuses on the foundations of FRBR, but it does not take into account the specificities and the complexity of the different implementations (e.g., FRBR-OO).

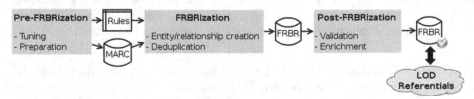

Fig. 1. Overview of the FRBRization process

4   Evaluating Pre-FRBRization

The pre-FRBRization aims at preparing the catalog to be FRBRized and the set of rules. The records of a catalog include bibliographic patterns [2], which complicate FRBRization. The different ways of cataloging and the set of rules are subject to potential issues. Since FRBRization may last hours or days, it seems interesting to be able to detect and solve potential problems prior to running the process (e.g., updating the set of rules or cleaning fields and values). To the best of our knowledge, this detection of problems is a novel contribution for digital libraries. Our pre-FRBRization metrics analyze a set of rules according to a catalog for estimating the records that will not be FRBRized correctly. Each metric computes a percentage defined as the number of records concerned with the given pattern/issue divided by the total number of records.

Library practice allows for the description of different entities and relationships according to the nature of the item described. These structures can be generalized into common patterns, but because the structure mainly is implicit such patterns are difficult to detect and FRBRize correctly [2,16]. The most frequent *core pattern* includes a Work, an Expression, a Manifestation and (mostly) the Agent creator of the Work. The *augmentation pattern* is defined as an additional content to an existing Work, with the assumption that the new content does not alter the main Work (e.g., illustrations, forewords). Several scenarios occur to FRBRize this pattern, for instance the creation of a new Work or a note for the original Work. The *derivation pattern* means that one Work is the modification of another Work (e.g., translations, imitations), and it usually implies the creation of Expression(s) under the same Work or relationships between Works. The *aggregation pattern* is commonly described as a whole-parts relationship (e.g., ensemble, aggregative work). The FRBRization of aggregations mainly results in the creation of relationships between Works and optionally new Agents. The *complementary works pattern* aims at modelling a relationship between Works which have the same importance (e.g., sequels, accompanying works). The FRBRization of complementary works mainly results in the creation of relationships between Works. The **metrics COR**, **AUG**, **AGG**, **DER** and **COW** respectively compute the percentage of records with a core pattern, an augmentation, an aggregation, a derivation and a complementary work.

In addition to bibliographic patterns, records may include cataloging errors or quality problems. Authors of the TelPlus project have established six requirements for FRBRization [13], that can be seen as errors in the initial records. We define the **metrics MID**, **MPD**, **MUT**, **MOT**, **MRC** and **MAR** which respectively compute the number of records that include the issues *missing record identifier*, *missing publication date*, *missing uniform title*, *missing original title*, *missing relator code*, and *missing authoritative responsibility*. We propose four new metrics related to cataloging issues. The **metric MTF** deals with *missing type and form of material*, which has an impact for correctly identifying Expressions (and sometimes Works). The **metrics TLE** and **RLE** relate to *title linkage error* and *responsibility linkage error*, which means that the unavailable related record (mainly in UNIMARC) has a negative impact in terms of completeness when FRBRizing. Finally, libraries make use of standards such as the International Standard Bibliographic Description (ISBD), widespread normalization of values (e.g., country codes) or codes specific to individual libraries (e.g., for a book category, value "r" corresponds to a roman). The **metric CPN** deals with these inconsistent *cataloging practices and norms*, which may contain useful information.

The set of rules has not been widely studied and mainly regarded as an artifact that needs to be tuned by librarians. Yet, this tuning has a crucial impact and its analysis can be exploited to improve FRBRization. In case of *missing rule*, a field cannot be processed by a rule, thus causing loss of information. Detecting these fields prior to FRBRization enables librarians to update the set of rules accordingly. A *not used rule* indicates that it is not useful for a given catalog. The *conflicting rules* issue occurs because the set of rules can be built using various techniques (e.g., written by librarians, merged from collected sets).

Actions associated to such rules can either be complementary or conflicting, which may degrade performance or quality during FRBRization. The **metrics MR, UR** and **CR** respectively compute the percentage of missing rules, not used rules and conflicting rules. The metric MR can be decomposed into more detailed metrics for a given pattern or issue, for instance **MR-AUG** to calculate the percentage of missing rules for detecting all augmentations. A formal notation of the pre-FRBRization metrics can be found in an online appendix [7].

5   Evaluating Post-FRBRization

When the process of FRBRization is finished, librarians typically need to check the FRBRized catalog produced by the tool. This evaluation is the most studied in the literature, because the resulting quality is currently an obstacle to the adoption of FRBR and because most FRBRization tools have demonstrated their capabilities through an experimental validation based on the analysis of the produced FRBR catalog [13,18]. In our context, we have chosen an evaluation based on expert FRBRized catalog. This is the most reliable evaluation, specifically because it directly assesses the quality of the FRBR catalog, including its complex relationships between entities. We have identified seven metrics to compare the FRBR catalog generated by a tool T and the expert FRBR catalog $\mathcal{E}$. These metrics are useful to understand the weak and strong points of a FRBRization tool, to estimate the manual effort which is needed to complete the FRBRization, or to provide an insight about the rules that should be added to improve the quality.

The first four metrics deal with data (entities, relationships and properties from the FRBR model). The **metric MD** is related to the missing data issue, i.e., data which appears in the expert catalog $\mathcal{E}$ is missing in the tool's catalog T. This metric computes the ratio between the number of missing data and the total number of data in the expert collection. It can be redefined for each type of data, i.e., MD-E for entities, MD-R for relationships and MD-P for properties. The **metric IAD** deals with incorrectly added data, i.e., duplicate data (e.g., a property which appears twice in an entity, because of a bad deduplication for instance) and incorrect data (e.g., an entity that should not have been created or a property with an unexpected value). It is defined as the number of incorrect data in T (which is not in $\mathcal{E}$) divided by the total number of data in T. Similarly to MD, the metric IAD can be redefined according to the data type. The **metric DLE** relates to errors in external link (e.g., to a referential or to the Linked Open Data). Either the link does not exist in $\mathcal{E}$ or it has a different value for the same external source. The metric calculates precision, i.e., the number of erroneous links in T divided by the total number of links in T. The **metric SMD** aims at computing semantic mismatch data, i.e., data which have a different semantics in both catalogs (e.g., a relationship *translated by* which appears as *contributed to*). The metric computes the amount of semantic mismatch data in T (compared to data in $\mathcal{E}$) with regards to the total number of data in T.

The next metrics deal with patterns. The detection of the pattern in a MARC record is crucial because it provides the FRBR structure. Yet, only part of a pattern may be incorrect and the evaluation should reflect this. Note that it is not possible to verify information about bibliographic patterns without annotations in the expert collection. The **metric MEND** (main entity not detected) relates to the detection of the main entity of a pattern (e.g., an Expression in the case of a translation). It measures the percentage of main entities from $\mathcal{E}$ that have been correctly detected in $\mathcal{T}$ among all main entities from $\mathcal{E}$. The **metric MRND** (main relationship not detected) checks whether the relationship associated to the main entity is correctly identified or not. For instance, an Expression is correctly identified but linked with a *"is a revision"* relationship rather than with a *"is a translation"* relationship. The metric MRND computes the percentage of main relationships from $\mathcal{E}$ that have been correctly detected in $\mathcal{T}$ among all main relationships from $\mathcal{E}$. Finally, the **metric ESE** deals with errors in secondary element(s) of the pattern, which means that the main entity and its relationship have been correctly detected, but other elements (e.g., the translator) are missing or incorrect. The metric ESE computes the percentage of correct secondary elements in $\mathcal{T}$ among all secondary elements from $\mathcal{E}$. A formal notation of the post-FRBRization metrics is given in an online appendix [7]. To use these metrics, it is necessary to have datasets with appropriate features.

6 Datasets

In BIB-R, two datasets allow the assessment of FRBRization tools. In our context, a **dataset** is a set of collections. Each **collection**, which contains records, is available in two input formats (MARC21 and UNIMARC) and it is associated with an expert FRBR collection. This expert collection has been manually created and verified by a librarian and three digital library researchers. All collections included in these datasets are based on the MARCXML and raw MARC formats. The records have been extracted from real-world catalogs, and modified when needed. The datasets are detailed in a report [8] and publicly available at http://bib-r.github.io/.

The first dataset **T42**[3] can be used for testing specific cases. In Sect. 4, we explained that a record has an inherent bibliographic pattern (e.g., core, augmentation) and it may include any number of issues (e.g., missing relator code, title linkage error). The objective of the dataset T42 is to check whether a FRBRization tool is able to handle each possible case. We define a **unit test** as the combination of a pattern and an optional issue. This dataset currently contains 42 meaningful tests which are crucial for testing specific aspects of FRBRization (a full list of combinations and statistics are available online).

The second dataset **BIB-RCAT**[4] simulates a real-world catalog in which various bibliographic patterns and issues may be found. It currently contains

---

[3] T42 is a reference to the novel *Hitchhikers Guide to the Galaxy*.

[4] BIB-RCAT is a recursive acronym that stands for *"BIB-RCAT Is Basically a Real-world CATalog"*.

three collections (MARC21 and UNIMARC formats, and the expert FRBR). It is mainly composed of records from various catalogs (e.g., a public French library, a public Swiss hospital). The size of this catalog (560 records) is smaller than catalogs found in cultural institutions, since the expert FRBR collection requires a tiresome effort to be manually produced and verified.

7 Experiments

In this section, we demonstrate the benefits of our benchmark BIB-R for the evaluation of FRBRization. Three tools, which are publicly available[5], have been used in these experiments: FRBR-ML, Extensible Catalog (XC) and Variations VFRBR. These tools are detailed in the Related Work (Sect. 2). The rest of this section describes three experiments using our benchmark: how to evaluate the strengths and weaknesses of FRBRization tools, how to compare tools in a real-world FRBRization scenario and how to facilitate the tuning of a tool. Due to page limit, only a few interesting results are presented, but all plots are publicly available in an online appendix [7].

7.1 Assessing Strengths and Weaknesses

This first experiment aims at demonstrating the benefit of the dataset T42 when it comes to evaluating the strengths and weaknesses of FRBRization tools. For the three tools, we have run each test from the dataset T42 and the evaluation is performed using post-FRBRization metrics. Note that we have not tuned the rules and rely on the set or rules provided with the tools, although they have been developed for different purposes. The first finding is about **missing data**. None of the tools completely FRBRize the data contained in the MARC records. Figure 2 illustrates this trend by showing the missings in terms of entities (MD-E), relationships (MD-R) and properties (MD-P) for various tests. With the core pattern (test 1.0), the tools may miss entities such as Concepts. Tools may also be implemented to merge some properties. For instance, XC merges the subtitle into the title, thus missing the subtitle property. The scores of VFRBR for missing data are strongly impacted by the fact that it does not create Work entities. The more complex the record becomes (tests 3.2 and 5.5), the more losses in the FRBRization. Secondly, only XC generates **incorrectly added data**, mainly in terms of properties and relationships (see online appendix). These additional data are in fact misplaced data, i.e., which should have been put in another entity or which should have linked other entities (e.g., the abstract is placed in the Work entity rather than in the Expression). Another study deals with the **detection of patterns**. Figure 3 depicts the scores obtained by the three tools for correctly detecting the bibliographic patterns without any cataloging issue (i.e., tests x.0). FRBR-ML obtains good results for detecting the core pattern (test 1.0). For other patterns (tests 3.0 and 5.0), it discovers half of the main entities (metric MEND) but it fails

---

[5] FRBR-ML (previously named marc2frbr), Extensible Catalog and Variations VFRBR (adjusted version, only to facilitate compilation).

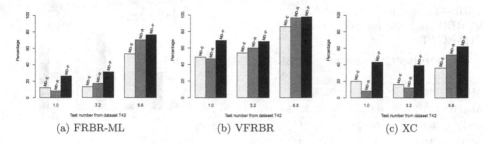

(a) FRBR-ML   (b) VFRBR   (c) XC

Fig. 2. Experiment results for evaluating missing data

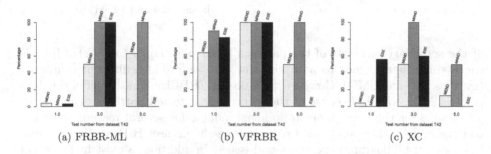

(a) FRBR-ML   (b) VFRBR   (c) XC

Fig. 3. Experiment results for evaluating bibliographic patterns

for the main relationship (metric MRND). VFRBR is not able to detect most patterns with its basic set of rules, even for the core pattern. This is mainly because this tool does not always create Works and Expressions. XC achieves good results with some patterns (core, complementary works) but not for derivations. Next, we note that all tools produce **semantic mismatch**, but only for the relationships (metric SMD-R). This issue occurs in 36 tests for FRBR-ML, 24 tests for VFRBR and 21 tests for XC (out of 42 tests), but the scores of the metric SMD-R are mostly below 10 %, thus indicating that less than 10 % of the relationships have a different semantics than in the collection annotated by experts. Since the more complex relationships are usually found in patterns, these results are also dependent on the ability of the tool to detect patterns. To summarize, our dataset T42 and post-FRBRization metrics are useful for understanding the failures of a tool.

7.2   Comparing Tools in Real-World Context

The objective of this second experiment is to compare FRBRization tools in a real-world context using post-FRBRization metrics. The post-FRBRization metric DLE is not presented, since the expert FRBRized collection cannot include a link for each existing authority files or knowledge bases. All tools rely on their basic set of rules (no tuning). Table 1 provides the results for the three tools. We note that they are able to identify only a few patterns (scores above 90 % for the metrics MEND and MRND). VFRBR is the only tool to FRBRize half

Table 1. Results of FRBR-ML, VFRBR and XC for the dataset BIB-RCAT

	FRBR-ML	VFRBR	XC	FRBR-ML tuned
MEND	94 %	98 %	94 %	1 %
MRND	100 %	100 %	100 %	29 %
ESE	99 %	55 %	100 %	21 %
MD	44 %	45 %	45 %	13 %
IAD	0 %	0 %	0 %	0 %
SMD	0 %	0 %	0 %	0 %

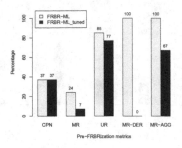

Fig. 4. Applying predictive metrics on BIB-RCAT for FRBR-ML basic rules and tuned rules

of the secondary elements of the patterns (ESE value equal to 55 %). All tools successfully manage not to add incorrect data or produce different semantics (metrics IAD and SMD). However, they do not FRBRize almost half of the data (metric MD), mainly because of the incorrectly detected patterns. These average results for the three tools are understandable for several reasons: contrary to dataset T42, these real-world records from the dataset BIB-RCAT can combine several bibliographic patterns and issues. In addition, almost half of them include cataloging practices, which complicate the interpretation of the records. Finally, some additional entities (e.g., Concept) are not processed and created. The basic set of rules are not sufficient for achieving an acceptable quality. To conclude, this experiment showed that our dataset BIB-RCAT and associated metrics are useful to compare tools in a real-world context.

7.3 Facilitating the Tuning

In this last experiment, we show how our pre-FRBRization metrics can help updating the set of rules. Only the FRBR-ML tool was used in this experiment, but the scenario could be applied to any tool. As shown in Table 1, the results of FRBR-ML for dataset BIB-RCAT could be improved. To provide insight to the expert, we compute predictive scores for the basic set of rules on the dataset BIB-RCAT. A subset of these scores is detailed in Fig. 4. The white bar stands for the results with the basic set of rules. For instance, we note that 37 % of the records contain cataloging practices (metric CPN). The basic set of rules contains many not used rules for the dataset BIB-RCAT (score of UR equal to 85 %) and it lacks 24 % of rules to take into account all fields from the dataset BIB-RCAT. Finally, the metrics for specific patterns indicate that 100 % of the rules are missing to tackle derivations (metric MR-DER) and aggregations (metric MR-AGG). Based on these predictive scores, an expert has enhanced the basic set of rules of FRBR-ML. This update took 4 h mainly for correcting minor changes (e.g., add rules for missing subfields) and implementing new templates to handle relator codes and missing concepts (e.g., augmentations, parent works). The enhanced set of rules has been tested with the prediction metrics (black bars in Fig. 4). Now, only 7% of the rules are missing to process all fields, and a few not used rules have been deleted (metric UR). The most significant enhancement deals with the pattern detection: all rules to identify derivations have been added, but the set still misses 67% of rules to process aggregations. Finally, FRBR-ML tuned with this enhanced set of rules was used to FRBRize the BIB-RCAT dataset. The results of this new FRBRization is shown in Table 1 (column *FRBR-ML tuned*). As expected, the quality of this enhanced FRBRization is better than with the basic set of rules, especially for the patterns. Adding relevant new rules enables us to reduce the amount of missing data, but 29% of relationships and 21% of secondary elements in the patterns are still missing. This experiment demonstrates how the predictive metrics help librarians update the set of rules and thus improve the quality of the FRBRization.

8   Conclusion

In this paper, we described BIB-R, the first benchmark for evaluating the interpretation of bibliographic records. It includes a set of metrics and two datasets (T42 and BIB-RCAT). Extensive experiments with our dataset T42 have been performed with three recent tools (FRBR-ML, Variations VFRBR and Extensible Catalog) to demonstrate the possibility to identify strengths and weaknesses. Our experimental validation is also the first to compare FRBRization tools with the same datasets and metrics. Finally, we showed how the pre-FRBRization metrics can be useful to help librarians update the set of rules. The release of this benchmark brings different perspectives. We plan to add more records in the real-world dataset BIB-RCAT. The main challenge is to update the FRBR expert collection. Next, we could enhance the benchmark to enable evaluation of ergonomics (quality of graphical user interfaces), performance (execution time) and quality of the semantic enrichment (for instance based on the Knowledge Base Population challenge[6]).

Acknowledgments. This work has been partially supported by the French Agency ANRT (www.anrt.asso.fr), the company PROGILONE (www.progilone.com/), a PHC Aurora funding (#34047VH) and a CNRS PICS funding (#PICS06945).

References

1. Aalberg, T.: A process and tool for the conversion of MARC records to a normalized FRBR implementation. In: Sugimoto, S., Hunter, J., Rauber, A., Morishima, A. (eds.) ICADL 2006. LNCS, vol. 4312, pp. 283–292. Springer, Heidelberg (2006)
2. Aalberg, T., Žumer, M.: The value of MARC data, or, challenges of FRBRisation. J. Documentation **69**, 851–872 (2013)
3. Alemu, G., Stevens, B., Ross, P., Chandler, J.: Linked data for libraries: benefits of a conceptual shift from library-specific record structures to RDF-based data models. New Libr. World **113**, 549–570 (2012)

---

[6] http://www.nist.gov/tac/2016/KBP/.

4. Bowen, J.: Moving library metadata toward linked data: opportunities provided by the eXtensible catalog. In: International Conference on Dublin Core and Metadata Applications (2010)
5. Buchanan, G.: FRBR: enriching and integrating digital libraries. In: Proceedings of Joint Conference on Digital Libraries, pp. 260–269 (2006)
6. Coyle, K.: FRBR, twenty years on. Cataloging Classif. Q. **53**, 1–21 (2014)
7. Decourselle, J., Duchateau, F., Aalberg, T., Takhirov, N., Lumineau, N.: Appendix: BIB-R: a benchmark for the interpretation of bibliographic records. Technical report, LIRIS, NTNU (2016). http://liris.cnrs.fr/~fduchate/docs/appendix/appendix-tpdl16.pdf
8. Decourselle, J., Duchateau, F., Aalberg, T., Takhirov, N., Lumineau, N.: Open datasets for evaluating the interpretation of bibliographic records. In: Proceedings of Joint Conference on Digital Libraries. ACM (2016)
9. Decourselle, J., Duchateau, F., Lumineau, N.: A survey of FRBRization techniques. In: Theory and Practice of Digital Libraries, pp. 185–196 (2015). https://hal.archives-ouvertes.fr/hal-01198487
10. Denton, W.: FRBR and the history of cataloging. In: Taylor, A.G. (ed.) Understanding FRBR: What it is and How it will Affect Our Retrieval Tools. Libraries Unlimited, Westport, Conn (2007)
11. Dickey, T.J.: FRBRization of a library catalog: better collocation of records, leading to enhanced search, retrieval, and display. Inf. Technol. Libr. **27**, 23–32 (2008)
12. Hickey, T.B., O'Neill, E.T.: FRBRizing OCLC's WorldCat. Cataloging Classif. Q. **39**, 239–251 (2005)
13. Manguinhas, H.M.A., Freire, N.M.A., Borbinha, J.L.B.: FRBRization of MARC records in multiple catalogs. In: Hunter, J., Lagoze, C., Giles, C.L., Li, Y.F. (eds.) JCDL, pp. 225–234. ACM (2010)
14. Notess, M., Dunn, J.W., Hardesty, J.L.: Scherzo: a FRBR-based music discovery system. In: International Conference on Dublin Core and Metadata Applications, pp. 182–183 (2011)
15. Riley, J.: Enhancing interoperability of FRBR-based metadata. In: International Conference on Dublin Core and Metadata Applications (2010)
16. Riva, P.: Mapping MARC 21 linking entry fields to FRBR and Tilletts taxonomy of bibliographic relationships. Libr. Resour. Tech. Serv. **48**(2), 130–143 (2013)
17. Riva, P., Žumer, M.: Introducing the FRBR Library Reference Model. IFLA Conferences (2015). http://library.ifla.org/1084/
18. Takhirov, N., Aalberg, T., Duchateau, F., Žumer, M.: FRBR-ML: a FRBR-based framework for semantic interoperability. Seman. Web J. **3**, 23–43 (2012)
19. Zhang, Y., Salaba, A.: Implementing FRBR in Libraries: Key Issues and Future Directions. Neal-Schuman Publishers, New York (2009)

Querying the Web of Data with SPARQL-LD

Pavlos Fafalios[1,2], Thanos Yannakis[1,2], and Yannis Tzitzikas[1,2(✉)]

[1] Computer Science Department, University of Crete, Heraklion, Greece
[2] Institute of Computer Science, FORTH-ICS, Heraklion, Greece
{fafalios,yannakis,tzitzik}@ics.forth.gr

Abstract. A constantly increasing number of data providers publish their data on the Web in the RDF format as Linked Data. SPARQL is the standard query language for retrieving and manipulating RDF data. However, the majority of SPARQL implementations requires the data to be available in advance (in main memory or in a repository), not exploiting thereby the real-time and dynamic nature of Linked Data. In this paper we present SPARQL-LD, an extension of SPARQL 1.1 Federated Query that allows to directly fetch and query RDF data from any Web source. Using SPARQL-LD, one can even query a dataset coming from the partial results of a query (i.e., discovered at query execution time), or RDF data that is dynamically created by Web Services. Such a functionality motivates Web publishers to adopt the Linked Data principles and enrich their digital contents and services with RDF, since their data is made directly accessible and exploitable via SPARQL (without needing to set up and maintain an endpoint). In this paper, we showcase the benefits offered by SPARQL-LD through an example related to the Europeana digital library, we report experimental results that demonstrate the feasibility of SPARQL-LD, and we introduce optimizations that improve its efficiency.

1 Introduction

While more and more structured data are published on the Web following the Linked Data principles [6], an important question is how one can efficiently access and query this constantly increasing body of knowledge. SPARQL [4] is the standard query language for retrieving and manipulating RDF data. However, the majority of SPARQL implementations requires the data to be available in advance, i.e., to exist in main memory or in a RDF repository accessible through a SPARQL endpoint. Nonetheless, Linked Data exists in the Web in various forms; even an HTML Web page can contain RDF data through RDFa [3], or RDF data may be dynamically created by Web Services.

In this paper we present SPARQL-LD, an extension (actually a generalization) of SPARQL 1.1 Federated Query [5] that allows to *directly* and *flexibly* exploit this wealth of data. SPARQL-LD extends the applicability of the SERVICE operator enabling to query any HTTP Web source containing RDF data. This extension does not require the named graphs to have been declared, thus one can even fetch and query a dataset returned by a portion of the query (i.e., whose URI is derived at query execution time).

© Springer International Publishing Switzerland 2016
N. Fuhr et al. (Eds.): TPDL 2016, LNCS 9819, pp. 175–187, 2016.
DOI: 10.1007/978-3-319-43997-6_14

Such a functionality can motivate Web publishers to enrich their documents and digital libraries with RDF since it makes their data directly accessible via SPARQL without needing to set up and maintain an endpoint (e.g., they can just publish RDF dumps). Actually, *availability* is the main bottleneck towards the success of the Semantic Web as a reliable technology. Buil-Aranda et al. [8] tested 427 public endpoints and found that their performance can vary by up to 3–4 orders of magnitude, while only 32.2 % of public endpoints can be expected to have monthly uptimes of 99–100%. Therefore, it may be more reliable to directly retrieve the triples of a dereferenceable URI than retrieving the same triples by invoking a query against a remote endpoint (considering of course that the query requirements are satisfied).

Figure 1 shows a query that can be answered by SPARQL-LD. The query first accesses Europeana's [13,15] SPARQL endpoint[1] for retrieving artists of works related to Renaissance (lines 2–3). Then, by querying the dereferenceable URI of each artist, the query retrieves and shows a description (in English) and an image of only those of Mannerist style (lines 4–6). Note that Europeana does not contain information about artist styles. Notice also that the artist URIs are derived at query execution time. One could also integrate in the same query data from any Web resource or Web Service that offers its data in a standard RDF format. As an example, consider that an online bookstore service exports its search results in RDF. Using SPARQL-LD one can directly access this service through SPARQL and find books about the artists returned by the two SERVICE patterns in the query of Fig. 1. Likewise, in the same query one could exploit a video service and find links of YouTube videos related to some of the artists.

Consequently, the functionality offered by SPARQL-LD can overcome the limitations of digital libraries (and information sources in general) related to information integration, enrichment and exploitation.

```
1  SELECT DISTINCT ?creator ?descr ?photo WHERE {
2    SERVICE <http://europeana.ontotext.com/sparql> {
3      ?work dc:subject dbr:Renaissance ; dc:creator ?creator }
4    SERVICE ?creator {
5      ?creator dct:subject dbc:Mannerist_painters ;
6             dbo:abstract ?descr ; foaf:depiction ?photo FILTER(lang(?descr)="en") } }
```

Fig. 1. An example of a SPARQL query that can be answered by SPARQL-LD.

SPARQL-LD was first demonstrated in a short (demo) paper [10]. With respect to that paper, in this paper: (i) we provide examples that illustrate the benefits offered by SPARQL-LD, (ii) we identify factors that affect efficiency and propose optimizations, and (iii) we extensively evaluate the efficiency of SPARQL-LD and the effect of the proposed optimization techniques. In addition, this paper provides a more detailed related work. The rest of this paper is organized as follows: Sect. 2 discusses related work, Sect. 3 introduces SPARQL-LD, Sect. 4 details optimization techniques, Sect. 5 presents evaluation results, and finally Sect. 6 concludes the paper.

---

[1] http://europeana.ontotext.com/sparql.

2 Related Work

The approach that we propose is considered a method to execute queries over the Web of Linked Data. Such approaches can be classified in three main categories: *query federation*, *data centralization*, and *link traversal*.

The idea of *query federation* is to provide integrated access to distributed sources on the Web. For example, the systems DARQ [23] and SemWIQ [17] provide access to distributed RDF data sources using a mediator service that transparently distributes the execution of queries to multiple SPARQL services. Given the need to address query federation, in 2013 the SPARQL W3C working group proposed a query federation extension for SPARQL 1.1 [5]. Buil-Aranda et al. [7] describe the syntax of that extension and formalize its semantics.

The idea of *data centralization* is to provide a query service over a collection of data copied (and probably transformed) from different sources on the Web. Such a collection is usually called "Warehouse". There are *domain independent* warehouses like SWSE [14], but also *domain specific* like the MarineTLO-based Warehouse [25]. In the same category falls the case of *digital libraries* containing descriptions and metadata about digital objects collected from multiple content providers (like Europeana [15,22]). Although such approaches require the data to exist in a single repository, they can significantly benefit from the functionality offered by SPARQL-LD. For instance, a query service over such a repository can support SPARQL-LD and offer the ability to also integrate (during query execution) data coming from online RDF sources (like in the example of Fig. 1).

Link traversal approaches exploit the Linked Data principles for discovering data related to URIs given in the query. For instance, the work in [11] discovers data that might be relevant for answering a query by following RDF links between data sources based on URIs in the query and in partial results. Diamond [19] is a similar in spirit query engine to evaluate SPARQL queries on distributed RDF data where, as a query is being evaluated, additional Linked Data can be identified by exploiting dereferenceable URIs. Finally, LDQL [12] is a declarative language to query Linked Data which is also based on link traversal. LDQL separates query components for selecting query-relevant regions of Linked Data, from components for specifying the query result.

SPARQL-LD actually *complements* the aforementioned approaches on query federation, data centralization and link traversal; it can be used in combination to such approaches. Works that focus on optimizing the execution of SPARQL federated queries, and that can be also applied in our case, are discussed in Sect. 4.

3 SPARQL-LD: Functionality and Examples

Motivation. Although the majority of SPARQL implementations requires the data to be available in advance (in main memory or in a repository), the specification of SPARQL allows to directly query a RDF dataset accessible on the Web (in a standard format) and identifiable by an URI through the operators FROM/FROM NAMED and GRAPH. However, this has an important limitation: it requires knowing *in advance* the URI of the dataset and having declared it in the FROM NAMED clause. Thus, a URI coming from partial results (that get bound after executing an initial query fragment) cannot be used in the GRAPH operator as the dataset to run a portion of the query. Furthermore, although RDFa [3] and JSON-LD [1] are W3C standards that are exploited by an ever-increasing number of publishers, we have not managed to find a SPARQL implementation that can directly query such RDF data. In addition, using the SERVICE operator of SPARQL 1.1 Federated Query [5], we can invoke a portion of a query against a remote RDF repository. However, SERVICE requires the URI to be the address of a SPARQL endpoint, thus one cannot exploit this operator for querying RDF data accessible on the Web but not available through an endpoint.

Extended SERVICE Definition. The SPARQL 1.1's SERVICE operator (SERVICE a P) is defined (in [7]) as a graph pattern P evaluated in the SPARQL endpoint specified by the URI a, while (SERVICE $?X$ P) is defined by assigning to the variable $?X$ all the URIs (of endpoints) coming from partial results, i.e. that get bound after executing an initial query fragment. The idea behind SPARQL-LD is to enable the evaluation of a graph pattern P not absolutely in a SPARQL endpoint a, but generally in a RDF graph G_r specified by a Web Resource r. Thus, now a URI given to the SERVICE operator can also be the dereferenceable URI of a resource, the Web page of an entity (e.g., of a person), an ontology (OWL), Turtle, or N3 file, etc. In case the URI is not the address of a SPARQL endpoint, the RDF data that may exist in the resource are fetched at real-time and queried for the graph pattern P.

SPARQL-LD is a generalization of SPARQL in the sense that every query that can be answered by the original SPARQL can be also answered by SPARQL-LD. Specifically, if the URI given to the SERVICE operator corresponds to a SPARQL endpoint, then it works exactly as the original SPARQL (the remote endpoint evaluates the query and returns the result). Otherwise, instead of returning an error (and no bindings), it tries to fetch and query the triples that may exist in the given resource.

Implementation. SPARQL-LD has been implemented using Apache Jena [2]. Jena is an open source Java framework for building Semantic Web applications. Specifically, we have extended Jena 2.13 ARQ component. ARQ is a query engine for Jena that supports SPARQL 1.1. The implementation is available as open source[2]. An endpoint that supports SPARQL-LD is publicly available[3].

The implementation can be described through the following process: we first check if the URI corresponds to a SPARQL endpoint by submitting the ASK query "ASK {?x ?y ?z}". In case we get a valid answer, we continue just like the default query federation approach, i.e. the corresponding graph pattern (query) is submitted to the endpoint. In case we do not get a valid answer, it means that the URI is not the address of an endpoint. Then, we read the *content type* header field of the URI by opening an HTTP connection and setting the value

---

[2] https://github.com/fafalios/sparql-ld.
[3] http://users.ics.forth.gr/~fafalios/sparql-ld-endpoint.

`application/rdf+xml` to the `ACCEPT` request property (we do that for handling also the case of RDFa). Now, according to the returned content type, we fetch and query the corresponding triples. For the case of HTML Web pages (the content type is `text/html` or `application/xhtml+xml`), we try to fetch and query the RDF triples that may be embedded in the Web page as RDFa. If the Web page does not contain any RDF data, the query returns no bindings. For reading possible RDF triples in a Web page, we exploit the `Semargl` framework (https://github.com/levkhomich/semargl) which also offers an integration with Jena. The implementation allows also reading and querying JSON-LD files.

Query Examples. Here we give two example queries that demonstrate the functionality offered by `SPARQL-LD`. More examples are available at the endpoint given in Footnote 3.

Querying Dynamically-Created RDF Data. `X-Link` [9] is a Linked Data-based Named Entity Extraction (NEE) framework which can export the result of the NEE process in RDF using the Open NEE model [9]. An `X-Link` Web service configured for the artistic domain is publicly available at http://83.212.107.202/ x-link-art. This service can identify names of several types of entities in a given Web document and link them to Web resources (URIs). For instance, we can request to perform NEE with *painters* and *countries* as the entities of interest at the Web page "https://en.wikipedia.org/wiki/Mannerism" and get the results in the default RDF/XML format, with the following request: http://83.212.107.202/x-link-art/api?categories=painter;country&url=https:// en.wikipedia.org/wiki/Mannerism.

Using the proposed extension, one can exploit the APIs of such services directly through SPARQL. For instance, Fig. 2 depicts a query that *parameterizes* and *calls* the above annotation service *at query execution time* (the namespaces have been omitted to save space). The query first retrieves Web pages related to *Mona Lisa* by querying its dereferenceable DBpedia URI (lines 2–3). Then, it calls the `X-Link` service for identifying names of *painters* and *countries* in the retrieved Web pages (lines 4–6), and for each detected entity the query retrieves (and shows) its name, its category and its number of occurrences in the Web pages (lines 7–9). Finally, the entities are ordered by the number of occurrences in descending order (line 10).

```
1  SELECT DISTINCT ?detectedEntity ?categoryName (count(?position) as ?NumOfOccurrences) WHERE {
2    SERVICE <http://dbpedia.org/resource/Mona_Lisa> {
3      dbr:Mona_Lisa dbo:wikiPageExternalLink ?page }
4    VALUES ?templ { <http://83.212.107.202/x-link-art/api?categories=painter;country&url=PAGE> }
5    BIND(REPLACE(str(?templ), "PAGE", str(?page), "i") as ?x) BIND(URI(?x) as ?service)
6    SERVICE ?service {
7      ?annot oa:hasBody ?ent .
8      ?ent oae:regardsEntityName ?detectedEntity ; oae:position ?position .
9      ?ent oae:belongsTo ?category . ?category rdfs:label ?categoryName }
10 } GROUP BY ?detectedEntity ?categoryName ORDER BY DESC(?NumOfOccurrences)
```

Fig. 2. Example of a SPARQL query that parameterizes and calls an annotation service at query execution time.

```
1  SELECT DISTINCT ?authorName ?paper WHERE {
2    SERVICE <http://users.ics.forth.gr/~fafalios/> {
3      ?p <http://purl.org/dc/terms/creator> ?author
4      FILTER(?author != <http://dblp.l3s.de/d2r/resource/authors/Pavlos_Fafalios>) }
5    SERVICE ?author {
6      ?author <http://xmlns.com/foaf/0.1/name> ?authorName .
7      ?paper <http://purl.org/dc/elements/1.1/creator> ?author } }
```

Fig. 3. Example of a SPARQL query that reads and queries RDF data embedded in a Web page (as RDFa) at query execution time.

Querying RDFa. The first author of this paper has enriched his personal Web page (http://users.ics.forth.gr/~fafalios) with RDFa describing information about his publications. Using the proposed extension, such RDF data embedded in Web pages is directly available through SPARQL. For example, the query in Fig. 3 returns all his co-authors together with their publications. The list of co-authors is obtained by querying the RDF data that is embedded in his personal Web page (lines 2–4), while their names and publications are obtained by querying the dereferenceable URI of each co-author (lines 5–7). Notice that the author URIs are derived at *query execution time*.

4   Optimizations

Several approaches have been proposed in the literature that aim at optimizing the execution of SPARQL federated queries, e.g., by reordering triple patterns based on cost estimation [24], by optimizing the evaluation of the OPTIONAL operator [7] (which is the most costly operator in SPARQL), by planning SERVICE queries against multiple endpoints based on the expected number of returned triples [20], or by parallelizing the execution of joins and union operators [24]. In addition, several caching approaches have been proposed that aim to improve the performance on answering SPARQL queries [16,18]. Obviously, all of them are beneficial for SPARQL-LD too. However, SPARQL-LD has the following extra requirements (points that need attention) that are not satisfied by existing works: (a) to *reduce the ASK queries that check whether a URI corresponds or not to a SPARQL endpoint*, and (b) to *avoid fetching remote resources that have been already fetched in the context of a single query execution*.

Below we describe optimizations that cope with the above requirements.

4.1   Index of Known SPARQL Endpoints

We have seen that, compared to the original SERVICE operator, the only additional cost is the time to run an ASK query (as we will see in Sect. 5, this cost is about 200 ms in average). To eliminate this cost, we can keep a small index with the URIs of known endpoints (like DBpedia's and Europeana's) as well as the URIs of endpoints that have been already checked. Thereby, if the SERVICE URI exists in the index, the query is directly forwarded to the endpoint, otherwise an ASK query is first submitted.

```
1  SELECT DISTINCT ?painter ?work  WHERE {
2    SERVICE <http://dbpedia.org/resource/Category:Greek_painters> {
3      ?painter <http://purl.org/dc/terms/subject> ?greekPainter }
4    SERVICE <http://europeana.ontotext.com/sparql> {
5      ?objectInfo <http://purl.org/dc/elements/1.1/creator> ?painter .
6      ?objectInfo <http://www.openarchives.org/ore/terms/proxyFor> ?work } }
```

Fig. 4. Example of a SPARQL query that calls the same remote SPARQL endpoint multiple times.

For example, consider the query of Fig. 4. The query first retrieves Greek painters from the dereferenceable URI of the corresponding DBpedia category (lines 2–3), and then it queries Europeana' SPARQL endpoint for retrieving works of these painters (lines 4–6). However, if the number of painter URIs returned by the first SERVICE invocation is n, the query will call the remote endpoint n times (one for each painter URI), which in turn requires to run n ASK queries. Thus, in case we do not use the proposed index of known endpoints, the expected cost for running n ASK queries is about $n \times 200$ ms.

4.2 Request-Scope Caching of Fetched Datasets

A SPARQL query may contain multiple SERVICE invocations against the same Web resource. Consider for example the query of Fig. 3. In case the same co-author exists in more than one publications, the corresponding RDF triples (of co-author's URI) will be redundantly fetched multiple times.

In such cases, fetching and loading repeatedly the same resource triples costs both in time, computer resources and traffic load. To avoid this, for a submitted query we can use a *request-scope* cache (usable only in the context of a submitted query) of datasets that have been already fetched. Thereby, in each new SERVICE invocation, we first check if the corresponding URI exists in the cache in order to avoid re-fetching its triples. The cache can be cleared after query execution. Of course, one could instead apply a caching policy that will keep the fetched resources in cache after query execution for serving future queries (for a period of time and according to the available main memory), e.g., a combination of static and dynamic caching as it is used by web search engines [21].

5   Evaluation

We have seen that using SPARQL-LD, one can run queries which are more expressive than those supported by SPARQL 1.1. Nevertheless, here we evaluate the efficiency of the extended SERVICE operator for several querying scenarios, examining also the cost of each task of query execution. We first evaluate the time for retrieving the properties of several randomly selected resources (URIs) using different access methods (Sect. 5.1). This can also reveal which RDF format offers the lower average query time. Then, we examine the case of querying very large Web resources, i.e. resources containing even millions of triples (Sect. 5.2). This allows us to inspect the scalability of our implementation. Finally, we evaluate the effect of the proposed optimizations (Sect. 5.3).

The experiments were carried out using an ordinary computer with processor Intel Core i7 @ 3.4 GHz CPU, 8 GB RAM and running Windows 7 (64 bit). The implementation is in Java 1.7. All data used in the experiments (URIs, queries, etc.), as well as the full results, are publicly available at http://users.ics.forth. gr/~fafalios/sparql-ld/Eval.zip.

5.1   Query Execution Time

We run experiments for 1,000 randomly selected DBpedia URIs belonging to the following 10 (randomly selected) DBpedia resource classes: *Artist, Painter, Scientist, Region, River, Fish, Athlete, BasketballPlayer, SportTeam, Chemical-Compound*. Notice that DBpedia publishes its data following the Linked Data principles (i.e., the URIs are dereferenceable) and also offers the data of a URI (properties and related entities) online in various formats including N3 and RDF/XML. Moreover, DBpedia has a publicly available SPARQL endpoint (http://dbpedia.org/sparql). We measured the total time that is required for retrieving the *outgoing properties* of each URI using the following 4 access methods: (1) by querying its dereferenceable URI, (2) by querying its RDF/XML file, (3) by querying its N3 file, and (4) by querying DBpedia's SPARQL endpoint.

Figure 5 depicts a boxplot of the results. We notice that in all cases the average query time is in the scale of milliseconds. Specifically, the average time is: 654 ms when querying the dereferenceable URIs (and the mean 636 ms), 335 ms when accessing the RDF/XML files (and the mean 297 ms), 323 ms when accessing the N3 files (and the mean 293 ms), and 305 ms when querying DBpedia's endpoint (and the mean 288 ms). We notice that querying the RDF/XML or N3 files has almost the same performance as querying the endpoint (the difference is only a few milliseconds), although querying the endpoint does not require

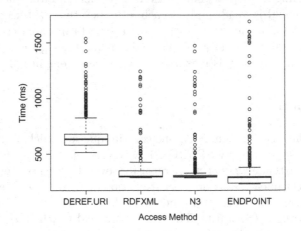

Fig. 5. Query execution time for several access methods.

checking the URI content type as well as reading and loading the corresponding RDF triples (since the query is directly evaluated by the remote endpoint and the result is returned). Moreover, we see that querying the dereferenceable URIs is more costly compared to the other approaches. This is maybe due to the fact that DBpedia first checks the value of the ACCEPT request property for returning the URI contents in the RDF/XML format (and not in HTML).

We should stress here that the results are highly affected by the network status at the time of query execution and by the status of the server hosting the remote resource. This is evident by noticing the several outlier cases in the boxplot. For example, for the case of N3 files, although the query execution time was about 300 ms for the majority of URIs, some of them required more that one second. By inspecting the contents of these URIs, we noticed that their number of triples does not differ compared to the average case, thus either the network or the remote server (DBpedia's server) is responsible for this delay.

We also examined the time required by the main subtasks of query execution, specifically: (a) the time to check if the URI given to the SERVICE operator corresponds to an endpoint, (b) the time to get the URI content type (in case it is not an endpoint), and (c) the time to fetch and load the RDF statements that correspond to the given URI (in case it is not an endpoint). Figure 6 depicts the results. The average time is 194 ms for (a) (and the mean 160 ms) and 174 ms for (b) (and the mean 146 ms). Here also we notice some outliers due to network delay. As regards (c), the average time is 289 ms (and the mean 283 ms) for the case of dereferenceable URIs, 148 ms (and the mean 121 ms) for the case of RDF/XML, and 138 ms (and the mean 118 ms) for the case of N3. We notice that accessing the N3 files is slightly more efficient. This can be justified by the fact that the N3 format is more compact and smaller in size than RDF/XML.

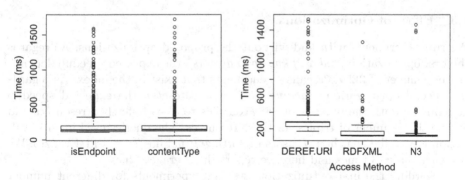

Fig. 6. Left: Time for (a) checking if the URI corresponds to a SPARQL endpoint, and (b) checking the URI content type. **Right:** Time for (c) fetching and loading the RDF statements corresponding to the given URI using different access methods.

5.2   Accessing Very Large Web Resources

We ran experiments for testing the case of accessing very large Web resources, i.e., resources containing a big number of RDF triples. We created four N3 files using real data coming from DBpedia. Specifically, we downloaded the triples of the English titles from the canonicalized 2014 dataset (single file containing about 11 millions triples). From this file, we created four files of 10^4, 10^5, 10^6, and 10^7 triples respectively, and we uploaded them in a Web accessible server. For each URI, we submitted a query that requests the properties of a particular resource (that exists in all files as subject in the triple).

Table 1 depicts the results. We notice that the total query execution time is less than 1 s in the case of 10^4 triples, while for 1 m triples the time is about 30 s. For bigger files, the time can be in the scale of minutes. However, we should stress here that, usually, the online RDF files are not very big in size because large files cannot be easily handled/exchanged. For instance, DBpedia publishes one file (of small size) for each named-entity. As regards the main subtasks of query execution, we notice that, as expected, fetching and loading the triples is the most time consuming task.

Table 1. Querying large online N3 files.

Num of triples	Total query time	Is endpoint	Get content type	Fetch triples
10^4	900 ms	318 ms	7 ms	449 ms
10^5	3.2 s	1.3 s	8 ms	1.8 s
10^6	31 s	11 s	8 ms	19 s
10^7	546 s	111 s	51 ms	433 s

5.3   Effect of Optimizations

We run experiments with and without the proposed optimizations. As regards the first optimization (index of known endpoints), the expected speedup depends on the number of SERVICE calls to endpoints that exist in the index. As regards the second optimization (caching of fetched datasets), the expected speedup depends on both the number of SERVICE calls to already-fetched resources and on the size (number of triples) of these resources. The queries used in this evaluation are available at http://users.ics.forth.gr/~fafalios/sparql-ld/Eval.zip. We run each query 3 times and here we report the average values.

Regarding the first optimization, we run experiments for different number of calls to "known" remote endpoints. Table 2 shows the speedup for each case. The speedup is calculated as the query execution time when the optimization is not applied divided by the optimized time. We notice that, using the proposed optimization method, the query execution time can be significantly improved (in our experiments, it is from 1.6 to 3.9 times faster).

As regards the second optimization, we run experiments for different number of calls to already-fetched resources and for different number of triples in these resources. Table 3 shows the results. As expected, this optimization can highly

Table 2. Effect of first optimization (*index of known endpoints*).

Query	Num of calls to indexed endpoints	Time without Opt.	Time with Opt.	Speedup
Q1	10	3.5 s	1.8 s	1.9×
Q2	10^2	27.2 s	16.5 s	1.6×
Q3	10^3	9.6 min	2.5 min	3.9×
Q4	10^4	44.8 min	24 min	1.9×

Table 3. Effect of second optimization (*caching of fetched datasets*)

Query	Num of calls to cached datasets	Num of triples	Time without Opt.	Time with Opt.	Speedup
Q5	**16**	10^3	11.9 s	1.4 s	8.5×
Q6	**16**	10^4	72.9 s	5.7 s	12.7×
Q7	**16**	10^5	10.3 min	39.8 s	15.5×
Q8	10	$\mathbf{10^2}$	11.8 s	9.6 s	1.2×
Q9	10^2	$\mathbf{10^2}$	35.9 s	10.7 s	3.4×
Q10	10^3	$\mathbf{10^2}$	4.8 min	11.6 s	24.6×

improve the efficiency of query execution (in our experiments, it is from 1.2 to 24.6 times faster), while it also reduces the transfer of data between local server and remote sources.

6   Conclusion

We have presented SPARQL-LD, a generalization of SPARQL 1.1 Federated Query that allows to directly fetch and query RDF data from any HTTP Web source. Using SPARQL-LD one can exploit and combine in the same SPARQL query: (i) data stored in the (local) repository, (ii) data coming from online RDF or JSON-LD files, (iii) data embedded in Web pages as RDFa, (iv) data coming from dereferenceable URIs, (v) data that is dynamically created by Web Services, and (vi) data coming by querying other SPARQL endpoints. A distinctive characteristic of this extension is that it enables to also query datasets coming from the partial results of a query (i.e., discovered at query execution time). We also identified factors that can affect the efficiency of SPARQL-LD and we proposed optimizations that manipulate such cases.

The functionality offered by SPARQL-LD motivates Web publishers to follow the Linked Data principles and expose their data in RDF without needing to set up and maintain a costly SPARQL endpoint. For instance, a museum can enrich its Web page with RDFa, or just put online a RDF dump, and thereby make its data directly accessible via SPARQL.

The conducted experiments showed that, as expected, the time for querying the triples of online RDF resources highly depends on the number of triples existing in the resource, on the status of the network between the local server and the remote server hosting the resource, and on the status of the remote server itself. Nevertheless, we saw that for common Web resources of normal size (less than 10^4 triples), and without using any caching method, the total query time is very low. We also saw that the performance of querying RDF/XML or N3 files is almost the same as querying an endpoint. Finally, as regards the proposed optimizations, experimental results showed that they can highly improve the query execution time. For instance, in our experiments, using a request-scope cache of fetched datasets, the execution time of a query reduced from about 5 min to only 12 s.

In future, we will study query planning approaches and more optimization and caching techniques appropriate for SPARQL-LD.

Acknowledgements. This work was partially supported by the BlueBRIDGE project (H2020 Research Infrastructures, 2015–2018, Project No: 675680).

References

1. A JSON-based Serialization for Linked Data. http://www.w3.org/TR/json-ld/
2. Apache Jena. http://jena.apache.org/
3. RDFa Core 1.1. http://www.w3.org/TR/2015/REC-rdfa-core-20150317/
4. SPARQL 1.1 Query Language (W3C). http://www.w3.org/TR/sparql11-query/
5. SPARQL Federat. Query. http://www.w3.org/TR/sparql11-federated-query/
6. Bizer, C., Heath, T., Berners-Lee, T.: Linked data-the story so far. Int. J. Semant. Web Inf. Syst. **5**(3), 1–22 (2009)
7. Buil-Aranda, C., Arenas, M., Corcho, O., Polleres, A.: Federating queries in SPARQL 1.1: syntax, semantics and evaluation. Web Semant.: Sci. Serv. Agents World Wide Web **18**(1), 1–17 (2013)
8. Buil-Aranda, C., Hogan, A., Umbrich, J., Vandenbussche, P.-Y.: SPARQL web-querying infrastructure: ready for action? In: Alani, H., et al. (eds.) ISWC 2013, Part II. LNCS, vol. 8219, pp. 277–293. Springer, Heidelberg (2013)
9. Fafalios, P., Baritakis, M., Tzitzikas, Y.: Exploiting linked data for open and configurable named entity extraction. Int. J. Artif. Intell. Tools **24**(02), 1540012-1–1540012-42 (2015)
10. Fafalios, P., Tzitzikas, Y.: SPARQL-LD: a SPARQL extension for fetching and querying linked data. In: The Semantic Web-ISWC 2015 (Posters & Demonstrations Track), Bethlehem, Pennsylvania, USA (2015)
11. Hartig, O.: SPARQL for a web of linked data: semantics and computability. In: Simperl, E., Cimiano, P., Polleres, A., Corcho, O., Presutti, V. (eds.) ESWC 2012. LNCS, vol. 7295, pp. 8–23. Springer, Heidelberg (2012)
12. Hartig, O., Pérez, J.: LDQL: a query language for the web of linked data. In: Arenas, M., et al. (eds.) ISWC 2015. LNCS, vol. 9366, pp. 73–91. Springer, Heidelberg (2015)
13. Haslhofer, B., Momeni Roochi, E., Schandl, B., Zander, S.: Europeana RDF store report (2011)

14. Hogan, A., Harth, A., Umbrich, J., Kinsella, S., Polleres, A., Decker, S.: Searching and browsing linked data with SWSE: the semantic web search engine. Web Semant.: Sci. Serv. Agents World Wide Web **9**(4), 365–401 (2011)
15. Isaac, A., Haslhofer, B.: Europeana linked open data-data. europeana. eu. Semant. Web **4**(3), 291–297 (2013)
16. Kjernsmo, K.: A survey of HTTP caching implementations on the open semantic web. In: Gandon, F., Sabou, M., Sack, H., d'Amato, C., Cudré-Mauroux, P., Zimmermann, A. (eds.) ESWC 2015. LNCS, vol. 9088, pp. 286–301. Springer, Heidelberg (2015)
17. Langegger, A., Wöß, W., Blöchl, M.: A semantic web middleware for virtual data integration on the web. In: Bechhofer, S., Hauswirth, M., Hoffmann, J., Koubarakis, M. (eds.) ESWC 2008. LNCS, vol. 5021, pp. 493–507. Springer, Heidelberg (2008)
18. Martin, M., Unbehauen, J., Auer, S.: Improving the performance of semantic web applications with SPARQL query caching. In: Aroyo, L., Antoniou, G., Hyvönen, E., Teije, A., Stuckenschmidt, H., Cabral, L., Tudorache, T. (eds.) ESWC 2010, Part II. LNCS, vol. 6089, pp. 304–318. Springer, Heidelberg (2010)
19. Miranker, D., Depena, R., Jung, H., Sequeda, J., Reyna, C.: Diamond: a SPARQL query engine, for linked data based on the rete match. In: AImWD 2012 (2012)
20. Montoya, G., Vidal, M.-E., Acosta, M.: A heuristic-based approach for planning federated SPARQL queries. In: COLD, vol. 905 (2012)
21. Papadakis, M., Tzitzikas, Y.: Answering keyword queries through cached subqueries in best match retrieval models. J. Intell. Inf. Syst. **44**(1), 67–106 (2015)
22. Purday, J.: Think culture: Europeana. eu from concept to construction. Electron. Library **27**(6), 919–937 (2009)
23. Quilitz, B., Leser, U.: Querying distributed RDF data sources with SPARQL. In: Bechhofer, S., Hauswirth, M., Hoffmann, J., Koubarakis, M. (eds.) ESWC 2008. LNCS, vol. 5021, pp. 524–538. Springer, Heidelberg (2008)
24. Schwarte, A., Haase, P., Hose, K., Schenkel, R., Schmidt, M.: FedX: optimization techniques for federated query processing on linked data. In: Aroyo, L., Welty, C., Alani, H., Taylor, J., Bernstein, A., Kagal, L., Noy, N., Blomqvist, E. (eds.) ISWC 2011, Part I. LNCS, vol. 7031, pp. 601–616. Springer, Heidelberg (2011)
25. Tzitzikas, Y., et al.: Integrating heterogeneous and distributed information about marine species through a top level ontology. In: Garoufallou, E., Greenberg, J. (eds.) MTSR 2013. CCIS, vol. 390, pp. 289–301. Springer, Heidelberg (2013)

A Scalable Approach to Incrementally Building Knowledge Graphs

Gleb Gawriljuk[1], Andreas Harth[1(✉)], Craig A. Knoblock[2], and Pedro Szekely[2]

[1] Institute of Applied Informatics and Formal Description Methods (AIFB),
Karlsruhe Institute of Technology, 76128 Karlsruhe, Germany
`harth@kit.edu`
[2] Information Sciences Institute, University of Southern California,
Marina Del Rey, CA 90292, USA

Abstract. We work on converting the metadata of 13 American art museums and archives into Linked Data, to be able to integrate and query the resulting data. While there are many good sources of artist data, no single source covers all artists. We thus address the challenge of building a comprehensive knowledge graph of artists that we can then use to link the data from each of the individual museums. We present a framework to construct and incrementally extend a knowledge graph, describe and evaluate techniques for efficiently building knowledge graphs through the use of the MinHash/LSH algorithm for generating candidate matches, and conduct an evaluation that demonstrates our approach can efficiently and accurately build a knowledge graph about artists.

1 Introduction

To be able to link the data about artists from the 13 museums and archives of the American Art Collaborative[1], we need a reference data set that contains all or most of these artists. Ideally, such a reference set would combine the known information from all of the high-quality sources available today. Given that such a reference set does not exist, we have to create the set from the input sources. Constructing a consolidated set is challenging as we need to link the common entities across large sources. Linking entities is challenging because the data is often inconsistent due to errors, misspellings, or out-of-date information.

We present an approach to building knowledge graphs by consolidating data from multiple sources. Constructing such a knowledge graph involves several challenges. First, when consolidating the data from various data sources, we had to align the data across available sources. Second, given the size of the data sources we had to develop a scalable approach that could link the data across sources with millions of entities. Finally, since the web is subject to a constant growth, we designed an approach that makes it easy to include new sources by extending an existing knowledge graph with the data of any new data sources that become available.

---

[1] http://americanartcollaborative.org/.

N. Fuhr et al. (Eds.): TPDL 2016, LNCS 9819, pp. 188–199, 2016.
DOI: 10.1007/978-3-319-43997-6_15

In our approach, each knowledge graph provides data about entities of a specific type. We define a knowledge graph as a set of typed properties between an entity and its property values. Given multiple datasets, we use an initial dataset to create an initial knowledge graph, and then incrementally consolidate further datasets into the knowledge graph. To consolidate data sources into the knowledge graph, we use a five-step approach. One step in the approach applies the MinHash/LSH (Locality-sensitive Hashing) algorithm [8] to produce candidates for rule-based matching functions. Therefore, we reduce the number of comparisons the matching functions must execute, and thus are able to perform entity linkage at scale.

In the remainder of the paper, we first present a motivating example of building a knowledge graph of artists. Second, we describe our scalable approach to incrementally building knowledge graphs. Third, we cover an experiment in which we construct a knowledge graph consolidating 161,465 artist entities from initially 17,539,125 entities, to demonstrate the scalability of the approach. We evaluate precision and recall on a manually built ground truth with 200 artist entities. Finally, we compare the work to other related work and present our conclusions.

2 Motivating Example

To illustrate our objective, we provide an example of consolidating data about the pop artist Roy Lichtenstein from four sources. We use the available SPARQL endpoints to access data from the Union List of Artist Names (ULAN), containing descriptions of 109,415 artists provided by the Getty[2], and the Smithsonian American Art Museum (SAAM), containing descriptions of 8,407 artists. We use the downloadable RDF files from DBpedia, which provides data about 1,176,759 people, and the Virtual International Authority File (VIAF), which provides data about 16,244,546 people[3]. Table 1 shows which data source provides what data about Roy Lichtenstein.

After all four sources are consolidated, the resulting knowledge graph holds the data about Roy Lichtenstein from all four source within one cluster. The cluster has links to the identifiers denoting Roy Lichtenstein from all four consolidated data sources, including provenance information. In case two or more sources provide the same value for a property, the knowledge graph mentions the value once and adds provenance for each source which includes the value.

3 Building and Extending a Knowledge Graph

We first give a general overview, and then discuss each step in detail. Our approach is designed to incrementally consolidate one data source at a time. The initial data source is used to create an initial knowledge graph. With each subsequent data source as input, we extend the knowledge graph with data of the subsequent data source as shown in Fig. 1.

---

[2] http://www.getty.edu/.

[3] Please note that not all of the people in DBpedia and VIAF are artists.

Table 1. List of properties expressed in the shared domain ontology by data source.

Property name	ULAN	SAAM	DBpedia	VIAF	Knowledge graph
name	X	X	X	X	X
alternateName		X		X	X
givenName		X	X	X	X
familyName		X	X	X	X
gender	X				X
nationality	X				X
birthDate	X	X	X	X	X
deathDate	X	X	X	X	X
birthPlace	X	X	X		X
deathPlace	X	X	X		X
description	X	X	X	X	X

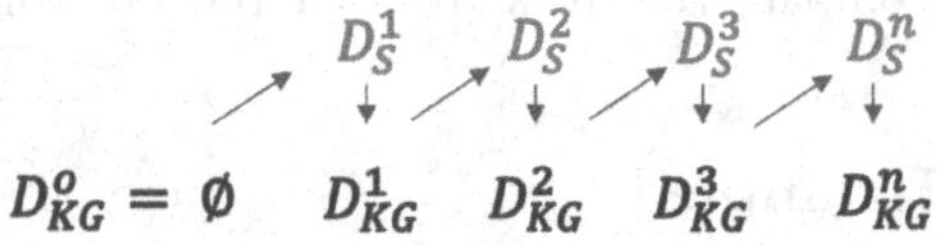

$$D_S^1 \quad D_S^2 \quad D_S^3 \quad D_S^n$$

$$D_{KG}^o = \emptyset \quad D_{KG}^1 \quad D_{KG}^2 \quad D_{KG}^3 \quad D_{KG}^n$$

Fig. 1. An illustration of the incremental growth of the knowledge graph.

Our approach consists of five steps as illustrated in Fig. 2. In the first step, the input data source is filtered to only select the entities relevant for the final knowledge graph. Then, the schema of the input data source is mapped to either the input knowledge graph or, if no input knowledge graph is provided, to the schema which should be used for the construction of the initial knowledge graph.

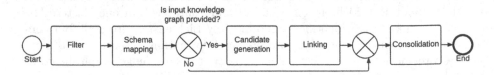

Fig. 2. The distinct steps of the approach.

After the input data source is filtered and mapped, there is a choice. In case no input knowledge graph is provided, the input data source is used to build an initial knowledge graph. For each entity in the initial data source, an entity in the knowledge graph is created in the consolidation step. In case an existing knowledge graph is provided next to the input data source, we continue with the candidate generation step, in which we apply the MinHash/LSH to generate candidates within the input knowledge graph for entities in the data source.

The linking step evaluates each generated candidate using rule-based matching functions to decide which candidate is a match. In the final step, the input data source is consolidated with the input knowledge graph based on the matching links selected by the matching functions.

The approach can be adjusted through parameters concerning the candidate generation and the linking steps. For the candidate generation, the user can specify MinHash/LSH parameters which define how strict the MinHash/LSH searches are for the candidates. For the linking step, the user can define the rule-based matching functions and similarity thresholds.

3.1 Filter

In case the input data source includes additional entity types, the filter step can be applied as an optional step to identify the entities of the target type in the data source. In the filter step, we execute the candidate generation step followed by the linking step for the input data source with the input knowledge graph. By applying the filtering step, we link entities in the data source to the knowledge graph and, thus, identify the linked entities as entities of the target type. Then, the approach proceeds to the next step only with the identified entities.

3.2 Schema Mapping

The schema of the input data source has to be mapped to the schema of the knowledge graph to be generated. Both the candidate generation and the linking step require to know which property in the data source semantically corresponds to a property in the knowledge graph. This is defined by the schema mapping. We use Karma [10], which automates the mapping from a source to an ontology, to build the source mappings for the datasets. For mapping the sources, we use schema:Person from the schema.org ontology.

3.3 Candidate Generation

The most time-consuming task in our approach is linking the entities from the data source to the entities in the knowledge graph. As mentioned by Doan et al. [4] hashing techniques can be applied to scale up the entity linkage. Therefore, to improve the scalability of our approach, we execute the candidate generation step to apply the MinHash/LSH hashing technique [8] before the entity linkage.

We apply the MinHash/LSH algorithm to find candidate entities in the knowledge graph for each entity in the data source which are likely to match, so that we only need to compare the found candidates. Hence, we significantly reduce the number of pair-wise comparisons required to execute and, therefore, speed up the approach.

MinHash/LSH operates over an n-gram representation of the name values[4]. The algorithm then hashes similar entities into the same cluster. The similarity of

---

[4] The 2-gram of the first name 'Roy' consists of {_R, Ro, oy, y_}.

two entities is defined by the Jaccard similarity between the two sets of n-grams representing the two entities[5].

How well the MinHash/LSH performs regarding recall and precision depends on the number of used minhashes m and the number of items in the generated hashes i. With m and i the LSH threshold t can be approximated as $t = \frac{1}{i}^{\frac{1}{m}}$. Thus, by adjusting m and i we can adjust the LSH threshold.

The LSH threshold t has the characteristic that t is the Jaccard-similarity between the entity e_1 in the knowledge graph and the entity e_2 in the data source so that e_1 has a 50 % chance to become a candidate. In other words, if the LSH threshold is set to 80 %, an entity e_1 has a 50 % chance to be selected as a candidate of entity e_2, when the Jaccard-similarity of e_1 and e_2 is 0.8. Hence, the threshold can be seen as a parameter to adjust the LSH trade-off between recall, precision, and performance. If a high recall is important then the threshold needs to be low and the candidate generation will require more time. If the performance of the MinHash/LSH and precision are important the threshold needs to be high.

To select the most effective LSH threshold for finding candidate based on the entity's name, we conduct extensive experiments of the MinHash/LSH. We assess the MinHash/LSH with either alternative m or alternative i which results in various LSH thresholds. Table 2 shows the resulting LSH thresholds with their recall and precision on the ground truth described in Sect. 4.1.

Table 2. LSH thresholds with their recall and precision.

t	34 %	40 %	46 %	52 %
Precision	14.12 %	15.69 %	23.12 %	26.96 %
Recall	100 %	99.50 %	99.50 %	98.01 %

The objective of the candidate generation step is to find as many matching candidates as possible. Hence, we need to keep the recall high. To have a high recall, we apply the MinHash/LSH with a low threshold of 46 %. A low threshold leads to a low precision because we find many false candidates. However, we tolerate a low precision of the candidate generation because the precision will be increased in the following linking step.

3.4   Linking

The linking step evaluates the clusters of candidates resulting from the Min-Hash/LSH algorithm. We apply rule-based matching functions to decide whether the candidate and the target entity in the knowledge graph are the same real-world entity. The objective for the matching functions is to increase the precision as much as possible while not decreasing the recall.

---

[5] The Jaccard similarity between sets S and T is defined as $\frac{|S \cap T|}{|S \cup T|}$.

When applying a matching function, the function iterates over each cluster. For each cluster, the function examines its matching rules for each candidate. If the candidate fulfills the matching rules, the candidate stays in the cluster; otherwise, the candidate is removed. As the clusters contain the data of each candidate and the target entity, the matching rules can leverage that data to decide whether a candidate is a match or not.

The matching functions can be exchanged with each other or chained to a sequence of functions, each evaluating a certain aspect of the candidate data. For example, when comparing people entities, we can define two matching functions to first check the similarity of the names and remove candidate with a low similarity and, then, further remove candidates with a different birth year.

3.5 Consolidation

In the final consolidation step, we either construct an initial knowledge graph from the initial data source or extend the input knowledge graph with the selected links of the input data source.

The data model of the knowledge graph illustrated in Fig. 3 is based on the schema:Role concept, which provides a way to represent multiple provenance information for one property value. Each distinct property value of an entity is modeled as a feature role holding the property value and its provenance. We use the properties of schema:Person to describe artists and the PROV-O[6] ontology to model the provenance properties. The provenance indicates for each property value the time of inclusion, the data source and the linking method used.

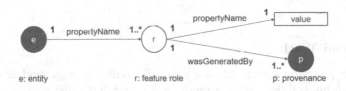

Fig. 3. Knowledge graph data model.

We construct the initial knowledge by creating a new entity for each entity in the data source, with a unique URI within the knowledge graph and a schema:sameAs property pointing to the original URI.

The knowledge graph is extended by adding only the properties and property values of the data source which do not exist yet for the relevant entity. Given the data of the existing knowledge graph D_{kg} and the data of the subsequent data source D_s, the extended knowledge graph D^e_{kg} holds the data from D_{kg} with data from D_s which was not there before $D^e_{kg} = D_{kg} \cup D_s$. Figure 4 shows how the consolidation of a subsequent data source would add an already existing name value (A), a new name value (B) and a new birth date value (C) to an

---

[6] http://www.w3.org/TR/prov-o/.

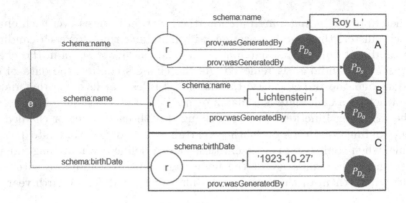

Fig. 4. Addition of a feature.

entity in the knowledge graph. The addition of the existing alternative name (A) is represented by the added provenance to the existing feature role.

After all matched candidates are processed, the entities from the subsequent data source that were not matched to any existing entity in the knowledge graph need to be consolidated. Hence, we iterate over the data source and include each unmatched entity as a new entity in the knowledge graph.

4   Evaluation

We evaluate our approach by assessing the quality of the results and the run-time performance.

4.1   Ground Truth

To quantify precision and recall, we manually build up a ground truth[7]. The ground truth holds links for the alphabetically first 200 artist entities which are represented in each of the four data sources. Because each entity exists in ULAN and SAAM, we know that the entities are artists even if they are not typed as an artist in DBpedia or VIAF. If a data source includes duplicates for an entity, these duplicates are included in the ground truth as well. Hence, links found by the approach that are not in the ground truth can be assumed to be wrong.

4.2   Quality Evaluation

To evaluate the quality of the results, we review the steps of the approach individually for each source, starting with the candidate generation (LSH). We apply two matching functions in sequence. First, we apply the Hybrid-Jaccard function to check the similarity of the name values between the candidate and the

---

[7] https://bitbucket.org/GlebGawriljuk/aifb-isi-knowledgegraphconstruction/raw/
168b6ec21654e1de01d546567f7232b77daaf1a2/groundTruth_final_2015.tsv.

target entity and only keep the candidates with a similarity above 0.5. Second, we apply the 'birth year' function to check whether the candidate has the same birth year as the target entity.

We start with the ULAN data and create a total of 109,414 entities in the initial knowledge graph. Consolidating the 8,406 SAAM entities, we link 3,443 entities and introduce 4,963 new entities, resulting in 114,377 knowledge graph entities. Then, we identify 58,944 DBpedia artist entities, link 34,838 of them, and introduce 24,106 as new entities, leading to 139,396 knowledge graph entities. Lastly, we identify 1,798,285 VIAF artist entities; we link 1,775,303 entities and introduce 22,982 as new. We merge many VIAF entities, as the VIAF data contains many duplicates and incomplete data (i.e., some records only have a last name) and our matching is solely based on the first and last name. The final knowledge graph holds 161,465 entities.

Table 3 shows the results of the ground truth evaluation based on the following quality measures:

- N: the number of ground truth entities which are introduced as new entities to the extended knowledge graph.
- T: the number of ground truth entities for which candidates are found.
- C: the number of candidates that are found for all ground truth entities.
- C_c: the number of correct candidates for all ground truth entities including duplicate candidates.
- C_t: the number of correct candidates for all ground truth entities excluding correct duplicate candidates.
- C_i: the number of incorrect candidates for all ground truth entities.

Table 3. Results of the ground truth quality evaluation.

		T	C	C_c	C_t	C_i	N	Recall	Precision
ULAN	Initial KG	-	-	-	-	-	202	-	-
SAAM	Candidate generation	200	1,841	201	200	1,640	0	100 %	10.87 %
	Linking Hybrid-Jaccard	200	217	201	200	16	0	100 %	92.59 %
	Birth year	187	188	188	187	0	13	93.50 %	100 %
DBpedia	Candidate generation	199	1,550	211	197	1,339	1	98.50 %	12.83 %
	Linking Hybrid-Jaccard	197	226	208	196	18	3	98.00 %	91.59 %
	Birth year	187	192	190	187	0	13	93.50 %	100 %
VIAF	Candidate generation	271	1,468	293	272	1,175	7	97.84 %	18.80 %
	Linking Hybrid-Jaccard	271	369	290	270	79	0	97.12 %	77.36 %
	birth year	262	276	266	260	10	9	93.53 %	96.30 %

When evaluating precision and recall we have to consider that the knowledge graph can hold duplicate entities which are all correct candidates for a target entity. Therefore, we differentiate between the number of correct candidates including the duplicates (C_c) and the number of correct candidates excluding the duplicates (C_t). Both precision and recall measures are based on C_t. The precision can be calculated based on C_c as well, however, the precision would be always higher and would only differ in at most 1 %.

In the candidate generation step we achieve a high recall and a low precision for all data sources. On the one hand, we miss only a few correct candidates due to a low LSH threshold, because the target's name differs from the candidate's name, due to a missing middle name, spelling errors or missing values. Therefore, both have a Jaccard similarity near or below our LSH threshold. Such candidates have only a chance of 50 % or less to be found by the MinHash/LSH. On the other hand, the MinHash/LSH also finds incorrect candidates due to similar name value between the candidate and the target entity.

We anticipate to reduce the number of candidates and increase the precision by applying the Hybrid-Jaccard function on the found candidates. We manage to increase the precision close to 90 % for all data sources, with a slight decrease in recall. This is achieved by removing incorrect candidates while keeping most of the correct candidates in the cluster. E.g., in the case of the SAAM consolidation, we decrease the number of candidates from 1,841 to 221 while keeping the number of correct candidates at 201.

We try to further increase the precision by applying the 'birth year'-function. Furthermore, we tolerate a decrease of the recall as a trade-off to the increased precision. We achieve to increase the precision for all data source consolidations. As foreseen, for all consolidation the recall decreases. The recall decreases because the 'birth year'-function removes correct candidates due to errors in the birth year values of the entities. An example is a correct candidate "Pietro Aquila" with "1592" as birth year value from ULAN[8] but "1650" from SAAM[9].

We do not achieve a 100 % precision because some incorrect candidates always remain, in case an incorrect candidate has a similar name and no birth year value. Due to a similar name both the LSH and the Hybrid-Jaccard keep the candidate in the cluster. The 'birth year'-function cannot evaluate the candidate because of the missing birth year and, thus, the incorrect candidate stays in the cluster.

The results show that we manage to achieve a high recall during the candidate generation as anticipated. Furthermore, we increase the precision by applying the Hybrid-Jaccard-function. Lastly, we succeed in further increasing the precision for the cost of a lower recall by executing the 'birth year'-function.

We tolerate a decreasing recall for an increasing precision because we argue that a higher precision is more important than a high recall. With a high recall, we can be sure to find all correct candidates for the cost of creating noise in form of schema:sameAs links between non-matching entities. In such a case, each data source added would lead to more noise, wrongly consolidating more entities into one entity in the knowledge graph. In contrast, by prioritizing a high precision we ensure that different entities are not consolidated into one entity, leading to less noise (but more potential duplicates) in the knowledge graph.

---

[8] http://vocab.getty.edu/ulan/500018769.
[9] http://edan.si.edu/saam/id/person-institution/121.

4.3   Performance Evaluation

For performance evaluation we measure the run-time for each step and review the run-time in relation to the data size. Experiments are conducted on an Ubuntu machine with 4 AMD Opteron 62xx class 2 GHz CPU cores and 32 GB RAM.

Table 4. Overall performance per data source in hours:min:sec.

Step	ULAN	SAAM	DBpedia	VIAF
Candidate generation	-	00:15:59	01:55:14	29:58:26
Linking	-	00:01:37	01:11:22	55:02:13
Consolidation	00:02:12	00:04:49	00:23:20	156:34:12
Total	00:02:12	00:22:25	03:29:56	229:00:39

Table 4 shows the run-time of each step of the approach for each data source. We find a long run-time for the Hybrid-Jaccard rule and the consolidation on the VIAF data source for two reasons. Firstly, the VIAF data source is significantly larger than the other data sources. Secondly, VIAF includes significantly more different name values for each entity in multiple languages and variations. Hence, the Hybrid-Jaccard rule must evaluate more values for each target entity when linking the VIAF entities. Overall we can observe an increasing run-time with each run as the size of both the knowledge graph and the data sources increases.

5   Related Work

The challenge of automatic knowledge graph construction is a trending topic in current research[10]. The body of work can be divided into three groups.

The first group, such as DIG [16], Knowledge Vault [5], CiteSeerX [1], Pujara et al. [12], the Knowledge Graph Identification [13] and NELL [3], builds knowledge graphs with a fixed schema from data retrieved using information extraction techniques. Similarly, we build up the knowledge graph with a fixed target schema. However, we do not apply information extraction techniques to retrieve data from the web. Our focus is on using existing structured web data sources, consolidating their data, and providing an integrated view on the data from multiple data sources within a single knowledge graph.

The second group, such as Reverb [6], OLLIE [11], and PRISMATIC [7], uses open information extraction without a fixed schema. Again, our approach differs in the objective: we integrate structured data rather than extract data. Also, we operate on a fixed schema, while the open information extraction approaches assume no fixed schema.

The third group consolidates data from structured data sources to build up the knowledge graph. We consider our work strongly related to LDIF [14], YAGO [15], YAGO2 [9], and Freebase [2], which we now discuss in detail.

---

[10] For example, the series of workshops on Automated Knowledge Base Construction (AKBC), http://www.akbc.ws/.

The Linked Data Integration Framework (LDIF) [14] designs a similar approach to construct knowledge graphs. Both LDIF and our approach consolidate web data into a knowledge graph by mapping input data to a target schema, linking entities, and consolidating the data based on the found links. The most significant difference is that LDIF uses a Hadoop implementation to parallelize the work of their transformation modules. In contrast, we scale by applying the MinHash/LSH algorithm. Furthermore, both approaches include provenance information in the knowledge graph. However, where LDIF includes only the origin of data, we additionally include the processing information.

YAGO [15] builds up a knowledge graph by defining entity classes from the conceptual Wikipedia categories and WordNet terms that do not represent an entity. However, entity data is included only from Wikipedia as all WordNet terms representing an entity are removed. Hence, no entity resolution among the entities in Wikipedia and WordNet is applied. YAGO2 [9] does apply entity resolution, but only on the location entities from Wikipedia and GeoNames[11] Similar to our approach, the entity resolution techniques can be seen as multiple rule-based matching functions. Their approach first matches the location entities based on their names. If multiple candidates are found, YAGO2 uses the data from the geo-coordinates of the entities to decide which candidate is a match. However, these rules are fixed part of the approach and cannot be exchanged.

Freebase [2] is a tuple database which is collaboratively created, structured, and maintained. Freebase provides an HTTP-/JSON-based API to maintain the data through read and write operations. However, entity linking is carried out in an interactive way with user input required for consolidation.

6   Conclusion

We have addressed the problem of efficiently building a consolidated knowledge graph of a specific entity type from data spread over multiple large data sources. Our approach works by aligning the data to a domain ontology, identifying the candidate links using the MinHash/LSH techniques, generating links with high precision using entity linking rules, and finally building the knowledge graph.

We have applied our approach to building a knowledge graph about artists containing 161,465 artists consolidated from four data sources. We have evaluated the resulting knowledge graph and showed that on a sample of data it has an average recall of 96,82 % and a precision of 99,17 %. We also have demonstrated that we are able to efficiently construct a knowledge graph of high quality, processing 17,539,125 entities in about 10 days on modest hardware.

References

1. Alexander, G., Ororbia, I., Wu, J., Giles, C.L: CiteSeerX: intelligent information extraction and knowledge creation from web-based data. In: Proceedings of the 4th Workshop on Automated Knowledge Base Construction at NIPS (2014)

---

[11] http://www.geonames.org/.

2. Bollacker, K., Evans, C., Paritosh, P., Sturge, T., Taylor, J.: Freebase: a collaboratively created graph database for structuring human knowledge. In: Proceedings of the ACM SIGMOD International Conference on Management of Data (2008)
3. Carlson, A., Betteridge, J., Kisiel, B., Settles, B., Hruschka, E.R.J., Mitchell, T.: Toward an architecture for never-ending language learning. In: Proceedings of the Twenty-Fourth Conference on Artificial Intelligence (AAAI) (2010)
4. Doan, A., Halevy, A., Ives, Z.: Principles of Data Integration. Elsevier, Amsterdam (2012)
5. Dong, X.L., Gabrilovich, E., Heitz, G., Horn, W., Lao, N., Murphy, K., Strohmann, T., Sun, S., Zhang, W.: Knowledge vault: a web-scale approach to probabilistic knowledge fusion. In: Proceedings of the 20th ACM SIGKDD International Conference on Knowledge Discovery and Data Mining (2014)
6. Fader, A., Soderland, S., Etzioni, O.: Identifying relations for open information extraction. In: Proceedings of the Conference on Empirical Methods in NLP and Computational Natural Language Learning (EMNLP) (2011)
7. Fan, J., Ferrucci, D., Gondek, D., Kalyanpur, A.: PRISMATIC: inducing knowledge from a large scale lexicalized relation resource. In: Proceedings of the 1st International Workshop on Formalisms and Methodology for Learning by Reading (2010)
8. Gionis, A., Indyk, P., Motwani, R.: Similarity search in high dimensions via hashing. In: Proceedings of the 25th International Conference on Very Large Data Bases (1999)
9. Hoffart, J., Suchanek, F.M., Berberich, K., Weikum, G.: YAGO2: a spatially and temporally enhanced knowledge base from wikipedia. Artif. Intell. J. **194**, 28–61 (2013)
10. Knoblock, C.A., Szekely, P., Ambite, J.L., Goel, A., Gupta, S., Lerman, K., Muslea, M., Taheriyan, M., Mallick, P.: Semi-automatically mapping structured sources into the semantic web. In: Simperl, E., Cimiano, P., Polleres, A., Corcho, O., Presutti, V. (eds.) ESWC 2012. LNCS, vol. 7295, pp. 375–390. Springer, Heidelberg (2012)
11. Mausam, Schmitz, M., Bart, R., Soderland, S., Etzioni, O.: Open language learning for information extraction. In: Proceedings of the Conference on Empirical Methods on NLP and Computational Natural Language Learning (EMNLP) (2012)
12. Pujara, J., Getoor, L.: Building dynamic knowledge graphs. In: Proceedings of the Knowledge Extraction Workshop at NAACL-HLT (2014)
13. Pujara, J., Miao, H., Getoor, L., Cohen, W.: Knowledge graph identification. In: Alani, H., Kagal, L., Fokoue, A., Groth, P., Biemann, C., Parreira, J.X., Aroyo, L., Noy, N., Welty, C., Janowicz, K. (eds.) ISWC 2013, Part I. LNCS, vol. 8218, pp. 542–557. Springer, Heidelberg (2013)
14. Schultz, A., Matteini, A., Isele, R., Mendes, P., Bizer, C., Becker, C.: LDIF - a framework for large-scale linked data integration graphs. In: Proceedings of 21st International Conference on World Wide Web (2012)
15. Suchanek, F., Kasneci, G., Weikum, G.: YAGO - a core of semantic knowledge. In: Proceedings of the 16th International Conference on World Wide Web (2007)
16. Szekely, P., et al.: Building and using a knowledge graph to combat human trafficking. In: Arenas, M., et al. (eds.) ISWC 2015. LNCS, vol. 9367, pp. 205–221. Springer, Heidelberg (2015)

Multimedia and Time Aspects

Sub-document Timestamping: A Study on the Content Creation Dynamics of Web Documents

Yue Zhao[✉] and Claudia Hauff

Delft University of Technology, Delft, The Netherlands
{y.zhao-1,c.hauff}@tudelft.nl

Abstract. The creation time of documents is an important kind of information in temporal information retrieval, especially for document clustering, timeline construction and search engine improvements. Considering the manner in which content on the Web is created, updated & deleted, the common assumption that each document has only one creation time is not suitable for Web documents. In this paper, we investigate to what extent this assumption is wrong. We introduce two methods to timestamp individual parts (sub-documents) of Web documents and analyze in detail the creation & update dynamics of three classes of Web documents.

Keywords: Timestamping · Sub-documents · Internet Archive

1 Introduction

Document timestamping is an important step in temporal information retrieval (T-IR) which determines the creation time of documents [4]. Depending on the type of document the creation time can either be extracted directly (e.g. news articles commonly list their creation date) or has to be inferred (as is the case for most Web pages). Such temporal knowledge is essential for a variety of tasks, including document clustering [5,9,16,17], timeline creation [10,22], and search engine adaptations for temporal queries [14,19].

Previous studies [5,9,11,15–17] on document timestamping usually employ a simplifying assumption that each document only has a single creation time. This assumption is suitable for historical documents and news documents, whose content is published at one point in time and is rarely or never updated. Web documents, however, are dynamic; content is added, removed and changed over time. Previous work [12] also shows that users prefer to know the creation time of contents rather than the evolution of Web pages. Therefore, we focus on inferring the creation time of content on Web pages in this paper.

We previously showed that a considerable fraction of Web documents (66.5 % of the explored sample) does indeed contain content created at two or more different points in time [23]. Importantly, the creation times of documents' so-called *sub-documents* (a sub-document can be a paragraph or a sentence) can vary

© Springer International Publishing Switzerland 2016
N. Fuhr et al. (Eds.): TPDL 2016, LNCS 9819, pp. 203–214, 2016.
DOI: 10.1007/978-3-319-43997-6_16

widely - content is not created within days, the median time between the oldest and most recent sub-document for the investigated sample of Web documents was 782.5 days. These findings though were derived from a very small set of high-quality Web documents (∼7000) only. Here, we take this work as a starting point and investigate sub-document timestamping on a much larger sample of Web documents (nearly half a million).

Following [15,16,23], crawl data from the Internet Archive[1] (IA) is leveraged to obtain ground-truth sub-document creation times. The IA has been archiving Web documents since 1996, and covers a significant but limited part of the Web.

Analyzing the content creation dynamics of Web documents though can only be the first step. Our ultimate goal is to reliably timestamp the sub-documents of all Web documents - independent of their availability in IA. Such fine-grained timestamping would enable large-scale investigations of information diffusion on the Web (e.g. how rumors or specific content spreads) as well as an in-depth exploration of temporal effects on retrieval models (studies of which have so far been restricted to small news corpora). To make this vision a reality, we develop a 2-stage machine learning approach which is not only based on features extracted from individual sub-documents (as done in [23]), but also leverages the relations among the sub-documents in the same Web document. We make the following contributions in this paper:

1. We explore the content creation dynamics of nearly half a million Web documents of varying quality.
2. We gain novel insights into the document factors that play a role in content creation over time.
3. We develop a two-stage machine learning approach to sub-document timestamping, significantly improving upon our previous work [23].

2 Related Work

Due to the importance of document creation times in T-IR, a number of studies have investigated creation time inference (based on the already outlined one-creation-time-per-document assumption). De Jong et al. [9] rely on temporal language models, built from news articles in different time ranges, to determine the most likely creation time range for non-timestamped documents. This approach was extended by Kanhabua et al. [15,16] and Kumar et al. [17] who introduce additional features, such as temporal entropy and the KL divergence between language models, to improve the accuracy of the temporal language models. Chambers et al. [5] and Ge et al. [11] take temporal expressions appearing in documents and knowledge about the relationships between news documents into account to improve inference accuracy.

The intuition of sub-document timestamping is based on research in Web dynamics which has largely focused on Web evolution [1,2,20] and Web crawling [6]. Research on the timestamping of Web documents that takes the Web dynamics into account is largely missing. Jatowt et al. [13] proposed a pipeline for

---

[1] https://archive.org/.

timestamping content based IA data; they however neither analyzed the content changes nor built inference tools to timestamp non-archived Web documents. In our previous work [23], we use a similar pipeline to [13] for IA-based sub-document timestamping. We made a first attempt at analyzing content dynamics (on a few thousand Web documents) and at inference. We now significantly extend our previous work, by analyzing a much larger set of Web documents with varying characteristics and leverage a new machine learning approach to improve the inference of sub-document timestamps.

3 Approach

We now introduce the timestamping pipeline to gain ground truth data and the machine learning approach to infer sub-document timestamps.

3.1 Timestamping Pipeline

Analogously to [23], our timestamping pipeline consists of 4 steps: (**S1**) historical versions extraction, (**S2**) sub-documents extraction, (**S3**) sub-documents timestamping and (**S4**) model training.

Let $D = \{d_1, d_2, ..d_n\}$ be the set of all Web documents d_i we *aim* to collect ground truth data for. For (**S1**) we first extract all historical versions $Hist$ of $d_i \in D$ based on their URLs from the IA. By our definition, two historical versions of a document d_i have to be different from each other, thus we skip archived versions without any content changes. Additionally, since not all Web documents may be available in the IA, we only process those Web documents with records in the IA ($D^{archived}$) in the next steps.

For (**S2**), all sub-documents are identified and extracted from each d_i in $D^{archived}$ and its corresponding historical versions $Hist_{d_i} = \{d_i^{h_1}, d_i^{h_2}, ..., d_i^{h_m}\}$ where $d_i^{h_1}$ is the earliest version of d_i on IA, and $d_i^{h_m}$ is the most recent version. Every Web document (original and historical versions alike) is then split into sub-documents based on its HTML markup: sub-documents are delimited by <p> and <div> tags. Only sub-documents with 50+ non-markup characters are considered, to ensure that each sub-document has sufficient content to be correctly matched in the next step.

For (**S3**), we compare the sub-documents of d_i with all $d_i^{h_i} \in Hist_{d_i}$. The comparison starts at the earliest version $d_i^{h_1}$ and continues in temporal order. We rely on approximate string matching [7] to detect the earliest appearance of each sub-document in d_i within $Hist_{d_i}$.

Finally, in (**S4**) we train our models to automatically estimate the timestamp of unlabelled sub-documents. For experimental purposes we split our datasets into three parts, train on two parts and test the accuracy of the models on the third part.

3.2   Features and Models

We first derive the same 21 features from *each* sub-document that were reported in [23]. These features fall into two categories: (i) term statistics of the sub-document and its sentences (e.g. position and length of the sub-document and its sentences), and, (ii) the values & locations of temporal expressions within the sub-document (e.g. the value of the temporal expression which is earliest or latest). Beyond that, we include two additional types of features: (iii) one numeric feature per year (for the years 1996 to 2012[2]) that expresses the number of temporal expressions containing the respective year in the sub-document, and, (iv) five numeric tense features that express the number of verbs in the respective tense appearing in the sub-document. Thus, in total we compute 44 features per sub-document. All temporal and part-of-speech based features were extracted using Stanford's coreNLP package[3].

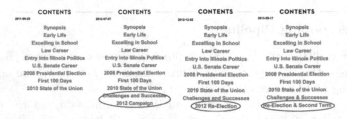

Fig. 1. Internet Archive based page updates for Barack Obama's biography page.

In [23], the initial 21 features were used in an ensemble learning setup (Random forests), that *classifies* sub-documents into different temporal categories. In this setup, the relations among the sub-documents (and their respective features) are ignored: for each sub-document the timestamp is estimated independently. Intuitively it makes sense to also consider the relations among the sub-documents as some may contain useful temporal features that could also benefit the inference of sub-documents' timestamps appearing in their neighbourhood. As a motivating example for the benefit of sub-document relations, consider Fig. 1, which shows how the section headings developed over time for the biography page of *Barack Obama*[4], based on the historical versions extracted from the IA. Generally, content is added or updated towards the end of the document and sections appearing in close spatial proximity are more likely to cover similar time periods compared to sections appearing far apart.

Based on this intuition, we propose a 2-stage model that incorporates the relations among sub-documents as shown in Fig. 2. In the first stage we also employ ensemble learning. In the second stage we leverage the predictions of the first stage and input those of spatially neighbouring sub-documents into a Conditional Random Field[5] [18] (CRF), a type of probabilistic graphical model

---

[2] This time range was chosen due to our experimental data, cf. Sect. 4.

[3] http://nlp.stanford.edu/software/corenlp.shtml.

[4] http://www.biography.com/people/barack-obama-12782369.

[5] CRF++: https://taku910.github.io/crfpp/.

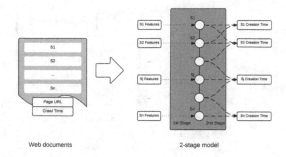

Web documents   2-stage model

Fig. 2. Overview of our 2-stage model.

widely used for sequential data labelling. The spatial neighbourhood of a sub-document is defined by its position in the HTML markup, instead of the rendered arrangement, to enable efficient processing.

4 Experiments

In our experiments, we leverage a subset of the publicly available ClueWeb12 corpus[6], a Web crawl of more than 700 million pages in early 2012. We investigate three types of Web documents:

Quality: This set includes all Web documents judged relevant to at least one of the 200 TREC topics released for ClueWeb12 that also appear in IA — 7,118 documents. This document set was employed in [23]. We consider the documents to be of high quality, as manual judges determined their usefulness to information needs (excluding spam and non-informative pages).

General: To counterbalance the Quality set, we randomly sampled documents from ClueWeb12[7], determined their existence in IA and crawled all historical versions available. Due to IA bandwidth limitations, we continued this process for six weeks, after which we had collected 433,082 ClueWeb12 documents with nearly 3 million (2,961,005) historic versions overall.

Seen: Lastly, we also sampled a set of "seen" (popular) Web documents, that is Web documents, that were of interest to at least some real users. Here, we were able to exploit the ClueWeb12 crawling strategy: added to the crawl frontier were not only URLs discovered during the standard crawling process, but also URLs that were mentioned in the public Twitter stream during the crawling period. These documents are marked as crawled from Twitter in the ClueWeb12 crawl and we sampled 23,077 of them that were also available in the IA (with 368,106 historic versions).

4.1 Exploratory Analysis

Do the Crawl Frequencies of Documents Differ in the IA? Efficient Web crawlers crawl some Web pages (or domains) more often than others, to

---

[6] http://www.lemurproject.org/clueweb12.php/.
[7] Specifically, we sampled from Disk1 of the ClueWeb12 corpus.

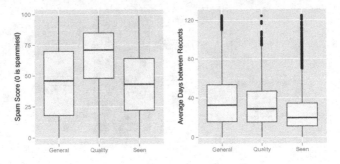

Fig. 3. On the left, the spam score distribution for `General`, `Quality` and `Seen` is shown. On the right, the IA crawling frequency is shown.

avoid re-crawling never changing documents and retaining up-to-date content for regularly changing documents. The IA crawler is no exception. In Fig. 3 (right) we plot for the three sets of documents the *average timespan* (in days) between subsequent IA crawled versions[8]. To make the comparison fair (and to remove the IA's changing technological abilities over the years from the comparison), we only consider documents whose first version appeared no earlier than January 2011 in IA and that were crawled at least three times. The results show that our set of `Seen` documents are crawled most frequently, while the set of `General` documents have the largest timespan between subsequent crawls.

To What Extent Do the Document Qualities Vary Across the Three Sets? Document quality can be measured in many ways, including readability, recency and trustworthiness. We take a practical view on quality and determine the amount of spam each document set includes. We rely on the pre-computed Web spam scores[9] released for ClueWeb12, and plot in Fig. 3 (left) the distribution of spam scores. Each document is assigned a spam score, and those scores vary between 0 (most likely to be spam) and 100 (least likely to be spam) — in practice, often documents with a score below 70 are considered to have at least some spam in them. Not unexpectedly, the `Quality` set is mostly spam-free, while the `Seen` documents and `General` documents have a similar amount of spam — indicating that through the public Twitter stream a significant amount of spam entered the dataset. A note of caution though: since the spam scores were derived automatically [8], in future work we will conduct a more qualitative analysis to further verify these findings.

The analyses that follow now are inspired by the questions raised in [23]. Recall though, that only the `Quality` set of documents was investigated before — we experiment with a much larger and more diverse set of data.

What Proportion of Web Documents is Created at Multiple Points in Time? Figure 4 shows the percentage of documents in each set that contains content (i.e. sub-documents) created at $\{1, 2, .., 10\}$ different points in time. More

---

[8] We mean here *all* versions available on IA, not just those with changed content.

[9] http://www.mansci.uwaterloo.ca/~msmucker/cw12spam/.

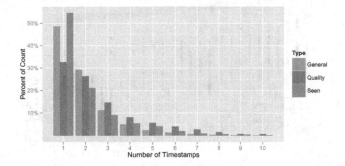

Fig. 4. Number of documents containing content created at different points in time.

than 95 % of all documents have less than 10 unique creation times. We also observe a marked difference between Quality and the other two document sets: less than 35 % of Quality documents have a single creation time, while this is the case for between 45–55% of documents in General and Seen. For Seen this difference can be attributed to an artifact in the data collection: 38 % of Seen documents were crawled by the ClueWeb12 crawler *before* they were first archived by the IA (a natural explanation being that people regularly tweet about newly created content on the Web). In these instance we assign the ClueWeb12 crawl time as their creation time. The same though cannot be said about Quality or General where this is the case in less than 7 % of all documents.

How Much Time Passes Between Content Updates? Having determined that there are indeed sufficient Web documents with multiple creation timestamps, we are now concerned with the time that has passed between the first and last creation timestamp of sub-documents in the same document. If changes were mostly made within a few days of the original creation of a document, there would be little need for sub-document timestamping in T-IR applications. In Fig. 5 we show for all documents (and all three document sets) with more than 1 creation timestamp how large the timespan between the earliest and latest sub-document creation time is. On average, we observe surprisingly large timespans: 350 days (Seen), 1881 days (General) and 1052 days (Quality) respectively. The considerably smaller timespan for Seen documents can be explained through Fig. 6: here we plot the distribution of earliest creation timestamps per document. All documents are crawled in 2012 and we observe that the earliest creation timestamps of documents distributed through Twitter are generally quite recent, most of which have been created in 2011 or 2012 — again pointing to the fact that users on Twitter tend to distribute recently created content.

What Proportion of Content Is Created over Time? Having observed that updates happen across one or more years, we are now concerned with the amount of content created at different points in time. If 99 % of a document's content were to be created at the earliest creation timestamp, it would be difficult to argue for adaptations of existing T-IR applications. For this experiment, we now focus on

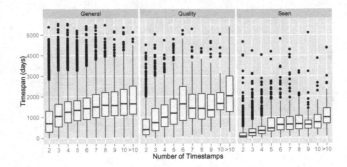

Fig. 5. The document set is partitioned according to the number of creation timestamps (documents with a single creation timestamp are ignored). Shown is the difference (in days) between the oldest and most recent creation timestamp.

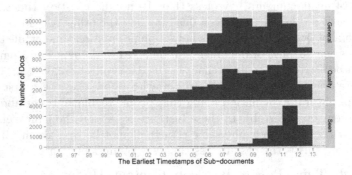

Fig. 6. The earliest sub-document timestamps of different type of Web documents.

Web documents with 2, 3 or 4 creation timestamps and compute the percentage of content present in each version. The results are shown in Fig. 7. Across the three document sets it holds that the more creation timestamps a document has, the less content is created initially. Quality documents have the highest percentage of initially created content across timestamps; one explanation for this observation is the high quality of the content: higher quality leads to more preservation of content over time.

Based on our experiments we conclude that across all 3 document sets, for a significant amount of documents the single-creation-time assumption is wrong.

4.2 Timestamp Inference

Having concluded our exploratory analysis, we now turn to the estimation of sub-document creation timestamps, a mechanism whose accuracy is essential to enable large-scale sub-document timestamping of the Web.

We treat sub-document timestamping as a classification task in line with [23] (ensuring a comparable baseline) and train & test separate models for Quality, Seen and General. Each of our datasets is split into 60 % training data, 20 %

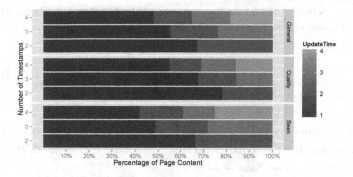

Fig. 7. Overview of content created at different points in time for documents with 2, 3 or 4 creation timestamps. Each bar shows the percentage of content available at each creation timestamp. *UpdateTime* indicates the timestamp if the content created.

validation data and 20 % test data. Our **baseline method** is the Random forest (RF) classifier with the same 21 features as in [23]. We first improve the RF with the enlarged feature set (44 features). Subsequently, we employ our 2-stage model (combining RF and CRFs) which employs the RF predictions as features in the CRF.

The algorithms' parameters are tuned by grid search. For the RF classifier, we tune the maximum number of features ($max\_features$) considered for the best split in generating each decision tree. Besides, the number of trees is fixed at 100 without any pruning methods based on some previous work on RF [3,21]. For CRF, the C-value is tuned from 1×10^{-6} to 1×10^{-4} to adjust the fit of the model.

Lastly, we explore how to exploit the relations between neighbouring sub-documents as part of our CRF models; CRF models that incorporate more neighbours lead to a better prediction, indicating that longer distance dependencies between sub-documents are more helpful than first-order dependency. To avoid an explosion in the number of features though, we only consider neighbourhoods of size four in our experiments.

Timestamping of Quality Sub-documents. Recall, that we aim to classify the timestamp of individual sub-documents (277,973 sub-documents in Quality in total), not individual documents. Due to the skewed nature of Quality with less than 1,000 sub-documents created between 1996 and 1998 and more than 69,000 created in 2012, we first balance our dataset in a 5-class setup. For each sub-document we determine the difference in days between the creation time of the sub-document (as found through our IA-based pipeline) and the crawl time of the sub-document as given in the ClueWeb12 metadata. Creating balanced classes yields the following five time intervals: *Class A* represents the interval $(0, 30]$, that is, the sub-documents were created no more than 30 days before they were crawled by the ClueWeb12 crawler. We similarly define the remaining four classes as $B = (30, 365]$, $C = (365, 1095]$, $D = (1095, 2190]$, and $E = (2190, \infty)$. We balance the dataset to test the effectiveness of different temporal inference models, as data imbalance is known to affect some models more than others.

Table 1. Sub-document timestamping inference. The baseline is RF with 21 features. Statistically significant changes over RF (all features) are marked ‡ ($p < 0.01$).

	Misclassified	F-measure/class				
		A	B	C	D	E
+++ Document set Quality +++						
Baseline method [23]	47.75 %	0.55	0.45	0.46	0.46	0.67
RF (44 features)	46.85 %	0.55	0.46	0.46	0.47	0.68
2-stage model (RF + CRF) ‡	44.64 %	0.59	0.47	0.49	0.50	0.70
+++ Document set Seen +++						
Baseline method	54.37 %	0.49	0.44	0.41	0.40	0.54
RF (44 features)	53.49 %	0.50	0.44	0.42	0.41	0.55
2-stage model (RF + CRF) ‡	50.30 %	0.52	0.49	0.44	0.44	0.60
+++ Document set General +++						
Baseline method	40.36 %	0.71	0.55	0.53	0.52	0.63
RF (44 features)	39.36 %	0.72	0.56	0.54	0.53	0.64
2-stage model (RF + CRF) ‡	36.70 %	0.72	0.59	0.57	0.56	0.69

Although these developed models cannot be employed as-is in the open Web setting (to timestamp all sub-documents of the Web), they allow us to focus on building sensible models first before tackling the next problem.

The results of these 3 classifier variations[10] on test data are shown in Table 1 (top). The 2-stage approach improves the classification accuracy significantly over our RF baselines[11]. When comparing the two RF classifiers, we observe a slight positive (and statistically significant) change when more features are employed. It means these features are helpful for improving the accuracy in each class, but they can only improve a little.

Timestamping of Seen Sub-documents. More than 50 % of sub-documents in Seen are timestamped by their ClueWeb12 crawl time, in correspondence with our finding in Sect. 4.1, that about 38 % of Seen documents were crawled by the ClueWeb12 crawler before they were picked up by the IA. Since those sub-documents are not useful for our purposes we ignore them here, and only consider the 306,210 sub-documents in Seen with historical version in IA before the ClueWeb12 crawl time.

We use the same balanced 5-class setup as in the previous experiment with $A = [0,0]$, $B = (0,28]$, $C = (28,152]$, $D = (152,444]$, and $E = (444,\infty)$. The results of three classifiers[12] in Table 1 (middle) show the same classification trends hold for Seen as for Quality (2-stage model outperforms the RF

---

[10] $max\_features$ is 3 and 6, C-value is 9×10^{-6}.

[11] McNemar's test was employed for statistical significance testing, with $p < 0.01$.

[12] $max\_features$ is 5 and 13, C-value is 9×10^{-5}.

significantly). However, the accuracies for Seen are all lower than Quality with the same setting. Since the timespans in each class except E of Seen are much smaller than Quality, the accuracy of Seen is acceptable.

Timestamping of General Sub-documents. The number of sub-documents in General exceeds six million, which is much larger than Quality and Seen. We first balance our dataset with $A = [0, 0]$, $B = (0, 367]$, $C = (367, 966]$, $D = (966, 1735]$, and $E = (1735, \infty)$. As shown in Table 1 (bottom), the classification accuracy of the three classifiers[13] in General is better than both Quality and Seen. One possible explanation is the larger amount of training data we have available for General. In future work, we will investigate in detail the reasons for this discrepancy.

We conclude that while features have to be selected with care, more relation-aware models (such as CRFs) improve the accuracy of the timestamping process significantly. We find that our trained classifiers do not perform equally well across all balanced classes. This indicates that our current features are more suitable for timestamp inference in a relatively coarse-grained setup, instead of high-resolution time intervals. Based on the observed mis-classification rates (between 37 % and 50 %) we conclude that we cannot yet employ our pipeline in any application that requires fine-grained and accurate sub-document timestamping.

5   Conclusions

In this work, we have presented a detailed analysis of the content dynamics of Web documents, with the ultimate goal to timestamp their individual sub-documents. We have added significantly to the existing corpus of work, analyzing a data set nearly two magnitudes larger than in previous research. Additionally, we contributed an improved sub-document timestamping inference model and showed its effectiveness across two different Web document sets.

Future work will focus on the improvement of the sub-document timestamping pipeline in order to be able to reliably timestamp all of the Web (or more realistically all of ClueWeb12), which will enable analyses in information diffusion, topic changes, content preservation and other areas.

References

1. Adar, E., Teevan, J., Dumais, S.T., Elsas, J.L.: The web changes everything: understanding the dynamics of web content. In: WSDM 2009, pp. 282–291 (2009)
2. Baeza-Yates, R., Pereira, Á., Ziviani, N.: Genealogical trees on the web: a search engine user perspective. In: WWW 2008, pp. 367–376. ACM (2008)
3. Bernard, S., Heutte, L., Adam, S.: Influence of hyperparameters on random forest accuracy. In: Benediktsson, J.A., Kittler, J., Roli, F. (eds.) MCS 2009. LNCS, vol. 5519, pp. 171–180. Springer, Heidelberg (2009)

---

[13] $max\_features$ is 7 and 11, C-value is 1×10^{-6}.

4. Campos, R., Dias, G., Jorge, A.M., Jatowt, A.: Survey of temporal information retrieval and related applications. ACM Comput. Surv. (CSUR) **47**(2), 15 (2015)
5. Chambers, N.: Labeling documents with timestamps: learning from their time expressions. In: ACL 2012, pp. 98–106 (2012)
6. Cho, J., Garcia-Molina, H.: The evolution of the web and implications for an incremental crawler (1999)
7. Cohen, W., Ravikumar, P., Fienberg, S.: A comparison of string metrics for matching names and records. In: KDD Workshop on Data Cleaning and Object Consolidation, vol. 3, pp. 73–78 (2003)
8. Cormack, G., Smucker, M., Clarke, C.: Efficient & effective spam filtering & re-ranking for large web datasets. Inf. Retrieval **14**(5), 441–465 (2011)
9. de Jong, F., Rode, H., Hiemstra, D.: Temporal language models for the disclosure of historical text. Royal Netherlands Academy of Arts and Sciences (2005)
10. Döhling, L., Leser, U.: Extracting and aggregating temporal events from text. In: WWW 2014, pp. 839–844 (2014)
11. Ge, T., Chang, B., Li, S., Sui, Z.: Event-based time label propagation for automatic dating of news articles. In: EMNLP 2013, pp. 1–11 (2013)
12. Jatowt, A., Kawai, Y., Ohshima, H., Tanaka, K.: What can history tell us?: towards different models of interaction with document histories. In: ACM HyperText 2008, pp. 5–14 (2008)
13. Jatowt, A., Kawai, Y., Tanaka, K.: Detecting age of page content. In: Proceedings of the 9th Annual ACM International Workshop on Web Information and Data Management, pp. 137–144. ACM (2007)
14. Jones, R., Diaz, F.: Temporal profiles of queries. ACM Trans. Inf. Syst. **25**(3), 14 (2007)
15. Kanhabua, N., Nørvåg, K.: Improving temporal language models for determining time of non-timestamped documents. In: Christensen-Dalsgaard, B., Castelli, D., Ammitzbøll Jurik, B., Lippincott, J. (eds.) ECDL 2008. LNCS, vol. 5173, pp. 358–370. Springer, Heidelberg (2008)
16. Kanhabua, N., Nørvåg, K.: Using temporal language models for document dating. In: Buntine, W., Grobelnik, M., Mladenić, D., Shawe-Taylor, J. (eds.) ECML PKDD 2009, Part II. LNCS, vol. 5782, pp. 738–741. Springer, Heidelberg (2009)
17. Kumar, A., Lease, M., Baldridge, J.: Supervised language modeling for temporal resolution of texts. In: CIKM 2011, pp. 2069–2072 (2011)
18. Lafferty, J., McCallum, A., Pereira, F.C.: Conditional random fields: probabilistic models for segmenting and labeling sequence data (2001)
19. Li, X., Croft, W.B.: Time-based language models. In: CIKM 2003, pp. 469–475 (2003)
20. Ntoulas, A., Cho, J., Olston, C.: What's new on the web?: the evolution of the web from a search engine perspective. In: WWW 2004, pp. 1–12 (2004)
21. Oshiro, T.M., Perez, P.S., Baranauskas, J.A.: How many trees in a random forest? In: Perner, P. (ed.) MLDM 2012. LNCS, vol. 7376, pp. 154–168. Springer, Heidelberg (2012)
22. Swan, R., Jensen, D.: Timemines: constructing timelines with statistical models of word usage. In: KDD Workshop on Text Mining, pp. 73–80 (2000)
23. Zhao, Y., Hauff, C.: Sub-document timestamping of web documents. In: SIGIR 2015, pp. 1023–1026 (2015)

Archiving Software Surrogates on the Web for Future Reference

Helge Holzmann[1]([✉]), Wolfram Sperber[2], and Mila Runnwerth[3]

[1] L3S Research Center, Appelstr. 9a, 30167 Hannover, Germany
`holzmann@L3S.de`
[2] zbMATH, FIZ Karlsruhe - Leibniz Institute for Information Infrastructure,
Franklinstr. 11, 10587 Berlin, Germany
`wolfram@zentralblatt-math.org`
[3] German National Library of Science and Technology (TIB),
Welfengarten 1b, 30167 Hannover, Germany
`Mila.Runnwerth@tib.eu`

Abstract. Software has long been established as an essential aspect of the scientific process in mathematics and other disciplines. However, reliably referencing software in scientific publications is still challenging for various reasons. A crucial factor is that software dynamics with temporal versions or states are difficult to capture over time. We propose to archive and reference surrogates instead, which can be found on the Web and reflect the actual software to a remarkable extent. Our study shows that about a half of the webpages of software are already archived with almost all of them including some kind of documentation.

Keywords: Scientific software management · Web archives · Analysis

1 Introduction

Software is used in science among all disciplines, from analysis software and supporting tools in the humanities, over controlling and visualization software in medicine to the extensive use of all kinds of software in computer science as well as mathematics. However, referencing software in scientific publications has always been challenging. One reason is that software alone is often not considered a scientific contribution and therefore, properly citable publications do not exist. This is particularly an issue if the software does not tackle a concrete research question, but was created as tool for various purposes or different kinds of research, such as standard software like *Microsoft Excel*. Another issue is, even if the software in question is well-known or even published, it undergoes dynamics and the version an article refers to might be different from the one currently available. Furthermore, a name like *Microsoft Excel* refers to a **product** rather than a concrete version of that software. Also publications typically deal with the innovation and benefits created by the software as a product rather than

This work is partly funded by the German Research Council under FID Math and the European Research Council under ALEXANDRIA (ERC 339233).

N. Fuhr et al. (Eds.): TPDL 2016, LNCS 9819, pp. 215–226, 2016.
DOI: 10.1007/978-3-319-43997-6_17

a concrete build, version or setup. However, this very specific **artifact** may be crucial to reconstruct a software instance as in the original experimental setup, to reproduce experiments and comprehend scientifically published results.

As an example consider the famous bug of *Excel 2007*, which produced the number 100,000 in a cell of which the underlying data equaled to 65,535[1]. To fix this, the appropriate patch was released shortly after[2], which resulted in an artifact with an updated minor version number, but of course did not update the major version *2007*. Actually, even though 2007 in this case is already more specific than just the product's name, it should be better considered a **sub-product** of the **product family**. *Excel* rather than a concrete version, since it does not refer to a concrete artifact. Therefore, to verify results of 100,000 in scientific experiments with *Excel 2007* involved, the precise version number referring to the exact artifact used in the experiment is required, but very unlikely to be mentioned in a publication.

In the context of our project *FID Math*, aiming for a mathematical information service infrastructure, we are facing the problem of referencing software with a focus on all kinds of mathematical applications, tools, as well as services. In the area of math, software is heavily used for various purposes, such as calculations, simulations, visualizations and more. Very often multiple are combined, while the critical task is performed by a script, which is software in itself, running inside an environment like MATLAB, Mathematica or Sage. Settings like these make it particularly challenging to reference a consistent state of the incorporated software. Further, the mix of open source and proprietary software introduces an additional challenge due to crucial differences in many aspects, such as code contributions, licensing, as well as the question for preservation. To address these difficulties in a universal manner, we propose Web archiving as a solution to preserve representations of software on the Web as surrogates for future reference.

2   Problem and Questions

In an ideal world, the results of every experiment conducted and published by scholars in their scientific work should be reproducible. This in turn implies that every software can be recovered in the exact state as used in their experiments. This either requires a detailed reference to the software's state and general access to software artifacts, or, alternatively, ways to freeze and preserve a software's state and provide it as attachment of a publication. Both seems unrealistic for practical as well as legal reasons. While open source projects often suffer a reliable release process with proper versioning, every committed state is usually precisely identified by a single hash, such as the *SHA* used by GIT[3]. This hash does not only encode the current state of the software, but refers to all previous

---

[1] http://blog.wolfram.com/2007/09/25/arithmetic-is-hard-to-get-right [from 25/09/2007].

[2] https://support.microsoft.com/en-us/kb/943075 [from 09/10/2007].

[3] https://git-scm.com.

commits comprising relevant meta data records. By contrast, structured meta data of proprietary software is often more explicit, with the author being the company behind the software and each bugfix or patch presumably increases the minor version number of the software. Accordingly, both types of software potentially allow referring to concrete artifacts. The challenge is to establish a unified representation and ways to recover the referenced software.

Open source licenses commonly allow redistribution, which facilitates sharing preservation copies with publications to replay experiments. For proprietary software this is usually considered piracy. Even more difficult to handle are Web APIs and services, where the user does not have access to the actual software, but only to the interface. However, by recovering in this context we do not necessarily mean to obtain a copy of the software, which is only required for replaying experiments and in many cases not usable without proper documentation anyway. Recovering can also mean to get an understanding of the software, for example through its *documentation, source code, related publications* or *change logs*. Already a brief description can be difficult to obtain though, as referenced software, after many years, might not even exist anymore.

Since the Web can be considered our primary source of all kinds of information today, our hypothesis is that most of the information listed above is available on the Web as well. Therefore, a snapshot of the corresponding websites of a software from when it was acquired for scholarly use, whether by downloading a copy or ordering in a shop, would constitute a representation of the software at that time. Although it might not include the artifact itself, it is the most comprehensive representation we can get, given the practical and legal restrictions. Hence, it can be considered a temporal surrogate of the actual software.

To realize a Web archiving solution for such a purpose, which enables reliably referencing software surrogates on the Web in scientific publications, we need to overcome a number of challenges. The system has to be aware of all relevant resources of a software and it has to ensure that these resources are preserved at the time of scholarly use. This will only work with manual intervention. Therefore, a concrete implementation idea remains for future work and is out of the scope of this paper. However, with existing Web archives we can already create a solution based on the publication dates of articles using software. In this study we ask the following questions to analyze the applicability of archiving software surrogates on the Web:

(**Q1**) How well is software represented by its surrogate on the Web?
(**Q2**) Which information of software is available on the Web?
(**Q3**) How many websites of mathematical software are archived?
(**Q4**) For how many of these can referenced versions from the past be recovered?

3 Related Work

The presented study touches different areas of interest from *scientific software management*, dealing with the scientific process and software citations, over the question of how to preserve artifacts in *software repositories*, to *Web archives* and their role in science.

3.1 Scientific Software Management

Since software has become an essential part of scientific work in the field of mathematics but also other disciplines, various initiatives for research data management have begun to focus on this aspect as well. Peng [1] addresses the need for reproducibility of computational research. In 2011 research was unthinkable without software-based experiments, its findings had no platform to be published properly. Hence, the claim of replicating experiments was technically possible but had no place in the publishing process. In 2012, Wilson et al. [2] published a study on scientific software. It discovered that the majority of scientists use software for their research. It offers best practices to software development for research purposes because most scientists have never been taught to do so.

The Software Sustainability Institute[4] (SSI) is a British facility with the objective to improve the role of scientific software. It is associated with the slogan *"Better software, better research"* coined by Goble [3]. The SSI discusses how software can be developed, archived and referenced in order to contribute to scientific knowledge. Closely associated is the Digital Curation Centre[5] with a more general mandate towards research data [4]. Funded by the German Research Foundation, SciForge[6] presents a concept to accompany software throughout the whole research process in a transparent and scientifically adequate manner.

Scientific standards for the handling of mathematical software are mentioned by Vogt [5]. On the one hand, researchers must be aware of how to develop software effectively. On the other hand, publishing habits must be adapted to archiving, citing and quality-approved software. As an example, the journal of *Mathematical Programming Computation* investigates software for its scientific impetus[7]. Running the software is part of their reviewing process. FAIRDOM[8] is a research data management initiative in systems biology. It proclaims a *fair* use of research data where *fair* is an acronym for findable, accessible, interoperable and reproducible [6]. A study with a focus on repeatability of computational research shows that standards are required to reconstruct published research on computational systems [7]. In 2012, Mike Jackson published two blog articles on how to cite and describe software[9,10]. Until today they influence the discussion about software citation. Psychology has adopted a citation standard in order to cite software in a standardized way [8]. For the humanities, the Modern Language Association of America (MLA) provides a styleguide to cite software [9]. Among many disciplines and especially in mathematics, *BibTeX* and *BibLaTeX* are widespread tools to manage bibliographies. Although they are adaptable, there is currently no approved standard to give a distinct reference to all varieties of software.

---

[4] http://www.software.ac.uk/.

[5] http://www.dcc.ac.uk/.

[6] http://www.sciforge-project.org/.

[7] http://mpc.zib.de/.

[8] http://fair-dom.org/.

[9] http://software.ac.uk/so-exactly-what-software-did-you-use.

[10] http://www.software.ac.uk/blog/2012-06-22-how-describe-software-you-used-your-research-top-ten-tips.

3.2   Software Repositories

There are various institutional and domain-specific repositories for research data. An overview is given by the Registry of Research Data Repositories[11] [10]. Many of these repositories are operated by universities or have been initiated by research institutes. The University of Edinburgh offers a research data repository, DataShare [11]. It is divided into domain-specific collections which can be searched separately or compositely. Harvard's DataVerse Project[12] supplies a web service to share, archive and cite research data. RADAR, funded by the German Research Foundation, pursues a service-oriented approach to provide infrastructure and services to host research data repositories [12]. However, currently, there is no designated repository for mathematical software beyond institutional level.

3.3   Web Archives

Preservation of Web resources has recently been of growing interest, resulting in a considerable number of publications around Web archives from different perspectives. First, Web archives have been gaining growing popularity as scholarly source [13] in disciplines like the humanities [14]. Second, Web archives have been of interest as subject of research themselves. In 2011, Ainsworth et al. [15] analyzed how much of the web is archived and found that for up to 90 % of the pages in the considered collections at least one archived version exists, however, only a few of them have a consistent coverage over time. Other works in this context focused on a particular subset of the Web, such as national domains [16,17], while our analysis has a very concentrated scope of mathematical software in Web archives. Beside the question of how much is archived, we also look at what resources are available, which falls in the area of profiling Web archive collections [18–20]. Finally, the goal of this study is to investigate the applicability of Web archiving to preserve representative surrogates of software. Even though this particular subject has not been tackled before, other researchers looked into Web archiving to create preservation copies of other types or Web resources, such as blogs [21] and social networks [22]. SalahEldeen and Nelson [23] found a nearly linear relationship between time and the percentage of lost social media resources.

Another challenge that we are facing is to find the resources related to a particular software in a Web archive. One approach in this direction is to use secondary data sources with temporally tagged URLs, like the social bookmarking platform Delicious. This can be used for temporal archive search as demonstrated by the Tempas system [24,25].

4   Data and Methodology

The primary source for this work was swMATH, an information service for mathematical software[13] ((M)SW). Based on the information of SW in this directory,

---

[11] http://www.re3data.org/.
[12] http://dataverse.org/.
[13] http://www.swmath.org.

we analyzed the linked URLs on the current Web as well as in a Web archive, provided by the *Internet Archive*[14].

4.1 swMATH in a Nutshell

swMATH is one of the most comprehensive information services for MSW [26]. It contains more than 12,000 records, each representing a SW **product** or **product-family** (cp. Sect. 1) with a unique identifier, as shown in Fig. 1. swMATH is based on the database of zbMATH[15], one of the most comprehensive collections of mathematical publications, with more 110,000 articles referring to MSW. The biggest challenge for a service like swMATH is to recognize these references. In many cases, only a name is mentioned, while a version or an explicit label as (M)SW is missing. swMATH tackles this with simple heuristics, by scanning titles, abstracts, as well as references of publications to detect typical terms, such as *solver*, *program*, or simply *software*, in combination with a name.

After new candidates have been detected, they are checked manually to ensure the high quality of the service. As part of this manual intervention step, additional meta data, such as the **URL** of a SW is added. Later on, websites are periodically checked and outdated URLs are removed or replaced. In case there is no permanent link that points to a website of the SW, the URLs of a corresponding repository record or a publication is used instead.

Another important feature for our analysis is the **publication list** for every SW on swMATH. Each article in this list is annotated with its publication year. The publications can be sorted chronologically or by the number of citations an article has received. In swMATH, publications also serve as source for additional information, such as related software and the keyword cloud shown in every record (s. Fig. 1).

Fig. 1. Mathematical software Singular on swMath

---
14 http://archive.org.
15 http://www.zbmath.org.

In order to enhance the functionality of swMATH, one goal is to capture the dynamics of (M)SW as reflected by the publications over time. We address this aspect by investigating which information of SW is available on the Web (s. Sect. 5.1) and can be recovered from Web archives (s. Sect. 5.1). Of importance for this study are **URLs** as well as the **publications** of a MSW, which are both available through swMATH.

4.2 Analysis

As an initial step of each analysis in our study (s. Sect. 5), we crawled the required datasets using Web2Warc[16], resulting in the following four collections, listed with the last time of crawl:

- **swMATH records** *(28/01/2016 - 14:14:53)*
  All 11,785 software pages available on swMATH at that time.
- **URLs** *(18/02/2016 - 18:55:03)*
  All webpages linked by the 11,125 URLs extracted from the swMATH records.
- **Publications** *(19/02/2016 - 08:30:15)*
  The top 100 user publications with respect to their number of citations received, for all swMATH records. These lists are dynamically loaded into the swMATH records and constitute therefore separate resource to be crawled.
- **Internet Archive** *(22/02/2016 - 19:38:55)*
  Meta data of the captures in the Internet Archive's Web archive for all URLs extracted from swMATH, using the *Wayback CDX Server API*[17]. For each URL we fetched the latest capture as well as the one closest to the time of the best publication of the corresponding software with respect to number of citations received.

To analyze these collections, which are small Web archives in themselves, we employed ArchiveSpark[18], a framework for Web archive analysis [27]. This way, we extracted the data of interest, such as URLs and publication dates from the swMATH records as well as linked URLs from the crawled webpages. Subsequently, URLs were analyzed further to identify what kind of resources they point to. We split the URLs by means of the following rules: 1. the host is split at dots [.], 2. the path is split at [/.-_], 3. the query string is split at [?=+&:-_]. As indicated by the segments we obtained through those splits, the URLs were classified as different resource type. The classes and segments are shown in Table 1. Segments in curly brackets denote whole URLs that match predefined URL patterns, such as GitHub URLs as denoted by {*github*}. These, for instance, are an indicator for available *source code*. Additionally, *documentation* and *artifacts* include all *publications* and *source code* respectively (bold in Table 1), since we consider publications to be documentation, and source code implicitly constitutes an artifact of software. Even though these heuristics are by no means

---

[16] https://github.com/helgeho/Web2Warc (*Last commit 73f0934 on Jan 29, 2016*).

[17] https://archive.org/help/wayback_api.php.

[18] https://github.com/helgeho/ArchiveSpark (*Last commit acc5a16 on Feb 17, 2016*).

Table 1. Segments extracted from software URLs and grouped into classes.

Class	Segments
Source code	*code, gpl, lgpl, {R-project}, {github}, {googlecode}, {sourceforge}, {cpc}, {gpl}, {bitbucket}, {gnu}*
Publications	*publications, papers, journals, publication, article, journal, doi, articles, library, bib, reports, {acm}, {springer}, {sciencedirect}, {wiley}, {cpc}, {arxiv}, {googlebooks}, {ieee}, {doi}, {manuscriptcentral}, {tandfonline}, {oxfordjournals}, {citeseerx}*
Updates	*changelog, history, news, blog*
Documentation	*doc, documentation, manual, api, reference, handbook, handbuch, referenz, doku, dokumentation, wiki, docs, readme,* **publications**
Artifacts	*exe, zip, gz, tar, download, tgz, files, downloads, ftp,* **source code**

complete, we tried to cover all cases that we observed by investigating the URLs manually and are convinced to convey a representative round-up of the available resources with this approach.

In addition to the listed datasets, we collected in-links from external websites to the URLs under consideration. These were extracted from the archived German part of the Web under the top-level domain *.de* from 1996 to 2013, which we had full local access to[19]. Due to scope of this dataset, we performed the in-link analysis only on URLs of domains ending in *.de* as well, assuming that these are better represented in the dataset than arbitrary URLs.

5   Analysis Results

The results of our study answer the questions introduced in Sect. 2. While **Q1** and **Q2** focus on software (SW) on the Web in general, **Q3** and **Q4** address its coverage by Web archives. In terms of terminology (cp. Sect. 1), the website of a SW, addressed in the first two questions, generally represents the **product** or **product-family**. At the same time, all information on the current state of the website usually refer to the current **artifact** of the SW, which may or may not be provided itself as download (s. Sect. 5.1). The versions available in a Web archive, addressed by the latter questions, are considered surrogates for the **artifact** that was prevailing at the time of capture (s. Sect. 5.2).

5.1   Software on the Web

The first objective of our analysis was to investigate whether the Web reflects real SW at all and how well webpages as potential surrogates represent actual SW (**Q1**). Our attempt to show such a relation involves the publications referring to a SW over time as well as the in-links to a SW's webpage, which we consider the equivalent to scientific citations on the Web. Figure 2a illustrates this remarkable

---

[19] This dataset has been provided to us by the Internet Archive in the context of *ALEXANDRIA* (http://alexandria-project.eu).

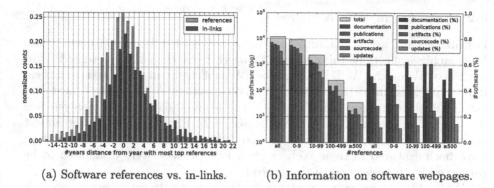

(a) Software references vs. in-links.   (b) Information on software webpages.

Fig. 2. Software surrogates on the Web.

correlation, with references slightly ahead of the in-links. The plot has been normalized by the highest number of publications and in-links for a given SW and aligned by the year with most publications for a SW at $x = 0$. It is based on the links extracted from *.de* pages as well as the number of articles in a year among the top 100 publications (cp. Sect. 4.2). Due to the available link data, only SW with URLs under *.de* was considered.

The fact that the Web lacks slightly behind references in literature suggest that the scientific use of SW leads to visibility and has a strong impact on its popularity. This motives our effort to archive SW's webpages at the time of publication or even the time of use for a publication. Without such a preservation copy, links that were created as a result of a publication may become stale as the SW and its website evolve and the original reference target cannot be recovered. The same applies to SW references in articles, too.

Next, we asked the question of which information can be obtained from the pages (**Q2**), based on the segments identified in the URLs (cp. Sect. 4.2). Figure 2b shows the total numbers as well as results separated by popularity with respect to the number of publications referencing a SW.

Interesting here is the finding that for around 60 % of the analyzed SW, the corresponding webpage links to some sort of documentation. In many of these cases, publications are available too, or comprise the documentation (cp. Sect. 4.2). However, this changes for more popular SW, where a larger fraction of their documentation is independent of publications. Also, as the plot shows, 50 % provide artifacts online, which is again even higher for popular SW. As a result, this trend suggests SW that is well represented on the Web is more prominent and more often used. Therefore, it was surprising to us that we only found source code for around 30 % throughout all popularities. Since only this minority enables a detailed tracing of the development process (s. Sect. 2), it is even more important to store temporal copies of the SW's webpages to get a sense of the development through the available information, such as documentation or update reports. Overall, we can conclude that SW webpages contain valuable information and indeed can serve as surrogates of the actual SW.

5.2   Software in Web Archives

For future work, we are planning to develop strategies and mechanisms to create Web archive collections tailored to SW, which cover the above presented information of SW on the Web. However, for SW published and referenced until then, it is valuable to investigate how well existing, generic Web archives have captured SW surrogates on the Web (**Q3**). The required data for this analysis was obtained from the Internet Archive (s. Sect. 4.2). Figure 3a shows that almost consistent over time the URLs of around 50 % of the SW under consideration have been captured in their Web archive. 10 % of these were disallowed to be preserved by the *robots.txt* of that website.

While 40 % of preserved contents is still a relatively satisfying number, it gets worse when looking at the fraction of archived captures at the time of the top publication referencing the SW, i.e., the article with most citations (s. *past* in Fig. 3a). Although we considered full years, only around 20 % of the SW pages were preserved in that period (**Q4**). However, due to the efforts at swMATH to replace outdated links (cp. Sect. 4.1), this number may actually be higher but not surfaced by our study. At the same time, this would mean that many of the original URLs have been outdated, which in turn suggests a certain development of the corresponding websites. Either way, our findings show the need for a sophisticated Web archiving infrastructure as part of the scientific SW management process, which assigns preserved material on the Web assigned to a specific SW or its reference in a publication, rather than URLs.

Another large portion that is not covered by the number of pages archived in the past are those webpages that were archived shortly before or after the year of the publication. The good news is, this gap is very small as shown in Fig. 3b. Most pages were captured in the exact year and the remaining were preserved closely around this time with decreasing numbers further away from the year of publication. Thus, by relaxing the time to identify a SW's surrogate in a Web archive to a couple of years around a publication, we can recover even more. Although the representativeness becomes less accurate this way, it can still be helpful to comprehend SW references or reproduce experimental results.

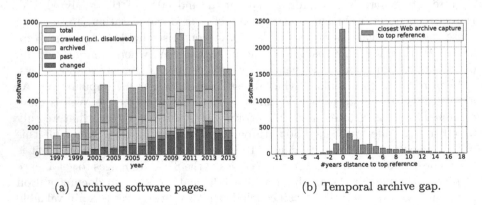

(a) Archived software pages.   (b) Temporal archive gap.

Fig. 3. Software surrogates in Web archives.

Not exactly surprising, but notable is the fact that almost all pages with archived captures in the past have changed according to the hash/digest of their archived record (s. Fig. 3a). This motivates to preserve those copies for future reference.

6 Conclusion and Outlook

As shown in our study, the Web reflects software to a remarkable extent, with documentation and artifacts on a considerable number of webpages. Hence, the Web can indeed serve as surrogate of actual software. However, only for about a half of the analyzed mathematical software, the corresponding webpages are preserved by an existing Web archive. To fix this in the future, we propose establishing new infrastructures to actively archive software surrogates on the Web at the time of use. The author of an article should be able to do this on demand and be provided with a handle to reliably reference the archived surrogate. As a first step, Web archives may be integrated with software directories, such as swMATH, to link software in existing scientific articles based on their publication date. Eventually, explicit references will improve the management of software in scientific publications.

References

1. Peng, R.D.: Reproducible research in computational science. Science (New York, NY) **334**, 1226–1227 (2011)
2. Wilson, G., Aruliah, D., Brown, C.T., Hong, N.P.C., Davis, M., Guy, R.T., Haddock, S.H., Huff, K.D., Mitchell, I.M., Plumbley, M.D., et al.: Best practices for scientific computing. PLoS Biol. **12**, e1001745 (2014)
3. Goble, C.: Better software, better research. Internet Comput. **18**, 4–8 (2014)
4. Rusbridge, C., Burnhill, P., Ross, S., Buneman, P., Giaretta, D., Lyon, L., Atkinson, M.: The digital curation centre: a vision for digital curation. In: Local to Global Data Interoperability - Challenges and Technologies (2005)
5. Vogt, T.: Software dokumentieren!. Mitt. Dtsch. Math.-Ver. **22**, 16–17 (2014)
6. Stanford, N.J., Wolstencroft, K., Golebiewski, M., Kania, R., Juty, N., Tomlinson, C., Owen, S., Butcher, S., Hermjakob, H., Le Novère, N., et al.: The evolution of standards and data management practices in systems biology. Mol. Syst. Biol. **11**, 851 (2015)
7. Collberg, C., Proebsting, T.A.: Repeatability in computer systems research. Commun. ACM **59**, 62–69 (2016)
8. A. P. Association: Publication Manual of the American Psychological Association, 6th edn. American Psychological Association, Washington (2009)
9. Gibaldi, J., Einsohn, A., Díaz, A., Uría, R., Rodríguez Sáenz, D., Labadie, J., Fontane, D., Floris, V., Chou, N.: MLA Style Manual and Guide to Scholarly Publishing, 3rd edn. Modern Language Association of America, New York (2008)
10. Pampel, H., Vierkant, P., Scholze, F., Bertelmann, R., Kindling, M., Klump, J., Goebelbecker, H.-J., Gundlach, J., Schirmbacher, P., Dierolf, U.: Making research data repositories visible: the re3data.org registry. PloS One **8**, e78080 (2013)

11. Macdonald, S.: Edinburgh DataShare - a DSpace data repository: achievements and aspirations. Presented at the Fedora-UK&I&EU Meeting, Oxford (2009)
12. Kraft, A., Razum, M., Potthoff, J., Porzel, A., Engel, T., Lange, F., van den Broek, K., Furtado, F.: The RADAR project - a service for research data archival and publication. ISPRS Int. J. Geo-Inf. **5**, 28 (2016)
13. Hockx-Yu, H.: Access and scholarly use of web archives. Alex. J. Natl. Int. Libr. Inf. Issues **25**, 113–127 (2014)
14. Gomes, D., Costa, M.: The importance of web archives for humanities. Int. J. Humanit. Arts Comput. **8**, 106–123 (2014)
15. Ainsworth, S.G., Alsum, A., SalahEldeen, H., Weigle, M.C., Nelson, M.L.: How much of the web is archived? In: JCDL (2011)
16. Alkwai, L.M., Nelson, M.L., Weigle, M.C.: How well are Arabic websites archived? In: JCDL (2015)
17. Holzmann, H., Nejdl, W., Anand, A.: The dawn of today's popular domains - a study of the archived german web over 18 years. In: JCDL (2016)
18. AlSum, A., Weigle, M.C., Nelson, M.L., Van de Sompel, H.: Profiling web archive coverage for top-level domain and content language. Int. J. Digit. Libr. **14**(3–4), 149–166 (2014)
19. Day, M., MacDonald, A., Pennock, M., Kimura, A.: Implementing digital preservation strategy: developing content collection profiles at the British library. In: JCDL (DL 2014) (2014)
20. Alam, S., Nelson, M.L., Van de Sompel, H., Balakireva, L.L., Shankar, H., Rosenthal, D.S.H.: Web archive profiling through CDX summarization. In: Kapidakis, S., et al. (eds.) TPDL 2015. LNCS, vol. 9316, pp. 3–14. Springer, Heidelberg (2015). doi:10.1007/978-3-319-24592-8_1
21. Kasioumis, N., Banos, V., Kalb, H.: Towards building a blog preservation platform. World Wide Web J. **17**, 799–825 (2014)
22. Marshall, C.C., Shipman, F.M.: An argument for archiving Facebook as a heterogeneous personal store. In: JCDL (DL 2014) (2014)
23. SalahEldeen, H.M., Nelson, M.L.: Losing my revolution: how many resources shared on social media have been lost? In: Zaphiris, P., Buchanan, G., Rasmussen, E., Loizides, F. (eds.) TPDL 2012. LNCS, vol. 7489, pp. 125–137. Springer, Heidelberg (2012)
24. Holzmann, H., Anand, A.: Tempas: temporal archive search based on tags. In: WWW (Demo) (2016)
25. Holzmann, H., Nejdl, W., Anand, A.: On the applicability of delicious for temporal search on web archives. In: SIGIR (Short) (2016)
26. Greuel, G.-M., Sperber, W.: swMATH – an information service for mathematical software. In: Hong, H., Yap, C. (eds.) ICMS 2014. LNCS, vol. 8592, pp. 691–701. Springer, Heidelberg (2014)
27. Holzmann, H., Goel, V., Anand, A.: Archivespark: efficient web archive access, extraction and derivation. In: JCDL (2016)

From Water Music to 'Underwater Music': Multimedia Soundtrack Retrieval with Social Mass Media Resources

Cynthia C.S. Liem[✉]

Multimedia Computing Group, Delft University of Technology,
Delft, The Netherlands
c.c.s.liem@tudelft.nl

Abstract. In creative media, visual imagery is often combined with music soundtracks. In the resulting artefacts, the consumption of isolated music or imagery will not be the main goal, but rather the combined multimedia experience. Through frequent combination of music with non-musical information resources and the corresponding public exposure, certain types of music will get associated to certain types of non-musical contexts. As a consequence, when dealing with the problem of soundtrack retrieval for non-musical media, it would be appropriate to not only address corresponding music search engines in music-technical terms, but to also exploit typical surrounding contextual and connotative associations. In this work, we make use of this information, and present and validate a search engine framework based on collaborative and social Web resources on mass media and corresponding music usage. Making use of the SRBench dataset, we show that employing social folksonomic descriptions in search indices is effective for multimedia soundtrack retrieval.

1 Introduction

Music is not just isolated sound, but also a continuously recurring element in our daily lives. We choose to listen to it as accompaniment to daily activities, such as studying, commuting and working out, or to get into or out of certain moods. We use it as an atmosphere-creating element during significant community events. We may (sometimes involuntarily) be confronted with it in public spaces, and through the multimedia we consume. If certain types of music tend to co-occur with certain types of non-musical context, the non-musical context may become an integral part of the meaning associated to the music.

Traditionally, digital music indexing has had a positivist orientation, focusing on direct descriptions of isolated musical content. However, considering the notions above, it would be interesting to see if music can also *indirectly* be indexed and retrieved based on typical non-musical contextual usage. This would imply that query vocabularies for music would reach beyond limited music-theoretical and mood-oriented vocabularies, or acoustically similar songs in case of a query-by-example scenario.

© Springer International Publishing Switzerland 2016
N. Fuhr et al. (Eds.): TPDL 2016, LNCS 9819, pp. 227–238, 2016.
DOI: 10.1007/978-3-319-43997-6_18

In previous work [8–10], we developed the notion of a search engine framework targeting such more subjective, contextual and identity-establishing effects of music. In [9], we considered the use case of a soundtrack retrieval system for user-generated video, and sketched the outline of a demonstrator system in which music search was based on vocabulary-unrestricted free-text query expressions of the intended narrative of the final multimedia result. To connect music to non-musical contexts, inspired by literature from the humanities, we employed folksonomic social collaborative resources on music usage in mass media as our prime data source of knowledge. In parallel, in [10], we investigated what free-text stories were generally associated to production music fragments, and showed that produced descriptions could be associated back to 'stimulus' fragments in a stable way.

The ideas from [9] were not validated yet. In [8], we offered preliminary insights into the capability of folksonomic social collaborative resources to connect musical and non-musical tags associated with soundtrack songs. In the current work, we now will present and validate the general system framework more thoroughly. In this, we will look in a more systematic way at effects of folksonomic tag frequency filtering, as well as various retrieval setups based on different information resource combinations for indexing.

An unusual feature of our approach is that it inherently allows for subjectivity in the query expression and relevance assessment. While this allows for creative media repurposing, this makes evaluation a non-trivial matter. In order to still perform an evaluation which is as unbiased as possible, we employ the SRbench dataset [15], which was meant as a benchmark for soundtrack recommendation systems for image collections.

The remainder of this paper is structured as follows. At first, we will discuss related work considering (aspects of) our problem from various disciplinary angles. Subsequently, we will discuss the general outline of our proposed framework. Then, we will introduce the benchmark dataset against which we will evaluate our approach, followed by a more detailed discussion of the evaluation setup. Evaluation results are then reported and discussed, after which we will finish the paper with a Conclusion and Outlook to future work.

2   Related Work

2.1   The Humanities Perspective: Music, Meaning and Mass Media

In this subsection, we will discuss several works from musicology and media studies in which musical meaning is considered in relation to (multi)media soundtrack usage. Our discussion is by no means a comprehensive review (that would be beyond the scope of this paper), but aims to point out representative thoughts.

Cohen [3] indicates that good synchronization between temporal visual development and a music soundtrack will lead to attributes associated in one domain to be transferred to the other, reporting how associated emotions to a musical fragment get attributed to an animated bouncing ball, as soon as the ball and the music are in sync. Cook [4] also considers transfer of associated meaning between music and multimedia, stressing the importance of mass media expressions. In his book, he relates and contrasts musical structure and development to events in associated media, proposing three basic models for analyzing musical multimedia which will have different interpretation effects: *conformance*, *complementation* and *contest* between music and the associated media.

Considering the function of soundtracks in film, Lissa [11] initially proposed a typology describing various reasons why a soundtrack would be a good match to displayed imagery. These reasons do not all require temporal development. Tagg and Clerida [16] ran an extensive study in which many free-text associations were obtained in response to played mass media title tunes, leading to a revision of Lissa's typology. In the context of our soundtrack retrieval work, in [10] we ran a crowdsourcing study in which production music fragments were played to online audiences, and free-text associated cinematic scenes were acquired. Considering self-reported reasons for cinematic associations, another revision of the original typology from Lissa, Tagg and Clerida was made, identifying 12 categories of reasons for non-musical connections to the played music. For the associated cinematic scene descriptions, consistency was found in the event structure of the provided user narratives, e.g. displaying agreement in whether a scene involves a goal, and whether this goal would be achieved during the scene.

2.2 Contextual Notions in Music Information Retrieval

In the digital domain, the field of music information retrieval traditionally has taken a positivist viewpoint on music, with major research outcomes (of which several examples can be found in the review by Casey et al. [2]) focusing on direct description of musical objects in terms of signal features, which then can be used for classification, similarity assessment, retrieval and recommendation. However, the notion of 'context' has been emerging. In Schedl et al. [12], distinction is made between user context and music context, with 'music context' considering contextual information related to music items, artist or performers, such as an artist's background, while 'user context' considers contextual information surrounding consumption by a user, such as the user's mood or activity. Music content is considered as an audio-centric phenomenon, and semantic labels and video clips are considered as parts of music context.

A slightly different terminology and conceptual division are used in the work of Kaminskas and Ricci [5]. Here, the authors distinguish between environment-related context (e.g. location, time, ambience), user-related context (activity, demographical information and emotional state) and multimedia context (text and images). Here, text includes lyrics and reviews; again, this paradigm implies that music fundamentally is considered as an auditory object.

In our current work, we bypass audio representations of music. Instead, we aim to demonstrate that *non-auditory, extra-musical associations to music can offer a valid alternative way to describe, index and retrieve music items*.

2.3   Automated Soundtrack Suggestion Approaches

Various works have been published aiming to automatically suggest soundtracks to videos or slideshows, many of them [6,7,13,18] considering emotion to be the main relevance criterion for matching music to video or slideshow imagery. Emotion-based feature representations are then proposed to be extracted from the audio and visual signals.

In Stupar and Michel [14], cinema is chosen as prime example of how soundtracks are used with imagery, and an approach is proposed in which soundtrack matches are learned in relation to imagery by comparing soundtracks from cinematic films to soundtracks in a music database. Cai et al. [1] propose a system to automatically suggest soundtracks as accompaniment to online consumed web content, again performing matching based on emotional information as inferred from the music and text features.

It is striking how emotion is omnipresent in these automated approaches, while it is not as prevalent in the soundtrack match typology found in our studies [10]. Furthermore, direct signal-to-signal matching employing features obtained from the signal will work on high-quality training data, but encounters quality issues when considered in the context of lower-end content such as user-generated video.

A final disadvantage of direct signal-to-signal matching approaches is that an absolute optimum is implied for the match from music to media item. This contrasts with the notion in multimedia production that the same footage can be used, transformed and repurposed in many radically different (and sometimes contrasting) ways, depending on the accompanying soundtrack.

3   Proposed Framework

Our proposed framework is particularly inspired by the earlier described notion in humanities literature that contextual extra-musical meaning associated to music has been established by the way in which music occurs as part of mass media. An interesting online social and collaborative resource holding such associations at scale is the Internet Movie Database (IMDb). In the IMDb, users enter various types of movie information, including plot summaries, keyword summaries, actor listings and reviews. The IMDb also has the possibility to add soundtrack listings to a movie. As a consequence, we can use this resource to associate film plot descriptions—describing the context in which soundtracks occur—to soundtrack listings. While the resource does not allow us to associate a soundtrack to a pinpointed event in the full plot description, our assumption is that *similar music will occur for similar plot descriptions*.

We now can associate film plot information to soundtrack song names. On top of this, we acquire a richer description for the songs by considering another major online social resource focused on music information: last.fm. This is a social music platform in which music listening behavior is stored and used for recommendation. On last.fm, users can describe songs in the form of free-text social tags. The public API of last.fm reveals this information, as well as associated information including popularity data.

The information from these two resources is connected by considering what IMDb movies have plot descriptions with soundtrack listings, and what soundtrack songs have social tag descriptions on `last.fm`. For our current work, we crawled 22,357 unique IMDb movies with plot descriptions, which had at least one soundtrack song with a `last.fm` tag associated to it. In total, considering the soundtracks of all these movies, 265,376 song tags could be found.

To model the connected information in a way that is useful for automated soundtrack retrieval, we encode it in the form of various search indices for information retrieval scenarios, as described in the following subsections.

3.1   An Information Retrieval Setup

We wish to automatically suggest music that would fit given non-musical contexts. For this, we first need to know what non-musical context is intended. Then, we need to match this information to knowledge from our crawled resources. We model this as an information retrieval problem, in which the desired context will form the query, expressed in vocabulary-unrestricted free textual form. Because of this allowed freedom, we can accommodate queries at many semantic levels, ranging from descriptions of multimedia items (e.g. a video or image), to more abstract narrative descriptions (e.g. a story or a situation).

Following the information retrieval paradigm, the query is used to retrieve relevant documents from one or more search indices, to be detailed below. Each search index considers a collection of documents. The document content is analyzed and matched against the query to assess relevance, while the document key represents the document in a short-hand way. For the creation of the search indices, which will be considered in various ways in our Evaluation Setup, we make use of the standard features of the Apache Lucene search engine library[1].

A full system setup consists of four steps:

1. Take a contextual free-text query description as input;
2. Match the query against a search index containing movie plot descriptions; return song tags for soundtracks belonging to the plot descriptions with highest relevance to the query;
3. Adapt the song tag collection by expanding it with the most frequently co-occurring song tags to the ones already in the collection;
4. Use the adapted song tag collection to query a music database (in which music items are described by song tags).

3.2   Mapping of Song Tags to Movie Plots

A movie m can be represented by a full-text plot description p_m, but also by a collection of soundtracks $S_m = \{s_1, s_2, s_3, ...\}$. Each soundtrack song $s_n \in S_m$ can in its turn be represented as a collection of corresponding social song tags: $s_n = \{t_1, t_2, ...\}$.

---

[1] https://lucene.apache.org.

We re-encode each social song tag by mapping it to a collection of movie plot descriptions: $t \mapsto \{p_m\} \quad \forall p_m : t \in s \in S_m$. The corresponding representation is treated as a document for a search engine, with song tags as document keys and the collection of movie plot descriptions as document content. These are indexed using Lucene's standard indexing options, which are based on the tf-idf weighting model. Because of this, stopwords and frequently occurring words across a corpus will naturally get lower weights than terms that are more representative of a particular document in a corpus.

Upon availability of the search index, it is possible to enter any narrative description as a query, upon which a ranking of matched 'documents' is returned in response to the query. Each of these documents will be represented by the document key. Therefore, upon entering an unrestricted free-text narrative description, a ranked list of song tags will be returned. In [8] we already visualized some interesting patterns emerging from this plot-based index; in the current work, we will consider it as part of our evaluations.

3.3   Mapping of Song Tags to Other Song Tags

In [9], we informally noted that retrieval results appeared to improve when a pseudo relevance feedback step was performed after the narrative querying step mentioned above. The reasoning behind such a step is as follows: if an initial search (for example on a movie plot search index) will yield relevant song tags, other song tags that co-occurred with the relevant song tags in a song may also be relevant. Therefore, it is good to retrieve the most important song tags that co-occur in songs with the tags we already have.

To acquire relevant co-occurring tags, we take a ranked list of suggested song tags, and use these to query an index in which song tags are mapped to other song tags with which they co-occur in a song. Put more formally, if we have a song $s_n = \{t_1, t_2, ...\}$, then $t_i \mapsto \{t_j\} \quad \forall t_i \in s_n, \forall t_j \in s_n,$ with $i \neq j$. When indexing this associated song tag representation in a similar way as we did before (taking the collection of co-occurring tags as 'document content', and the tag of interest as 'document key'), frequently occurring tags will only be considered as relevant terms when they do not dominate the full corpus.

3.4   Mapping Database Target Songs to Song Tags

When we ultimately want to associate narrative queries to songs, we need a database of candidate songs. Each song in the database should be represented in such a way that the steps above lead to a querying representation against which the song can be matched. Therefore, each song in the database should be represented as a collection of song tags. Unusually, our aim is not to restrict the vocabulary here to allow for as-flexible-as-possible matching. In [9], we employed tag vocabulary, song titles and song descriptions in a production music database as contributed by the songs' composers. In the current work, we do not restrict to production music, but make use of a general commercial song database.

4   The SRbench Benchmark Dataset

To evaluate the appropriateness of the retrieval mechanisms as outlined in Sect. 3, we need a dataset of songs, vocabulary-unrestricted queries, and corresponding relevance or recognition assessments. No standardized datasets exist for our problem. Furthermore, since our approach begins with a free-text, contextual expression of the music information need, the system allows for a high degree of flexibility and subjectivity in query vocabulary.

We could design a validation campaign through a user study, although an additional concern may be that free-text querying of music information is not commonly adopted yet in search engines, and therefore might need more detailed instruction or examples to test subjects, risking bias effects. Furthermore, the initial description does not have to be generated by a human per se; for example, we could envision the output of automatic image (or video) captioning systems (e.g. [17]) to also be usable as input to our framework.

As in the current work, we are not interested in the querying mechanism, but rather in the retrieval performance as a consequence of given queries, for our current evaluations, we will employ the existing SRbench dataset [15]. SRbench was established as an evaluation benchmark for soundtrack retrieval in connection to collections of still images. It consists of 25 query photo collections, which were crawled from popular categories in Picasa, as shown in Table 1. Next to this, it offers audio snippets of 470 popular commercial songs in various genres. Through locally and online conducted user assessments, for each of the photo collections, for each pair of songs taken from the dataset, six different people indicated how they rate the fit of each of the songs to the photo collection. The strength of their preference for one song over the other is expressed through a *preference strength* p_s, on an integer scale from 0 (no preference) to 5 (strong preference). A system can then be evaluated by taking (a part of) its output ranking in reponse to a query, and assessing to what extent rankings between pairs in the system-produced ranking match those indicated by the human assessors.

Stupar and Michel [14] measure performance with respect to a top-20 output, according to two measures:

1. Preference precision: $\frac{\text{\# correctly ordered pairs}}{\text{\# evaluated pairs}}$.
2. Weighted preference precision: $\frac{\sum p_s \text{of correctly ordered pairs}}{\sum p_s \text{of evaluated pairs}}$.

The measures are calculated considering ratings at three agreement levels: those ratings with 6/6, 5/6 and 4/6 assessor agreement, respectively, on which song in the assessment pair was a better fit to the given image collection. In our current work, we also make use of these measures, as explained in the following section.

Table 1. Query image collection categories in SRbench.

Aviation, Architectural, Cloudscape, Conservation, Cosplay, Digiscoping, Fashion, Fine art, Fire, Food, Glamour, Landscape, Miksang, Nature, Old-time, Portrait, Sports, Still-life, Street, Underwater, Vernacular, Panorama, War, Wedding, Wildlife

5   Evaluation Setup

In order to evaluate our approach against SRbench, a few steps should be performed in the evaluation setup. First of all, we should *translate the target song database to a song tag representation*. To this end, we cross-match artist and title names with `last.fm`. We managed to match 439 out of 470 songs. For classical music items, typically one particular rendition (not necessarily the one matched to our metadata) would have the most extensive tag vocabulary; we manually corrected the mapping such that this vocabulary would be used.

Another necessary step is to *translate the 25 image collection queries into textual queries to our search indices*. As can be noted from Table 1, the image categories are very general and abstract. We wanted to obtain a richer description without putting in personal bias. While one could run visual concept detectors for this, we retained the spirit of using social online knowledge resources, employing the API of the Flickr image service. Through the `tags.getRelated` API call, using SRBench category names as query tag input, a larger set of related tags could be obtained as commonly co-occuring in Flickr. As an illustration of the adequateness of this technique, in Fig. 1 we show the SRbench images for the category 'underwater' alongside the Flickr API response for `tags.getRelated('underwater')`.

As not all songs could be matched to `last.fm`, our set of 439 songs comprises 5847 song pairs, as opposed to the 6836 song pairs in SRbench for the original full set of 470 songs. Since SRbench does not include raw ranking outputs for earlier proposed systems, we cannot recompute performance for the approaches in [15] based on a smaller set of song pairs, so direct comparison of evaluation outcomes against those originally reported in [15] is impossible. To still get a good sense of performance, we will consider various configurations within our proposed framework, in the form of four possible setups which also are illustrated in Fig. 2:

1. Direct querying of the target song database index (*direct*). This approach is expected to give a baseline of performance of our system, as we do not expect many non-musical terms from the image query collection to coincide with social music tags.
2. Querying of an index associating song tags to movie plots (*plot*). Returned song tags are used as query to the target song database.
3. Querying of a song tag co-occurrence index (*cooc*). Returned song tags are used as query to the target song database.
4. Pseudo relevance feedback querying (*prf*): first query the index associating song tags to movie plots, use the results to query the song tag co-occurrence index, and use those results to query the target song database.

As a final, fully independent baseline, we also consider a *random* approach. For this, we generate 1000 random top-20 outcomes. For each of these outcomes, we calculate the evaluation metrics (unweighted and weighted performance, considering 6/6, 5/6 and 4/6 annotator agreement), averaging over all trials.

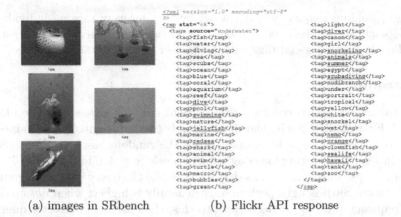

(a) images in SRbench   (b) Flickr API response

Fig. 1. Images of the 'underwater' category in SRbench, alongside Flickr API response output for `tags.getRelated('underwater')`.

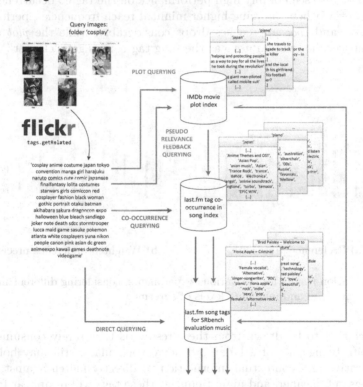

Fig. 2. Illustration of the various experiments performed for evaluation of our approach.

In reports on tag metadata (obtained through the `tag.getInfo` API call), `last.fm` distinguishes between 'reach' (the number of individual consumers of a given tag) and 'taggings' (the number of times a tag is used). To test the effect of 'cleanness' of social tags, we test three setups varying the reach frequency of the tags (avoiding frequency bias caused by very actively tagging individuals): (1) all social tags are indexed; (2) only social tags with a reach of at least 100 are indexed; (3) only social tags with a reach of at least 1000 are indexed.

6   Results

The obtained evaluation results for the various experimental setups are plotted in Fig. 3. First of all, there is clear distinction between the outcomes obtained for our various system setups in comparison to the random baseline. We generally notice that upon tolerating lower agreement levels (e.g. 4/6 instead of 6/6), performance drops. This is logical, as that would make the query more ambiguous to interpretation. Surprisingly, performance generally is highest when *not* applying reach frequency filtering to tags. Next to this, the baseline of directly querying the song database (*direct*) performs reasonably well, even outperforming movie plot index querying (*plot*) when considering all social tags for indexing. Under this setup, we also see notably high performance on the tag co-occurrence index (*cooc*). However, when enforcing higher minimal reach frequencies, performance on the *direct* and *cooc* approaches drops considerably, while the *plot* and *prf* querying approaches will improve as the song tag corpus gets 'cleaner'.

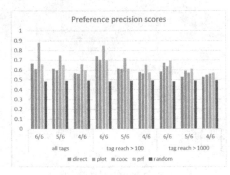
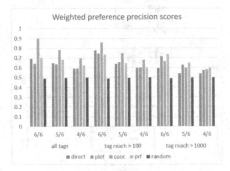

(a) Preference precision   (b) Weighted preference precision

Fig. 3. Evaluation results for our various experiments, considering different minimum tag reach frequencies and minimum levels of agreement.

A conclusion to be drawn from these results is that rarely consumed song tags, despite being part of a large and noisy vocabulary, still may hold some useful narrative and contextual information to directly match against. Yet as soon as tags get 'cleaner' and more frequent, these tags will disappear from the index. In that case, direct matching against song tags becomes more difficult, and a narrative matching step (*plot* and particularly *prf*) will make more sense. Applying the pseudo relevance feedback stage consistently outperforms querying of the movie plot only, and as such seems to hold the best of both worlds: focusing a song tag corpus based on co-occurrence, but also taking advantage of any relevant narrative contextual associations.

7 Conclusions and Future Opportunities

In order to associate music to non-musical narrative elements, we proposed to make use of socially established associations from online mass media. In our proposed approach, we start from an unrestricted free-text narrative query expressing the desired context. We modeled folksonomically expressed associations as obtained from the IMDb and `last.fm` as search indices in an information retrieval scenario and considered various search index setups.

Evaluating our approach against the SRbench benchmark dataset shows that social song tags turn out fairly adequate to match non-musical queries against—especially when 'filtered' through a tag co-occurrence index. Future work should examine if co-occurrence has particularly strong effects due to the songs having been used as movie soundtracks, or if this would hold for any corpus with social tagging. Employing matches against full-text movie plots especially performs well when combining this with co-occurrence filtering. Performance effects caused by restricting to higher minimum reach frequencies suggest that tags with low reach may actually hold relevant contextual information. At the same time, this is at the expense of having to deal with a larger and noisier index.

We believe that our work holds promise in several ways. From a disciplinary viewpoint, we made an important step in unifying data-driven approaches with more subjective and qualititative notions on music utility. While current online music service models tend to focus on on-demand consumption of known songs, employing an approach like ours can assist in transforming this into on-demand immersion in known experiences. We therefore hope our work can ultimately offer novice audiences new and serendipitous entrances to lesser-known genres.

Acknowledgements. The research leading to these results has received funding from the European Union Seventh Framework Programme FP7/2007–2013 through the PHENICX project under Grant Agreement No. 601166.

References

1. Cai, R., Zhang, C., Wang, C., Zhang, L., Ma, W.-Y.: MusicSense: contextual music recommendation using emotional allocation modeling. In: Proceedings of the 15th ACM International Conference on Multimedia (ACM MM), pp. 553–556, Augsburg, Germany (2007)
2. Casey, M., Veltkamp, R., Goto, M., Leman, M., Rhodes, C., Slaney, M.: Content-based music information retrieval: current directions and future challenges. Proc. IEEE **96**(4), 668–696 (2008)
3. Cohen, A.J.: How music influences the interpretation of film and video: approaches from experimental psychology. In: Kendall, R., Savage, R.W. (eds.) Selected Reports in Ethnomusicology: Perspectives in Systematic Musicology, vol. 12, pp. 15–36. Department of Ethnomusicology, University of California, Los Angeles (2005)
4. Cook, N.: Analysing Musical Multimedia. Oxford University Press, New York (1998)

5. Kaminskas, M., Ricci, F.: Contextual music information retrieval: state of the art and challenges. Comput. Sci. Rev. **6**(2–3), 89–119 (2012)
6. Kuo, F.-F., Chiang, M.-F., Shan, M.-K., Lee, S.-Y.: Emotion-based music recommendation by association discovery from film music. In: Proceedings of the 13th ACM International Conference on Multimedia (ACM MM), pp. 507–510. Singapore (2005)
7. Li, C.-T., Shan, M.-K.: Emotion-based impressionism slideshow with automatic music accompaniment. In: Proceedings of the 15th ACM International Conference on Multimedia (ACM MM), pp. 839–842, Augsburg, Germany (2007)
8. Liem, C.C.S.: Mass media musical meaning: opportunities from the collaborative web. In: Proceedings of the 11th International Symposium on Computer Music Multidisciplinary Research (CMMR), Plymouth, UK (2015)
9. Liem, C.C.S., Bazzica, A., Hanjalic, A.: MuseSync: standing on the shoulders of Hollywood. In: Proceedings of the 20th ACM International Conference on Multimedia (ACM MM), pp. 1383–1384, Nara, Japan. ACM (2012)
10. Liem, C.C.S., Larson, M.A., Hanjalic, A.: When music makes a scene — characterizing music in multimedia contexts via user scene descriptions. Int. J. Multimedia Inf. Retrieval **2**, 15–30 (2013)
11. Lissa, Z.: Ästhetik der Filmmusik. Henschelverlag, Berlin (1965)
12. Schedl, M., Gómez, E., Urbano, J.: Music information retrieval: recent developments and applications. Found. Trends Inf. Retrieval **8**(2–3), 127–261 (2014)
13. Shah, R.R., Yu, Y., Zimmermann, R.: ADVISOR: personalized video soundtrack recommendation by late fusion with heuristic rankings. In: Proceedings of the 22nd ACM International Conference on Multimedia (ACM MM), pp. 607–616, Orlando, Florida, USA (2014)
14. Stupar, A., Michel, S.: PICASSO — to sing you must close your eyes and draw. In: Proceedings of the 34th Annual ACM SIGIR Conference, Beijing, China (2011)
15. Stupar, A., Michel, S.: SRbench — a benchmark for soundtrack recommendation systems. In: Proceedings of the 22nd ACM International Conference on Information & Knowledge Management (CIKM), San Francisco, USA (2013)
16. Tagg, P., Clarida, B.: Ten Little Title Tunes — Towards a Musicology of the Mass Media. The Mass Media Scholar's Press, New York and Montreal (2003)
17. Vinyals, O., Toshev, A., Bengio, S., Erhan, D.: Show and tell: a neural image caption generator. In: Proceedings of the 28th IEEE Conference on Computer Vision and Pattern Recognition (CVPR), Boston, USA (2015)
18. Wang, J.-C., Yang, Y.-H., Jhuo, I.-H, Lin, Y.-Y., Wang, H.-M.: The acousticvisual emotion guassians model for automatic generation of music video. In: Proceedings of the 20th ACM International Conference on Multimedia (ACM MM), pp. 1379–1380, Nara, Japan. ACM (2012)

Digital Library Evaluation

The "Nomenclature of Multidimensionality" in the Digital Libraries Evaluation Domain

Leonidas Papachristopoulos[1,2], Giannis Tsakonas[3(✉)], Michalis Sfakakis[1],
Nikos Kleidis[4], and Christos Papatheodorou[1,2]

[1] Department of Archives, Library Science and Museology,
Ionian University, Corfu, Greece
{l11papa,sfakakis,papatheodor}@ionio.gr
[2] Digital Curation Unit, Institute for the Management of Information Systems,
'Athena' Research Centre, Athens, Greece
[3] Library and Information Center University of Patras, Patras, Greece
gtsak@upatras.gr
[4] Department of Informatics,
Athens University of Economics and Business, Athens, Greece
klidisnik@aueb.gr

Abstract. Digital libraries evaluation is characterised as an interdisciplinary and multidisciplinary domain posing a set of challenges to the research communities that intend to utilise and assess criteria, methods and tools. The amount of scientific production, which is published on the field, hinders and disorientates the researchers who are interested in the domain. The researchers need guidance in order to exploit the considerable amount of data and the diversity of methods effectively as well as to identify new research goals and develop their plans for future works. This paper proposes a methodological pathway to investigate the core topics of the digital library evaluation domain, author communities, their relationships, as well as the researchers who significantly contribute to major topics. The proposed methodology exploits topic modelling algorithms and network analysis on a corpus consisting of the digital library evaluation papers presented in JCDL,ECDL/TDPL and ICADL conferences in the period 2001–2013.

Keywords: Digital library evaluation · Research communities · Topic modelling · Network analysis

1 Introduction

The birth and rise of the digital libraries (DL) domain has been the result of the cooperation and engagement of a wide set of disciplines, which amalgamate concepts, terminologies, methods, tools and needs in a single research space. From the early era of DLs, Borgman has noted that interdisciplinarity and multidisciplinarity are its main characteristics and that increased participation and "cross-fertilization of ideas" form a fuzzy domain [1]. The level of complexity

© Springer International Publishing Switzerland 2016
N. Fuhr et al. (Eds.): TPDL 2016, LNCS 9819, pp. 241–252, 2016.
DOI: 10.1007/978-3-319-43997-6_19

is considerably increased when we put the field of DL evaluation under our spotlight. This particular field focuses on diverse criteria, such as effectiveness, quality, and performance, that can be viewed under different perspectives, such as user-centred or system-centred, and are used by non-monolithic assessment approaches to unveil patterns and to meaningfully interpret behaviours [2]. The cartography of the domain would be an important tool for the researchers who intent to study and contribute to a such a multidimensional field, i.e. the identification of the issues the domain confronts and the communities deal with. Actually this cartography aims to the development of a human network associated to the DL evaluation research directions.

This paper aims to reveal the "nomenclature of multidimensionality"[1], which in fact presents an approach to detect communities or researchers that play an important role in the DL evaluation field due to their ability to contribute to different topics. These members of the human network enrich the already rich diversity of the field. Their possible absence would be a loss for the DL evaluation as the inventory of methods, the points of view, tools, terms etc. would be reduced, abolishing at the same time the main characteristic of the domain, the "cross-fertilisation of ideas" [1]. According to literature, the nodes of a network that connect clusters are called "bridgers" and "provide valuable opportunities for innovation, growth, and impact because they have access to perspectives, ideas, and networks that are otherwise unknown to most network members" [3]. The detection of researchers with "bridging" capabilities would act as a recommendation tool for collaboration [4], or for assigning papers for review [5] and literature research.

The analysis of the researcher's activity in such a complex domain can be confronted with combinational approaches. On one hand text mining techniques, like topic modelling, can provide probabilistic assumptions about the topical orientation of research documents. This work applies Latent Dirichlet Allocation (LDA) in a corpus consisting of the full and short papers presented at the JCDL, ECDL/TPDL and ICADL[2] conferences during the period 2001–2013 and deal with DL evaluation. On the other hand Network Analysis offers a set of methods, tools and techniques for gaining awareness of the structure, relationships and information flows between individuals, groups, organisations, URLs and various other knowledge entities. In this paper the interactions between the emerged topics are investigated as well as co-authorship information is analysed for revealing bipartite relationships among the topics and the author groups. The results of the study will contribute to the creation of a topic map of DL evaluation domain and to the discovery of author communities having impact to several topics. Hence it goes a step beyond co-authorship graphs and it augments the expressiveness of the human network by attributing topical identification to agents and communities. The outcome of the work will both fulfil topic orientation needs of candidate researchers and will also raise the awareness of the

---

[1] Nomenclature: a system of names for things especially in science from Merriam-Webster.com.

[2] ICADL proceedings are available since 2002.

expertise distribution within the research community. Actually in this paper we attempt to answer the following research questions:

1. Which are the most prominent topics emerged in DL evaluation and how they interact? Are there cohesive groups of topics that usually co-exist in research efforts? Could these groups of topics formulate super-topics (semantic categories) that chart the research efforts of the DL evaluation domain?
2. Which are the researchers' communities in the DL evaluation domain that share same scientific interests and which is the performance of the research activity in each topic?
3. How much multidimensional is the behaviour of the researchers in the field? Are there groups of researchers who bridge two or more topics and are considered highly significant factors of the human network as they connect the knowledge of large parts of the network?

The remaining sections of this article are organised as follows: Sect. 2 presents the related work in the field of topic modelling and network analysis; Sect. 3 demonstrates the steps for topic interpretation and the generation of a graph that presents interactions between topics and between authors and topics. Section 4 shows light on the findings from our analysis and finally Sect. 5 epitomises the basic outcomes of the study and raises some issues over future works.

2 Related Work

The multidisciplinarity of the field of DL has been the main motivation for enormous literature activity on the specification of the topics that compose the 'topical ecosystem' - a network of interconnected and interacting topics - of the domain. Some attempts have been made on an abstract level attempting to investigate the nature of collaboration in information science under a philosophical lens, leaving though unanswered the extent and contribution of various disciplines [6]. Attempting to highlight precisely the topics that have been emerged in the DL field, [7] used a traditional method like content analysis on conference proceedings (JCDL, ECDL/TPDL, ICADL) from the period 1990–2010. The study identified 21 core topics and 1,015 subtopics having although acknowledged the fact that the temporal aspects of their appearance had been neglected. On the other hand there are initiatives that attempted to decompose the DL domain using more machinery and sophisticated methods. The LDA method was also applied on 3,121 doctoral dissertations completed between 1930 and 2009 at North American Library and Information Science programs in order to indicate the dominant topics of the field [8]. The study revealed that the topics have significantly changed through time, but topics such as 'information retrieval' and 'information use' present a constant behaviour.

Recently, the detection of special communities and significant members within diverse domains has been drawn to the centre of the attention. A study based on graph decomposition, analysed the collaboration between scientists and between institutions in the field of physics to get meaningful inferences about the funding that they received for a 10 years period [9]. The work underlines the fact that successful funding is based on three factors: the number of collaborations of a scientist, the number of collaborations in her/his University and the ability of the scientist to interact with different research teams during time.

Singh et al. combined keyword analysis with the examination of national and international research networks in the field of hardware architecture in order to facilitate the characterisation of specific institutions and the determination of the collaborations established in the specific field for the years 1997 to 2011 [10]. The knowledge of the specialisation of the core members of a wide community such as the hardware architecture would be beneficial for decision making regarding collaborations for the development of new products. The need to unveil the community that focuses on a specific methodology was stimulant for the analysis of 131 articles (reports of health technology assessment agencies) published during the period 1989 to 2011 [11]. The main objective of the authors was the understanding the maturity and growth of Cost-Effectiveness Analysis - CEA methodology and the reveal of collaborative patterns between authors of the surveyed literature. In [12] a method of topic identification and community detection techniques was attempted for the information retrieval field. The study underlines the fact that "cognitive and social perspectives" are considered as of equal importance, which are dynamically connected and a change to one would cause a change to the other.

In the field of DL the need for identification of the most prestigious actors in the community was evident from its early stages. A study based on the JCDL conference series for the period 1994–2004 created a co-authorship network which was exploited through centrality measures in order to highlight the most centred participators in the field [13]. Additionally, the rankings derived from centrality measures were validated through a matching with JCDL committee member who are considered a priori prestigious actors. Another work [14] assessed the performance of a PageRank-like algorithm on the ArnetMiner Dataset and the UvT Expert Collection in order to validate query-dependent expert ranking tasks in digital libraries. This paper attempts to extent a previous study on topic detection in the most significant DL conferences (ECDL/TPDL, JCDL and ICADL) and monitoring the topic's drifts along the time [15]. The study concluded that JCDL eavesdrops the alternant interests of researchers changing consequently its agenda, while on the other hand the rest follow a less fluctuating behaviour. This paper seeks an answer regarding the topics interaction in the research literature. Moreover it integrates co-authorship related data to the aforementioned study's results and unveils researchers' communities and agents playing a central role within the DL evaluation domain.

3   Methodology

3.1   Corpus Composition and Topic Extraction

The role of conference proceedings has been upgraded within the scholarly communication environment as they constitute a credible channel for fast dissemination of the outcomes of scientific production. Accordingly, in the DL evaluation domain a large part of this production is presented at the JCDL, ECDL/TPDL and ICADL conferences. The corpus of current experiment is composed by evaluation studies published during the period 2001–2013 and was formed by a two-phased procedure. During selection phase, a Naïve Bayes classifier was applied on 2,001 full and short papers (866 from JCDL, 597 from ECDL/TPDL, 538 from ICADL) and 395 papers were classified as DL evaluation papers and formulated the corpus. To evaluate our classifier three domain experts, having achieved a high inter-raters's agreement score, validated the DL evaluation identity of the corpus. The validation criteria were the papers' title, abstract and author keywords [16].

The first step towards the topic extraction was the conversion of the documents to text format for better manipulation. Subsequently, the texts were tokenised in order to construct the necessary 'bag of words', which would constitute the base of the topic modelling process. That 'bag of words' was cross-checked to exclude the 'negative dictionary' [17] and to remove the frequent and rare words (words appeared over 2,100 or under 5 times). The aforementioned process resulted to a vocabulary of 38,298 unique terms and 742,224 tokens. Token is a word which is meaningful on a linguistic and methodological level. The majority of tokens derived from JCDL's papers and on average each paper contributes 1,879 tokens. The preprocessing phase was followed by the implementation of the LDA algorithm to our corpus. Via topic modelling we intended to categorise each paper to one or more topics. Topic modelling presumes that a topic is a probability distribution over a collection of words and each document is composed by a number of topics. A topic derived from topic modelling is composed by a number of words and each interpreter takes into consideration the most significant - according to topic modelling score - in order to entitle a topic. A tab-delimited text file was imported to Mimno's jsLDA (javascript LDA) web tool [18] and we run 1,000 training iterations in order to achieve a stable structure of topics. Additionally, we executed a number of tests asking for various numbers of topics (25, 20, 15, 14, 13, 11, 9 topics) in order to specify the optimal interpretable number. Three domain experts examined the word structure of each topic and concluded that the optimal interpretable number of topics is 13.

3.2   Topic Correlation and Centralities

Apart from the typical features that a topic modelling tool offers, jsLDA offers a topic correlation functionality based on the Pointwise Mutual Information (PMI) [19]. PMI compares the probability of co-occurring two topics in a document with the independent existence of each one within the same document. More specifically, if $P(a)$ and $P(b)$ are the probabilities of the topics a and b to occur in document D respectively, and $P(a, b)$ is the join probability of the two topics to occur in D, then their mutual information $PMI(a, b)$ is defined as: $PMI(a, b) = log_2 \frac{P(a,b)}{P(a)P(b)}$.

The undirected graph in Fig. 1 - constructed with Gephi[3] as all the graphs in this paper - presents the 13 topics (nodes) of the corpus and the 36 relationships (edges) that exist between them. An edge correlates two topics if their PMI is greater than a cutoff of 0.002. The thickness of the edges represents the PMI values and hence the magnitude of correlation between the topics [15].

The revelation of the topics that play an important role in the domain is a focal point of our analysis and therefore three measures of centrality were applied: *degree centrality, betweenness centrality* and *closeness centrality* [20]. The *degree centrality* (C_d) of a node of a graph measures the numbers of adjacencies of a node in direct contact. The exploration of topic *degree centrality* aims to underline potential interactions between the topics of our graph. Topics with high degree centrality are supposed to have semantic communication. On the other hand, an opposite behavior will indicate topic isolation and therefore papers that belong to the specific category are probably monopolized by a single or a limited set of topics.

Betweenness centrality (B_c) constitutes an alternative point of view for centrality measuring the capacity of a node to connect a pair of others. An intermediate position of a topic implies its dynamic role to affect the starting and the ending points of a path. Each topic having high *betweenness centrality* rate is considered to have a high transitional role as it plays a communicative bond between two different topics.

Closeness centrality (C_c) is the average shortest paths from a focal node to all other nodes of a graph. A low closeness centrality indicates that a node is more likely to be directly connected to most others in the graph. In our topic graph a high number of intermediary nodes implies a high heterogeneity between the reference topic and the other of the graph. The minimum length between two topics refers to a close semantic relationship.

Apart from the centrality measures that have been used to identify the centric nodes of the topic graph, we also used in our analysis the *clustering coefficient* (C_I) measure in order to localise neighbourhoods of the graphs having strong relationships. The *clustering coefficient* CI of a node i is the number of triangles connected to node i divided by the number of triangles centered around node i. According to Mislov et al. "the clustering coefficient of a node with N neighbours is defined as the number of directed links that exist between the node's N neighbours, divided by the number of possible directed links that could exist between the node's neighbors" [21]. Regarding the topics graph, we consider any topic having a high clustering coefficient degree as an 'entry point' to a set of topics having a close relationship between them.

3.3   Author – Topic Correlation

Furthermore we extracted the names of the authors of each paper of our corpus, as well as we associated each name to the topics of the papers they have written.

---

[3] Gephi: The Open Graph Viz Platform, https://gephi.org/.

Totally 905 unique authors were identified. A graph $G(V,E)$ was created as follows:

The set of nodes V is the union of the set of topics T and the set of author names N ($V = T \cup N$, where T and N are disjoint sets), while the set of edges E is the union of the sets NT and CA ($E = NT \cup CA$), where: NT is the set of pairs (n_i, t_j), where n_i is the name of the author who has written a paper that belongs to the topic t_j with a probability greater than a threshold equal to 0.12 and CA is the set of pairs (n_i, n_c), where n_c is the name of a co-author of the author with name n_i.

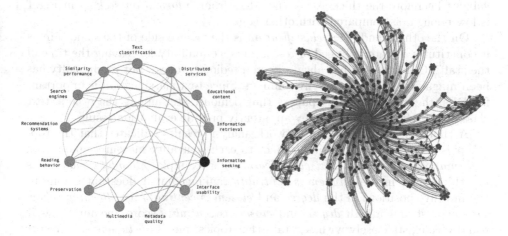

Fig. 1. Topic correlation   **Fig. 2.** The community of the information seeking topic

Hence the graph G represents both co-authorship relationships and author – topic relationships. The edges of the set NT represent the author – topic relationships, while the edges of the set CA represent co-authorships. Actually the graph G reveals the community of each topic of the DL evaluation domain. Figure 2 presents a fragment of the graph G, which corresponds to the research community of the topic *Information Seeking*, the topic with the largest community and the greatest number of papers. The red edges correspond to co-authorship relationships (CA set). The co-authors of a paper form a fully connected component in the graph. There exist authors that connect fully connected components of the graph and we can infer that they participate in more than one author groups.

4 Findings

4.1 Topic Interactions

All the centrality measures presented in Table 1 reveal *Information seeking* and *Text classification* as the core topics of DL evaluation domain. The high frequency of appearance of *Information seeking* in the corpus indicates the significant role of the topic in the DL evaluation domain. Actually, the study of information seeking behaviour is vital for the evaluation of DL systems and, thus, the interest is focused on several levels of human activity, such as task accomplishment, system support, etc. The fact that *Information seeking* was observed in a high proportion of the documents implies that the probability to occur in a document is also high. According to the PMI definition, this high probability decreases the PMI values of the *Information seeking* with the other topics observed in the same document. Other pairs of topics that co-exist in the same document and have low individual probabilities, have increased PMI values. Therefore the thickness of the edges from *Information seeking* in Fig. 1 is low enough, as compared with other edges.

On the other hand, *Text classification* is the inverse side of the same coin as it constitutes an internal need of researchers' community for 'saving the time of the staff'. Classification has always been a tedious process, but its difficulty has been magnified in the digital environment. Since text classification is important for DLs, the majority of the papers that demonstrate, among the others, text classification functionalities, they apparently provide evaluation results.

A thorough examination of network analysis results indicates that the top-5 topics are not the same with respect to each centrality measure, apart from *Information seeking* and *Text classification*. For example, *Metadata quality* has the third place in the *betweeness centrality* ranking, but it does not belong to the first five positions in the *degree* and *closeness centrality* list. *Reading behavior* demonstrates a high *degree* and *closeness centrality* score, meaning that it usually co-exists closely with several other topics, but its *betweeness centrality* value indicates that the topic cannot be considered as a communicative bond between two different topics.

All three centrality measures indicate that *Information retrieval* is among the core topics proving the unappeased interest for the advancement of content discovery. This result highlights *Information retrieval* as one of the main concerns of the domain. Furthermore *Search Engines* is a topic that usually appears in the same papers with *Information retrieval* as indicated by their PMI value (Figure 1), while its *betweeness centrality* has also a high value.

Considering Figure 1 it is worth observing the cohesive topical neighbourhoods that exist in the DL evaluation environment. The neighbourhood of *Educational content* seems to be the most cohesive. This is validated by its *Clustering coefficient* value, which is the highest in the topics graph ($C_I = 0.833$). Almost all of the neighbours of *Educational content* interact each other, underlying that there is a semantic completion among the topics: *Reading behavior, Information seeking, Interface usability, Metadata quality* and *Educational content*. Actually, in the DL evaluation papers, it is very common to use educational content as a basis for the examination of users reading and information seeking behaviour or for the quality of a user interface. Similarly, the pair *Recommendation systems* and *Similarity performance* share the same *Clustering coefficient* value ($C_I = 0.7$) and generates another cohesive group that consist of the topics *Information retrieval, Search engines, Text classification, Similarity performance, Recommendation systems* and *Information seeking*. The cohesive subgroups could be considered as 'super-topics' that reduce the width of the topic set and create a connected perspective of it.

4.2 Author Activities in DL Evaluation

Our DL evaluation corpus consists of 395 papers and includes 905 unique authors. Given that an author could participate to more than one paper, the total number of author-participations to the papers equals 1,335. Hence the average number of authors per paper is 3.38, while the average author participation in the papers is 1.47. Table 1 presents the number of authors, the number of papers and the average number of co-authorships for each topic. The majority of the papers were classified to *Information seeking*, fact that affects the number of authors that are involved in the specific topic. However *Preservation* and *Multimedia* demonstrate low productivity. Regarding the number of authors needed to accomplish an evaluation process, *Educational content* seems to be demanding as 4.395 authors contribute averagely in a paper of this topic. On the other hand *Reading behavior* is the topic that exhibits the lowest co-authorship rate.

One of the most productive researcher in the DL evaluation domain is Dion Hoe-Lian Goh, who has been hyperactive in ICADL and JCDL conferences and is the author of 20 papers. The names of Yin-Leng Theng (19 papers), Edward A. Fox (13 papers), Ee-Peng Lim (11 papers), Hussein Suleman (9 papers) are accomplishing the list of the most productive authors in the specific domain.

Table 1. Productivity results and centralities across topics

Topic	Paper productivity measures			Network analysis measures			
	#Authors	Average co-authorship	#papers	Degree	Closeness centrality	Betweeness centrality	Clustering coefficient
Distributed services	140	3.578	45	5	1.583	2.750	0.200
Educational content	129	4.395	43	4	1.667	0.333	0.833
Information retrieval	223	3.011	87	6	1.500	2.083	0.600
Information seeking	803	3.369	352	11	1.083	19.917	0.364
Interface usability	248	3.438	96	5	1.583	1.000	0.700
Multimedia	75	3.414	29	4	1.667	1.000	0.667
Metadata quality	205	3.824	68	5	1.583	3.033	0.400
Preservation	76	2.933	30	4	1.667	0.450	0.667
Reading behaviour	126	2.885	52	6	1.500	2.167	0.600
Recommendation systems	161	3.273	55	5	1.583	0.783	0.700
Search engines	35	3.192	52	5	1.583	2.950	0.400
Similarity performance	107	3.45	40	5	1.583	1.167	0.700
Text classification	306	3.016	129	7	1.417	4.367	0.524

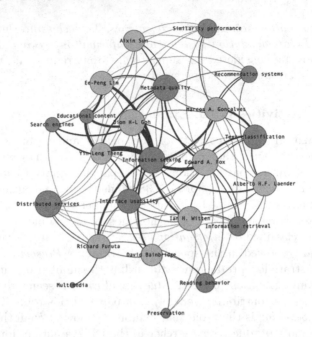

Fig. 3. 'Multidimensional' researchers for the DL evaluation domain

Dion Hoe-Lian Goh, Yin-Leng Theng and Ee-Peng Lim could be characterised as *Educational Content* oriented researchers since they have written on the topic 8, 10 and 8 papers respectively. Edward A. Fox has actively participated in two topics, *Text classification* and *Metadata quality*, and Hussein Suleman has focused on *Distributed services*, having written 8 papers on the topic.

The bipartite graph of Fig. 3 reveals the top-10 researchers who have contributed to more than one topics. These researchers could be considered 'multidimensional', since they have contributed to most topics of DL evaluation and they combine knowledge from different domains in order to accomplish their tasks. The blue nodes of the graph correspond to the authors, while the red correspond to the topics. The size of the blue nodes corresponds to the degree of each node, i.e. the number of topics that each researcher covers. The edges of the graph connect the 10 authors with the topics they have contributed. The thickness of the edges of the graph denotes the number of papers each author has written on a topic. Two authors, Dion Hoe-Lian Goh and Edward A. Fox are the most multidimensional as their works is related to 10 different topics. The list of the ten most multidimensional authors of the domain is completed with the names of Ee-Peng Lim, Marcos Andre Goncalves, Richard Furuta, Ian H. Witten, Alberto H. F. Laender who deal with 9 topics and Yin-Leng Theng, David Bainbridge, Aixin Sun who cover 8 topics.

A high proportion of the authors, 382 (42 %), contribute to three topics, whereas 207 researchers have dealt with two different topics. Finally, from the entire body of 905 authors, only 37 can not be characterised as 'bridgers', as their work has been assigned to only one topic. According to Fig. 3, the *Preservation* and *Multimedia* topics are not among the interests of the 'mulitdimensional' authors. In contrast, *Information seeking, Text classification, Metadata quality, Distributed services* and *Interface usability* are in the centre of their attention.

5 Conclusions

Regarding the first research question the examination of the centrality measures indicate that among the 13 topics, *Information seeking* and *Text classification* have been in the centre of the attention of DL community. Content analysis models, such as LDA and PMI, indicate that DL evaluation main interest is the amelioration of services which facilitate the users seeking behaviour. The investigation of the second research question, which focuses on the interaction of the topics and was based on PMI and *clustering coefficient*, revealed that *Reading behavior, Information seeking, Interface usability, Metadata quality* and *Educational content* create a cohesive subgraph. Another cohesive subgroup consists of *Information retrieval, Search engines, Text classification, Similarity performance, Recommendation systems* and *Information seeking*. These groups imply high probabilities of coexistence between the particular topics and confirm the multidimensionality of DL evaluation domain. They also confirm the existence of super-topics, that operate as latent semantic categories and cover complementary issues.

Finally, the results of the study indicate a significant multidimensionality in the behaviour of key researchers of the domain, since their majority tend to contribute to more than two topics. The most significant of them have achieved to introduce knowledge and methods even from ten different topics signifying the community inclination as well as their personal ability to combine different perspectives.

References

1. Borgman, C.L.: What are digital libraries? Competing Vis. Inf. Process. Manag. **35**, 227–243 (1999)
2. Saracevic, T.: Digital library evaluation: toward evolution of concepts. Libr. Trends. **49**, 350–369 (2000)
3. Hoppe, B., Reinelt, C.: Social network analysis and the evaluation of leadership networks. Leadersh. Q. **21**, 600–619 (2010)
4. Sun, J., Xu, W., Ma, J., Sun, J.: Leverage RAF to find domain experts on research social network services: a big data analytics methodology with MapReduce framework. Int. J. Prod. Econ. **165**, 185–193 (2015)
5. Karimzadehgan, M., Zhai, C., Belford, G.: Multi-aspect expertise matching for review assignment. In: Proceeding of the 17th ACM Conference on Information and knowledge Mining - CIKM 2008, pp. 1113–1122. ACM, New York (2008)
6. Holland, G.A.: Information science: an interdisciplinary effort? J. Doc. **64**, 7–23 (2008)

7. Nguyen, S.H., Chowdhury, G.: Digital library research (1990–2010): a knowledge map of core topics and subtopics. In: Airong, J. (ed.) ICADL 2011. LNCS, vol. 7008, pp. 368–372. Springer, Heidelberg (2011)

8. Finlay, C.S., Sugimoto, C.R., Li, D., Russell, T.G.: LIS dissertation titles and abstracts (1930–2009): where have all the Librar* gone? Libr. Q. **82**, 29–46 (2012)

9. Bellotti, E.: Getting funded. multi-level network of physicists in Italy. Soc. Netw. **34**, 215–229 (2012)

10. Singh, V., Perdigones, A., Garcia, J.L., Cañas-Guerroro, I., Mazarrón, F.R.: Analyzing worldwide research in hardware architecture, 1997–2011. Commun. ACM **58**, 76–85 (2014)

11. Catala-López, F., Alonso-Arroyo, A., Aleixandre-Benavent, R., Ridao, M., Bolanos, M., García-Altes, A., Sanfelix-Gimeno, G., Peiro, S.: Coauthorship and institutional collaborations on cost-effectiveness analyses: a systematic network analysis. PLoS One **7**, e38012 (2012)

12. Yan, E., Ding, Y., Milojević, S., Sugimoto, C.R.: Topics in dynamic research communities: an exploratory study for the field of information retrieval. J. Informetr. **6**, 140–153 (2012)

13. Liu, X., Bollen, J., Nelson, M.L., de Sompel, H.: Co-authorship networks in the digital library research community. Inf. Process. Manag. **41**, 1462–1480 (2005)

14. Gollapalli, S. Das Mitra, P., Giles, C.L.: Ranking authors in digital libraries. In: 11th Annual International ACM/IEEE Joint Conference on Digital Libraries - JCDL 2011, pp. 251–254. ACM, New York (2011)

15. Papachristopoulos, L., Kleidis, N., Sfakakis, M., Tsakonas, G., Papatheodorou, C.: Discovering the topical evolution of the digital library evaluation community. In: Garoufallou, E., et al. (eds.) MTSR 2015. CCIS, vol. 544, pp. 101–112. Springer, Heidelberg (2015). doi:10.1007/978-3-319-24129-6_9

16. Afiontzi, E., Kazadeis, G., Papachristopoulos, L., Sfakakis, M., Tsakonas, G., Papatheodorou, C.: Charting the digital library evaluation domain with a semantically enhanced mining methodology. In: 13th Annual International ACM/IEEE Joint Conference on Digital Libraries - JCDL 2013, pp. 125–134. ACM, New York (2013)

17. Fox, C.: A stop list for general text. SIGIR Forum **24**, 19–21 (1989)

18. Mimno, D.: jsLDA: an implementation of latent Dirichlet allocation in javascript. https://github.com/mimno/jsLDA

19. Church, K.W., Hanks, P.: Word association norms, mutual information, and lexicography. Comput. Linguist. **16**, 22–29 (1990)

20. Freeman, L.C.: Centrality in social networks conceptual clarification. Soc. Netw. **1**(3), 215–239 (1978)

21. Mislove, A., Marcon, M., Gummadi, K.P., Druschel, P., Bhattacharjee, B.: Measurement and analysis of online social networks. In: Proceedings of the 7th ACM SIGCOMM Conference on Internet Measurement - IMC 2007, pp. 29–42. ACM (2007)

Dissecting a Scholar Popularity Ranking into Different Knowledge Areas

Gabriel Pacheco, Pablo Figueira$^{(\boxtimes)}$, Jussara M. Almeida,
and Marcos A. Gonçalves

Computer Science Department, Universidade Federal de Minas Gerais,
Belo Horizonte, Brazil
{gabriel.pacheco,pabfigueira,jussara,mgoncalv}@dcc.ufmg.br

Abstract. In this paper, we analyze a ranking of the most "popular" scholars working in Brazilian institutions. The ranking was built by first sorting scholars according to their h-index (based on Google scholar) and then by their total citation count. In our study, we correlate the positions of these top scholars with various academic features such as number of publications, years after doctorate, number of supervised students, as well as other popularity metrics. Moreover, we separate scholars by knowledge area so as to assess how each area is represented in the ranking as well as the importance of the academic features on ranking position across different areas. Our analyses help to dissect the ranking into each area, uncovering similarities and differences as to the relative importance of each feature to scholar popularity as well as the correlations between popularity metrics across knowledge areas.

Keywords: Citation analysis · Scholar popularity · Academic features

1 Introduction

Taking successful cases as reference is an effective way of having a successful career in several areas, including research. Knowing the characteristics of successful scholars, such as the venues (journals or conferences) in which they publish or their supervising patterns, may bring valuable insights for young scholars, helping them in decision making and career management. Indeed, insights into how to maximize the impact of the research is valuable even for experienced scholars.

However, the most influential characteristics in the success and visibility of a scholar may differ across different knowledge areas, since each area has its particular idiosyncrasies. Thus, in order to better understand such characteristics, one should be careful and perform a separate analysis in a per area basis. Indeed, we have recently analyzed the relative importance of several academic features (e.g., numbers of conference and journal publications, number of supervised students) on scholar popularity (estimated by total citation count), considering scholars working on different knowledge areas in two major Brazilian institutions [9].

In this paper, we take a step further, building on our prior work [9]. We perform a characterization of a set of the most popular scholars in Brazil, classifying

© Springer International Publishing Switzerland 2016
N. Fuhr et al. (Eds.): TPDL 2016, LNCS 9819, pp. 253–265, 2016.
DOI: 10.1007/978-3-319-43997-6_20

and separating them by knowledge area, and identifying, within these subgroups, which academic features are most impacting throughout their careers. Compared to [7], we focus on a *ranking* of the most popular scholars of the country, analyzing how different areas are spread over it. We also cover a 6 times larger set of scholars and multiple popularity metrics (not only citation count, as in [9]).

Our research explores a ranking produced by a study sponsored by the ACUMEN project[1] and published in the "Webometrics Ranking of World Universities"[2]. The ranking is composed only by scholars working on Brazilian institutions and contains the top 6,003 most popular scholars, from an universe of 33,000 scholars, collected from Google Scholar in the 4th week of April, 2015. The ranking is organized in a descending order according to the scholar's h-index, with the total number of citations received by the scholar being used as tie breaker. We aim at characterizing the scholar ranking in terms of how each knowledge area is distributed and how different academic features are correlated with the scholar's ranking position, for scholars in different areas.

Our study is important to both young and experienced scholars, since it provides insights into effective strategies to maximize their work's impact. It may also help institutions formulate policies to improve the productivity and impact of the scholars' research. Understanding the influence of different academic features on scholar popularity can also benefit the design of more effective popularity prediction models (e.g., [2,5]), which in turn can help improving digital library services such as search and expert or collaboration recommendation.

Our main contributions are thus twofold. First, our results complement prior work by showing some significant differences in terms of how scholars of different knowledge areas are spread over the ranking and how their popularity correlates with different academic features. We find that the top ranking positions are dominated by two areas – Physics and Health – with most scholars from other areas, notably Humanities, occupying the lower ranking positions. Second, some surprising results are revealed such as the strong positive correlation between h-index and i-10 index (i.e., number of papers of a scholar with ten or more citations), even stronger than the correlation with the total number of publications. This suggests that, rather than the total number of publications, the number of publications with a minimum of citations (10) is more important to achieve higher popularity. Another important contribution of our work is the methodology used to explore the scholar ranking, which is independent of data source, encouraging future comparisons with scholar rankings of other countries.

The remainder of this paper is organized as follows. Section 2 discusses related work. Our dataset is described in Sect. 3. Section 4 presents our analyses and main results. Conclusions and future work are offered in Sect. 5.

2   Related Work

Several prior studies aim at assessing the popularity or productivity of scholars (e.g., [4,7,11]), but most of them focus on a single knowledge area. Works

---

[1] http://research-acumen.eu/.

[2] http://www.webometrics.info/en/node/102.

covering multiple areas are very rare and focus mostly on the popularity of publication venues [4,11], not scholars. On the other hand, there is also an increasing interest in ranking scholars and academic institutions, possibly because institutions tend to use these rankings as promotion tools or a certificate of excellence which may bring advantages in terms of human and economic resources [3]. Given this interest, several rankings are released yearly using distinct factors and criteria. Such rankings are also an important tool for bibliometric studies, including characterizations of institutions/scholars in the top rank positions, comparisons of rankings, and investigations of metrics used to produce them [6,8].

Our work does not aim at pointing out the "best" metrics for ranking generation but rather evaluating the importance of several academic features to popularity across knowledge areas, considering the most "popular" scholars in a country (in our case, Brazil). In [1], the authors correlate scholar productivity with number of citations, finding a positive correlation between these two factors. In here, we look at a larger set of features, correlating them with the positions in the popularity ranking, for various popularity metrics.

In this sense, our current study extends a prior work of our own [10] aimed at assessing the importance of several academic features on scholar popularity, but limited to a single area (Computer Science) as well as a more recent work [9] that performs such investigation across different knowledge areas. Both studies focus on a single popularity metric, namely total citation count. In contrast, our present study analyzes how different knowledge areas are represented in different portions of a *national scholar popularity ranking*, and study how different *academic features* correlate to different metrics of popularity.

3   Datasets

Our study is focused on a scholar ranking published by the Webometrics project[3] in the 4th week of April, 2015. The original purpose of such rankings is to promote the academic research on the Web, supporting Open Access initiatives, aiming at enhancing scientific and cultural knowledge transfer to society. Towards that goal, the publication of rankings is a powerful tool to suggest and/or consolidate changes in academic procedures and behaviors, improving "best practices".

The choice of Webometrics was supported by its high coverage (larger than other rankings, according to information in the project's website)[4]. Moreover, aligned with our research goals, it is one of the few rankings focused on individual scholars in a nationwide level. In each country ranking, scholars are sorted first according to their h-index, and then by their total citation count (tie breaker), both extracted from Google Scholar at the date of data gathering. Google Scholar[5] is a free very large bibliometric repository that provides the number and list of citations received by each author. It reportedly covers over 160 million unique documents [12], and provides different citation indices for

---

[3] http://www.webometrics.info/en/node/116.

[4] http://www.webometrics.info/en/Objetives.

[5] http://scholar.google.com.

both scholars and individual publications. In particular, for each scholar, Google Scholar offers the citation count, h-index and i10-index[6].

We here choose to analyze the Webometrics ranking of Brazilian scholars, which includes a total of 6,003 scholars of various knowledge areas. However, we note that the methodology we adopt to analyze the ranking is independent of data source, and may be replicated on other rankings. As such, this methodology by itself is a contribution of our work. We also note that the Brazilian ranking is representative, in terms of number of scholars and range of scholar popularity, of other rankings produced by Webometrics. For example, the h-index of the 6,003 scholars in the Brazilian ranking varies between 12 and 106. The Chilean ranking, with 1,410 scholars, has a range of h-index of 5 to 98, while the h-index of the Turkish ranking, with 4,999 scholars, ranges from 6 to 109.

With the goal of characterizing the set of scholars in the ranking, we collected the following set of variables associated with each scholar: total citation count, h-index, i10-index, number of publications in journals, number of publications in conferences, number of supervised students and years after receiving the doctorate degree. The first three variables represent different metrics of scholar popularity. The last four variables capture different features associated with the academic career of the scholar. These variables were gathered from two main sources: Google Scholar and the Lattes platform[7].

The main source of information for Webometrics is the Google Scholar platform. Accordingly, we also used the same platform to collect the total citation count, h-index and i10-index of each scholar[8]. Surprisingly, we were not able to find 32 of the 6,003 scholars of the ranking in Google Scholar, even after manually searching for variations of their names (e.g., only initial of the first name, etc.). For each of the 5,971 scholars found, we also collected the scholar's main research area, provided by the scholar himself in his Google Scholar profile.

The academic features – numbers of publications in journals and in conferences, number of supervised students and years after doctorate – of each scholar were collected from Lattes[9], a platform developed by the Brazilian National Council for Scientific and Technological Development (CNPq) to collect and publish curriculum vitae (CV) information of all researchers in Brazil. The researchers must frequently update their CVs on Lattes in order to apply for funding and other types of promotions. The Lattes platform is considered a powerful and reliable tool, since its is fed by the researchers themselves, thus containing trustful and updated information. We were able to collect the features of all 5,971 scholars. We note that, in the present study, the numbers of publications in journals and conferences encompass only full papers (and not

---

[6] An index created by Google Scholar to account for the number of publication of a scholar with at least 10 citations.

[7] http://lattes.cnpq.br.

[8] Although the platform provides both total and partial (for the past 5 years) indices, we here focus on the indices covering the entire career of each scholar.

[9] Though the number of journal and conference publications could be extracted from Google Scholar by parsing and counting them in each category, we chose to gather them from Lattes, which already provides such statistics at a much lower cost.

extended abstract). Books and book chapters are not included either. Moreover, the number of students consists of all PhD and Master supervisions completed by the scholar until the date of data gathering.

After collecting the popularity metrics and academic features from Google Scholar and Lattes, we grouped scholars into seven major knowledge areas: Biological Sciences (including, for example, Pharmacy, Cell Biology), Natural Sciences (e.g., Chemistry, Agronomy), Social Sciences, Physics, Humanities (e.g., History, Literature), Health (e.g., Odontology, Medicine), and Technology (e.g., Engineering and Computer Science). Scholar grouping was performed manually, using the main research area collected from the Google Scholar profiles, and clustering related areas together into the aforementioned seven knowledge areas[10].

Table 1 shows the distribution of scholars in the ranking across the knowledge areas (2nd and 3rd columns). As we can notice, there is a large dominance of scholars from Health (around 33 % of all scholars), followed by Natural Sciences (23 %) and, basically tied, Biological Sciences (13 %) and Technology (12 %).

To make a comparison with the general population of Brazilian scholars, we also collected the total amount of Brazilian researchers with doctorate degrees from Lattes, grouping them into the same knowledge areas. We were able to collect a total of 270,593 scholars. Given the large number of scholars, the clustering into areas was done automatically by applying filters to the search service available on Lattes[11]. The distribution of these scholars across the seven knowledge

Table 1. Distributions of scholars across knowledge areas in the ranking (sample) and in the Lattes platform (population).

Knowledge area	# Scholars in ranking (sample)	% Scholars in ranking (sample)	# Scholars in Lattes (population)	% Scholars in Lattes (population)	% population covered by ranking
Biological sciences	756	13 %	39, 706	15 %	2 %
Natural sciences	1, 399	23 %	27, 859	10 %	5 %
Social sciences	507	8 %	50, 906	19 %	1 %
Physics	511	9 %	11, 197	4 %	4.5 %
Health	1, 984	33 %	47, 389	17 %	4 %
Humanities	120	2 %	51, 422	19 %	0.2 %
Technology	694	12 %	42, 114	16 %	1.6 %
Total	5, 971	100 %	270, 593	100 %	2.2 %

---

[10] This clustering procedure was performed by two people independently, and a third person acted as tie-breaker in case the same area was categorized differently.

[11] Some discrepancies may exist between the manual categorization of the ranking and the automatic categorization available in Lattes. Yet, if present, they should not affect our analyses, since we use the scholar population extracted from Lattes simply as a first-cut estimate of the representativeness of each area in the ranking.

areas are also shown in Table 1 (4th and 5th columns). Note that, although the population of scholars in Humanities is relatively large (19 % of the total population of scholars), this is the area with the smallest number of scholars in the top 6,000, according to the analyzed ranking.

On the other hand, 5 %, 4.5 % and 4 % of the total population of scholars in Natural Sciences, Physics and Health, respectively, are represented in the ranking, as shown in the rightmost column of Table 1. These proportions are much larger than those observed for the other areas. For example, Biological Sciences, the most represented area out of the other four, has only half of this fraction of scholars in the ranking. We note the particular case of Physics: despite being the most underrepresented area in the global population (with only 4 % of all scholars in Lattes), it is the second largest area in the ranking in terms of representativeness: 4.5 % of the Brazilian physicists (in Lattes) are among the top 6,000 most popular scholars of the country. This may be due to the tradition and quality of the Physics programs in Brazil. In fact, in a recent evaluation (2013) of all Graduate programs in Brazil, conducted by the Ministry of Education, about 25 % of the Physics programs received the highest grades (6 or 7), in a 3–7 scale, meaning a quality comparable to the best Physics programs in the world.

4   Analyses

In this section we analyze the Brazilian scholar ranking by focusing on two main dimensions. First, we analyze how researchers of the different knowledge areas are spread over the top 6,000 ranking[12]. Next, we assess how different academic features associated with the scholars may influence their positions in the ranking.

We start our discussion by analyzing how each knowledge area is represented in different portions of the ranking. Table 2 shows, for each area, the number of scholars present in different portions of the ranking. We first note a large representation of Physics scholars in the top 10 positions (8 out 10). One fact that

Table 2. Numbers of scholars present in specific portions of the ranking.

Knowledge area	Top 10	Top 100	Top 1,000	Top 6,000
Biological sciences	-	6	102	756
Natural sciences	-	19	182	1,399
Social sciences	-	2	62	507
Physics	8	29	100	511
Health	1	42	459	1,984
Humanities	-	-	10	120
Technology	1	1	76	694

---

[12] We note that the ranking indeed contains only 5,971 scholars (as discussed in Sect. 3). Yet, for simplicity, we refer to it as top 6,000.

might contribute to such dominance is that most of those 8 physicists work on Particle Physics and are linked to the experiments of the Large Hadron Collider, which led to the discovery of a new particle with the characteristics of the Higgs boson. This theory is capable of explaining how all particles and forces in nature behave, and thus is expected to attract a lot of attention over the world.

Moving forward in the ranking, we find that scholars from Health start dominating it: almost half of the top 100 and top 1,000 scholars are from that area. Indeed, around 20 % and 23 % of all scholars in Physics and Health in the ranking are among the top 1,000 scholars. In contrast, only 8 % of the scholars in Humanities present in the ranking occupy positions in the same range, but none are among the top 100. The fractions of scholars of the other areas in the top 1,000 fall between 11 % and 13 %. In the particular case of Technology, we find that only one scholar is among the top 10 and top 100 most popular ones.

Delving further into this analysis, Fig. 1 shows how each knowledge area is represented (percent-wise) in two regions of the ranking defined by the top n positions for n varying between 10 and 100 (Fig. 1(a)) and between 1,000 and 6,000 (Fig. 1(b). Each bar contains the fractions of scholars from each area who are among the top n of the ranking (thus adding up to 100 %).

Note the decrease in the representativeness of Physics as n increases between 10 and 100. Such decrease, which is expected given the comparatively small number of physicists among the top 6000, attests the dominance of Physics in the top positions. Also note how the representativeness of Health increases with n in the same region of the ranking. Indeed, the fraction of Health scholars among the top n surpasses 35 % reaching 42 % for $30 \leq n \leq 100$. For $n \geq 1000$, the fractions of both Health and Physics scholars tend to decrease with n, leaving room to scholars from Natural and Biological Sciences, Technology and Humanities.

Table 1 and Fig. 1 attest the dominance of Physics and Health as the areas with the most popular scholars, according to the analyzed ranking. *How is such popularity related to different academic features, such as number of publications in different venues, number of supervised students, and experience (estimated by*

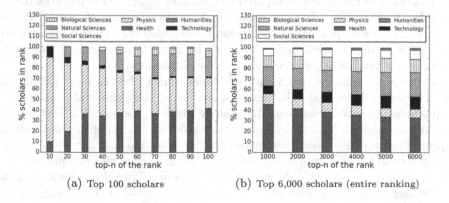

(a) Top 100 scholars   (b) Top 6,000 scholars (entire ranking)

Fig. 1. Distribution of knowledge areas across different regions of ranking.

Table 3. Averages of popularity and academic features for scholars in each area.

Knowledge area	Popularity metric			Academic feature			
	Citation count	i10-index	h-index	# Conference papers	# Journal articles	# Student supervisions	Years after doctorate
Biological sciences	1,684.9	38.3	19.9	12.8	72.9	21.6	18.7
Natural sciences	1,635.6	37.6	19.4	30.1	65.5	23.9	19.5
Social sciences	1,887.4	32.6	18.7	39.4	46.8	32.6	21.5
Physics	2,953.9	48.9	22.6	21.0	83.0	14.1	21.0
Health	2,173.6	46.7	21.8	9.3	88.6	21.8	18.8
Humanities	1,956.7	27.9	17.6	21.1	50.8	34.8	23.7
Technology	1,759.2	36.1	18.7	89.1	43.0	27.3	19.3

the number of years after doctorate degree)? How do different rankings, produced according to different popularity metrics, correlate with each other?

To answer these questions, we calculate the average of each academic feature and popularity metric for scholars in each area. Results are in Table 3. Similarly, Fig. 2 shows the cumulative distribution functions (CDFs) for some popularity metrics and academic features. A CDF shows the fraction of scholars (y-axis) for which the given variable is equal to or smaller than the value in the x-axis.

We note that, both on average and for the complete distribution, physicists tend to be the most popular scholars in the ranking. This holds in terms of not only h-index but also total citation count and i10-index. According to Fig. 2, 10 % of the physicists in the ranking (i.e., 51) have total citation count and h-index exceeding 5,367 and 34, respectively. Similarly, Health is the second most popular area, in all three popularity metrics. Both areas also contain scholars with the largest number of journal publications. Interestingly, despite being very popular, Health scholars tend to publish fewer conference papers, while physicists tend to supervise fewer students, compared to the other areas. These patterns reflect possible idiosyncrasies of those areas which relate to the relative importance of these features to popularity (see below). Note also that Health scholars are among the youngest ones, while Physicists have an "academic age" similar to scholars in other less popular areas (e.g., Social Sciences).

In the other extreme, Humanities scholars tend to be the least popular in terms of both h-index and i10-index, despite having more citations (on average) than scholars in other areas (e.g., Technology). Indeed, compared to Technology, Humanities scholars tend to supervise more students and publish more journal papers. They are also more experienced (on average). On the other hand, Technology scholars tend to publish a very large number of conference papers: this number is more than twice as large as in any other area, and also twice as large as the number of journal articles published by the same scholars (on average). Consistently with results in [9], these observations reflect the importance given by Technology scholars to conference papers (as opposed to journals). We note also

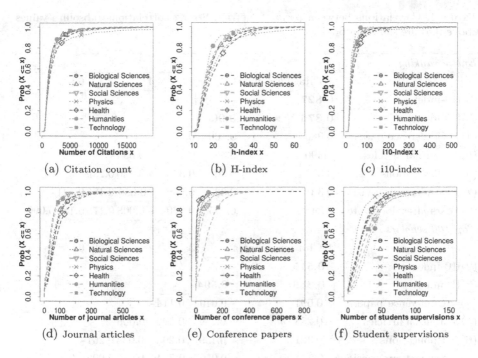

Fig. 2. CDFs of popularity and academic factors for scholars in each area.

that in terms of overall number of publications, scholars in Technology have the largest average numbers: 30 % larger than the average for physicists, for example. Yet, they fall way behind them in terms of all three popularity metrics.

A final observation about Fig. 2 is that scholars in Natural Sciences tend to publish fewer papers than scholars in any other area. Moreover, they also tend to have smaller citation counts. However, their i10-index and h-index are, at least on average, above than those in other areas, including Technology.

These suggest important cross-area differences as well as along different ranking regions. We go further to analyze such differences by assessing the relative importance of different academic features to scholar popularity, both globally and for specific areas. We do so by building different scholar rankings, one for each academic feature and popularity metric. In addition to the numbers of conference and journal publications, we also consider a scholar ranking based on the total number of publications (sum of the two). In each ranking, scholars are sorted in descending order of the considered variable, using their position in the original (Webometrics) ranking as tie breaker. We then compute the Spearman's rank correlation coefficient[13] between ranking pairs to assess how the two considered variables are related.

Table 4 shows the correlations computed for the entire ranking (top) and for the top 100 scholars (bottom). We assign an ID to each metric (e.g., 1 for citation

---

[13] A non-parametric measure of statistical dependence between two ranked variables.

Table 4. Correlations between pairs of metrics. Strong correlations (absolute values above 0.5) are shown in bold.

Entire ranking

	1(Citation count)	2	3	4	5	6	7
(2)i10-index	**0.827**	-					
(3)H-index	**0.876**	**0.906**	-				
(4)# Conference papers	0.086	0.100	0.121	-			
(5)# Journal articles	0.390	0.401	0.413	−0.034	-		
(6)# Publications	0.352	0.361	0.375	−0.020	0.229	-	
(7)# Student supervisions	0.331	0.330	0.356	0.039	0.186	0.141	-
(8)Years after doctorate	0.340	0.335	0.370	−0.008	0.173	0.133	0.090

Top 100 scholars

	1(Citation count)	2	3	4	5	6	7
(2)i10-index	**0.515**	-					
(3)H-index	**0.886**	**0.594**	-				
(4)# Conference papers	0.091	−0.019	0.114	-			
(5)# Journal articles	0.288	0.259	0.322	−0.006	-		
(6)# Publications	0.321	0.139	0.294	0.033	0.261	-	
(7)# Student supervisions	−0.234	−0.056	−0.254	0.010	0.002	0.016	-
(8)Years after doctorate	−0.002	−0.029	−0.013	−0.047	0.117	0.051	−0.054

count, 2 for i10-index), and use them to identify the metrics in the columns and rows of the table. For example the correlation between citation count (1) and years after doctorate (8) computed for the entire ranking is 0.340 (9th line and 2nd column). Correlations with absolute values above 0.5 are shown in bold.

Focusing on the entire ranking, we note very high correlations among all three popularity metrics. While the strong correlation between h-index (3) and citation count (1) is somewhat expected, i.e., scholars with more citations tend to have higher h-index values, the very strong correlation (0.906) between i10-index (2) and h-index is surprising. Such correlation is even stronger than the correlation of any popularity metric with the number of publications (6), a feature previously associated with large citation counts [9,10]. Although we cannot make causality claims, these correlations suggest that, while having a large number of publications, which may increase the scholar's visibility, may help popularity, having many popular publications (in the case, 10 or more citations) seems more important to reach high h-index and large citation counts.

Another interesting observation is the low-to-moderate correlation between popularity (any metric) and years after doctorate (8). This is one of the major h-index criticisms: older scholars have more time to accumulate more citations in a few highly influential papers when compared to younger scholars. Our analysis shows that, at least in the Brazilian ranking, there is not a strong correlation between these two factors. Indeed, for the top 100 scholars, the correlations are nearly 0 (bottom of table). We also note stronger correlations between number of journal articles and popularity (any metric) compared to the correlations with number of conference papers or even total number of publications. Thus, overall, journal articles seem more important to popularity than conference papers. Yet, once again such correlations decrease for the top 100 scholars.

In general, the correlations between any academic feature and popularity metric analyzed are at most moderate over the entire ranking, and seem to decrease among the top 100 scholars. Indeed, for the top 100 scholars, the number of student supervisions, for example, has low and negative correlation with both citation count and h-index, suggesting that having more students over their careers does not translate into higher popularity (at least for those top scholars).

These low-to-moderate correlations observed between the academic features and popularity metrics suggest that: (1) variations across knowledge areas may affect the global result, and/or (2) other factors may be more important to popularity, especially for top scholars in a country. Towards tackling (1), we analyze the correlations between the rankings for each area. Each such ranking was built by restricting to scholars of a given area and sorting them according to the original (entire) rankings. Table 5 shows the correlations for Physics, Technology and Humanities. Other results are omitted due to space constraints.

The same strong correlations between all pairs of popularity metrics, in particular i10-index and h-index, are observed for all areas. Indeed, the latter is above 0.87 in all areas. Focusing on Physics, the number of journal articles is clearly the most important academic feature (out of those analyzed) to popularity (in any metric): the correlations vary between 0.43 and 0.46. In contrast, the number of conference papers is the least important one. These results reflect the importance given by the area on journal publications (see Table 3). The numbers of students and years after doctorate have only low influence on popularity, especially in terms of citation count (corroborating findings in [9]).

Compared to physicists, Health scholars exhibit similar patterns (omitted), with the following differences. The correlations between the popularity metrics are even stronger. Number of students and years after doctorate seem to play stronger roles on popularity (correlations above 0.4). According to Table 3 and Fig. 2, these scholars tend to supervise more students than physicists, though, on average, they have fewer years of experience. Moreover, despite publishing fewer papers in conferences this type of publication seems to be more important to the popularity (especially h-index) of Health scholars (correlation of 0.31).

Turning to Technology, Table 5 reveals the slightly stronger importance of conference papers to popularity, compared to physicists. Yet also in Technology, the number of journal articles is still the most important feature, for any popularity metric considered. This observation, consistent with [9], comes despite the much greater focus given by Technology scholars on publishing conference papers (twice as much). Scholars in Natural Science have, in general, similar patterns (omitted) to those in Technology, except for a somewhat greater importance of both numbers of students and years after doctorate (correlations falling in the

Table 5. Correlations between pairs of metrics for specific knowledge areas.

Physics

	1(Citation count)	2	3	4	5	6	7
(2)i10-index	**0.799**	-					
(3)H-index	**0.856**	**0.901**	-				
(4)# Conference papers	0.169	0.253	0.194	-			
(5)# Journal articles	0.433	0.438	0.460	0.046	-		
(6)# Publications	0.361	0.404	0.414	0.005	0.272	-	
(7)# Student supervisions	0.284	0.312	0.286	0.067	0.173	0.154	-
(8)Years after doctorate	0.268	0.307	0.328	0.103	0.198	0.154	0.163

Technology

	1(Citation count)	2	3	4	5	6	7
(2)i10-index	**0.805**	-					
(3)H-index	**0.855**	**0.882**	-				
(4)# Conference papers	0.205	0.224	0.211	-			
(5)# Journal articles	0.339	0.401	0.343	0.141	-		
(6)# Publications	0.320	0.293	0.353	0.151	0.151	-	
(7)# Student supervisions	0.310	0.278	0.308	0.142	0.136	0.118	-
(8)Years after doctorate	0.283	0.326	0.320	0.100	0.085	0.056	0.070

Humanities

	1(Citation count)	2	3	4	5	6	7
(2)i10-index	**0.727**	-					
(3)H-index	**0.799**	**0.899**	-				
(4)# Conference papers	0.152	0.171	0.183	-			
(5)# Journal articles	0.114	0.167	0.181	0.103	-		
(6)# Publications	0.136	0.170	0.182	−0.056	−0.016	-	
(7)# Student supervisions	0.260	0.242	0.352	0.061	0.019	0.072	-
(8)Years after doctorate	0.379	0.397	0.422	−0.027	0.096	0.081	0.270

ranges of 0.356–0.393 and 0.336–0.371, respectively) to popularity, as well as a lower importance of conference papers (correlations between 0.154 and 0.188).

Finally, as shown in Table 5, both journal and conference publications are very weakly correlated with popularity (in any metric) for scholars in Humanities, while "years after doctorate" appears as the most important factor, out of those analyzed, especially for h-index. These results suggest that, particularly for this area, follow-up studies should consider other academic factors with potentially higher importance (e.g., number of books, short papers, etc.).

In sum, we indeed found some interesting idiosyncratic aspects related to the different areas, though there are also surprising similarities. We also acknowledge that we need more investigation to bring to light aspects responsible for these scholars' success, not revealed in this study (e.g., for scholars in Humanities).

5 Conclusions and Future Work

We analyzed the importance of different academic features on the ranking of the most popular Brazilian scholars across different knowledge areas. Some of our most interesting findings include: (i) the dominance of Physicists and Health scholars in the top ranking positions; (ii) the low representativeness of Humanities in the entire ranking, dominated by the two aforementioned areas and Natural Sciences; (iii) the higher importance of journal publications on popularity (citation count, h-index and i10-index) in all areas, including Technology, in which conference publications are dominant; (iv) the strong correlation between the h- and i-10 indices, an unexpected result. In the future, we plan to expand our study to include other academic features (e.g., other types of publications) and countries, contrasting our findings in a cross-area and cross-country manner, as well as analyse aspects related to author name ambiguity in the ranking.

Acknowledgments. This work was partially funded by projects InWeb (grant MCT/CNPq 573871/2008- 6) and MASWeb (grant FAPEMIG/PRONEX APQ-01400-14), and by the authors' individual grants from CNPq, CAPES and FAPEMIG.

References

1. Abramo, G., Cicero, T., D'Angelo, C.A.: Are the authors of highly cited articles also the most productive ones? J. Informetr. **8**(1), 89–97 (2014)
2. Acuna, D.E., Allesina, S., Kording, K.P.: Future impact: predicting scientific success. Nature **489**(7415), 201–202 (2012)
3. Aguillo, I.F., Bar-Ilan, J., Levene, M., Priego, J.L.O.: Comparing university rankings. Scientometrics **85**(1), 243–256 (2010)
4. Bollen, J., Rodriquez, M.A., de Sompel, H.V.: Journal status. Scientometrics **69**(2), 669–687 (2006)
5. Chakraborty, T., Kumar, S., Goyal, P., Ganguly, N., Mukherjee, A.: Towards a stratified learning approach to predict future citation counts. In: JCDL (2014)
6. Cronin, B., Meho, L.I.: Using the h-index to rank influential information scientists. JASIST **57**(9), 1275–1278 (2006)
7. Ding, Y., Cronin, B.: Popular and/or prestigious? measures of scholarly esteem. IP&M **47**(1), 80–96 (2011)
8. Dorogovtsev, S., Mendes, J.: Ranking scientists. CoRR, abs/1511.01545 (2015)
9. Figueira, P., Pacheco, G., Almeida, J.M., Gonçalves, M.A.: On the impact ofacademic factors on scholar popularity: a cross-area study. In: Kapidakis, S., Mazurek, C., Werla, M. (eds.) TPDL 2015. LNCS, vol. 9316, pp. 139–152. Springer, Heidelberg (2015)
10. Gonçalves, G., Figueiredo, F., Almeida, J.M., Gonçalves, M.A.: Characterizingscholar popularity: a case study in the computer science research community. In: JCDL (2014)
11. Leydesdorff, L.: How are new citation-based journal indicators adding to the bibliometric toolbox? JASIST **60**(7), 1327–1336 (2009)
12. Orduña-Malea, E., Ayllon, J.M., Martín-Martín, A., López-Cózar, E.D.: Methods for estimating the size of google scholar. Scientometrics **104**(3), 931–949 (2015)

Exploring Comparative Evaluation of Semantic Enrichment Tools for Cultural Heritage Metadata

Hugo Manguinhas[1], Nuno Freire[1], Antoine Isaac[1,6(✉)],
Juliane Stiller[2], Valentine Charles[1], Aitor Soroa[3], Rainer Simon[4],
and Vladimir Alexiev[5]

[1] Europeana Foundation, The Hague, The Netherlands
{hugo.manguinhas,antoine.isaac,
valentine.charles}@europeana.eu,
nuno.freire@theeuropeanlibrary.org
[2] Humboldt-Universität zu Berlin, Berlin, Germany
juliane.stiller@ibi.hu-berlin.de
[3] University of the Basque Country, Bizkaia, Spain
a.soroa@ehu.eus
[4] Austrian Institute of Technology, Vienna, Austria
rainer.simon@ait.ac.at
[5] Ontotext Corp, Sofia, Bulgaria
vladimir.alexiev@ontotext.com
[6] Vrije Universiteit Amsterdam, Amsterdam, The Netherlands

Abstract. Semantic enrichment of metadata is an important and difficult problem for digital heritage efforts such as Europeana. This paper gives motivations and presents the work of a recently completed Task Force that addressed the topic of evaluation of semantic enrichment. We especially report on the design and the results of a comparative evaluation experiment, where we have assessed the enrichments of seven tools (or configurations thereof) on a sample benchmark dataset from Europeana.

Keywords: Semantic enrichment · Metadata · Cultural heritage · Evaluation · Europeana

1 Introduction

Improving metadata quality is crucial for Europeana, the platform for accessing digitized Cultural Heritage (CH) in Europe, and many projects that have similar goals in more specific areas. A key technique is semantic enrichment, which puts objects in context by linking them to relevant entities (people, places, object types, etc.). In the CH domain, semantic enrichment provides richer context to items and allows systems to add information to existing metadata [2, 3]. Systems can indeed later obtain "semantic" descriptions for the related entities when these are published, e.g., using Linked Open Data technology. Semantic enrichment is a useful technique with a variety of applications. For instance, enriching flat documents with instances of

© Springer International Publishing Switzerland 2016
N. Fuhr et al. (Eds.): TPDL 2016, LNCS 9819, pp. 266–278, 2016.
DOI: 10.1007/978-3-319-43997-6_21

structured knowledge can be used in search, where results for named entity queries will include facts about the entities involved [1]. A variety of tools has been recently developed – or adapted from other domains – to enrich objects by exploiting the existing metadata.

Enrichment of CH metadata is however a very difficult task due to: (1) bewildering variety of objects; (2) differing provider practices for providing extra information; (3) data normalization issues; (4) information in a variety of languages, without the appropriate language tags to determine the language [2]. Adding to this difficulty, the importance of these factors may change between one dataset or application and another. Moreover, the various tools and approaches available can of course perform very differently on data with different characteristics. This makes it difficult to identify and apply the right enrichment approach or tool (including using the right parameters when an approach can be tuned). To help practitioners, especially those from the Europeana family of projects, the following R&D questions required specific effort:

- perform concrete evaluations and comparisons of enrichment tools currently developed and used in the Europeana context;
- identify methodologies for evaluating enrichment in CH, specifically in Europeana, by making sure that evaluation methods (i) are realistic wrt. the amount of resources to be employed, (ii) can be applied even when enrichment tools are not trivially comparable (i.e. when they link objects with different target datasets) (iii) facilitate the measure of enrichment progress over time;
- build a reference set of metadata and enrichments that can be used for future comparative evaluations.

To gain insight on these points, a group of experts from the Europeana community has gathered as a Task Force and undertook an evaluation campaign for representative enrichment tools[1]. In this paper we explain the phases of the evaluation, covering methodology, results and analysis. We conclude with a summary of the group's lessons learned and recommendations. We refer the reader interested in more detail to browse the technical report of the Task Force, where advanced explanations on the evaluation method and results can be found [4].

2   Related Work

For information access systems, several initiatives exist for conducted structured evaluation experiments, e.g. CLEF[2]. In the Semantic Web community, the Ontology Alignment Initiative assesses ontology alignment systems[3]. For enriching data, no well-established benchmark exists, although the number of tools and enriched datasets is growing constantly. A notable exception in the Linked Data community is GERBIL [5], a framework for benchmarking systems on various annotation tasks, including

---

[1] http://pro.europeana.eu/taskforce/evaluation-and-enrichments.

[2] http://www.clef-initiative.eu/.

[3] http://oaei.ontologymatching.org/.

named entity recognition and named entity linking. GERBIL enables implementers to compare results of tools on the same datasets, in a principled, reproducible way. Using GERBIL, Usbeck et al. compared more than 15 systems on 20 different datasets. GERBIL can be used with systems and datasets from any domain. However, these datasets do not include multilingual CH metadata. Evaluating enrichment tools against them would not bring the insight needed to answer our research questions[4].

Within the digital CH domain, numerous studies have evaluated automatic semantic enrichments. The DM2E[5] project performed sample evaluation of alignment of (local) places and agents to DBpedia where 150 random agents and 150 random places were selected. The sample was based on the amount of agents/places each collection contains. The results showed that 18 % of the agents and 60 % of places are linked. From these, 83 % of the agent links and 85 % of the place links are good.

The Paths Project[6] developed functionalities for information access in large-scale digital libraries, focusing on metadata enrichment to let users better discover and explore CH material. Evaluation of the Paths prototype focused on assessments with focus groups within laboratory settings. Enrichments were tested indirectly following a methodology of Interactive Information Retrieval in a laboratory setting: users performed tasks and logs, screen recordings and observer notes were collected [6].

In [3], the authors explore the feasibility of linking CH items to Wikipedia articles. They develop a small dataset comprising 400 objects from Europeana and manually link them to Wikipedia whenever there exists an article that exactly describes the same object as the CH object. This dataset is then used to evaluate two systems, which yielded relatively poor performances.

OpenRefine[7] has also been used to perform evaluations. One was an evaluation of structured field reconciliation[8]. The other was an evaluation of named-entity recognition on unstructured fields (performed with OpenRefine and a plugin): Both have been evaluated on concrete datasets with a manual validation.

Stiller et al. in [2] evaluated enrichments in Europeana looking at the intrinsic relationship between enrichments and objects but also taking extrinsic factors such as queries into account. The results for the extrinsic evaluation were subjective as the choice of queries for the evaluation influenced the results. Nevertheless, if a representative query sample is chosen the approach can give insights about the likelihood of a user encountering beneficial enrichments or incorrect ones. A previous evaluation of the enrichment in Europeana qualitatively assessed 200 enrichments for the four different types: time, persons, location and concepts [7].

As enrichments impact the search performance and are often implemented to improve search across several languages, all evaluations targeting the search performance are also relevant as enrichment evaluations. Within the cultural heritage domain,

---

[4] In fact a possible outcome of our work could be to contribute datasets and gold standards to GERBIL so as to make it a more suitable platform for future evaluations in our domain.

[5] http://dm2e.eu/.

[6] http://www.paths-project.eu/.

[7] http://openrefine.org/.

[8] http://freeyourmetadata.org/publications/freeyourmetadata.pdf.

search evaluation was performed [8] as well as retrieval experiments bases on data from cultural heritage portals [9]. Additionally, user-centric studies aimed at improving usability of these services [10].

Although not specific to cultural heritage, the evaluation studies conducted in the area of ontology alignment are relevant, as CH makes extensive use of ontologies in the data, and many enrichment tools enrich data by targeting ontologies. Work in this area has found similar difficulties in evaluation as our case, such as defining a gold standard and reaching a high level of rater agreement [11].

3   Evaluation Setting

The Task Force conducted the evaluation of selected enrichments tools fulfilling the following steps which are more detailed in Fig. 1: select a sample dataset for enrichment, apply different tools to enrich the sample dataset, manually create reference annotated corpus and compare automatic annotations with the manual ones to determine performance of enrichment tools.

A proper evaluation requires a dataset representative of the multilingual and cross-domain diversity of Europeana's metadata. We selected data from The European Library[9] (TEL). TEL is the biggest data aggregator for Europeana; it has the widest coverage of countries and languages. For each of the 19 countries, we randomly selected a maximum of 1000 records. In order to have varied data in the evaluation dataset, we partitioned these larger datasets in 1000 sequential parts and blindly selected one record from each partition. In total the evaluation dataset contains 17,300 metadata records in 10 languages (see more info, incl. the full list of countries and languages, in our extended document and archive [12]). TEL records are expressed in the Europeana Data Model[10]. Figure 2 lists the properties used in the evaluation dataset.

Albeit coming from libraries, our dataset is quite heterogeneous, as TEL includes books, prints, and maps. However, at a later stage of the evaluation, we detected a bias towards scientific materials, which as key TEL resources are more frequently represented than in Europeana. Still, these documents belong to varied domains of science: mathematics, biology, agriculture, etc.

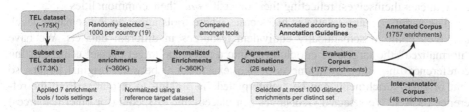

Fig. 1. Evaluation workflow.

[9] http://www.theeuropeanlibrary.org/.

[10] http://pro.europeana.eu/edm-documentation.

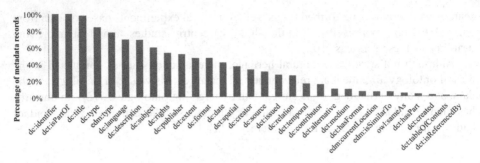

Fig. 2. Frequency of properties found within the evaluation dataset.

4 Enrichment Results Obtained from the Participants

Within our Task Force, the following participants have applied their tools to enrich the evaluation dataset: the Europeana Foundation (EF); TEL; the LoCloud project[11] (with two different tools); the Pelagios project[12] and Ontotext[13] (with two different settings for determining language of metadata). Table 1 lists the tools, methods and target datasets each participant used. Participants sent their enrichment results using an agreed format containing (i) the identifier of the enriched object; (ii) the enriched property (e.g., dcterms:spatial); (iii) the identifier of the target entity (e.g., a DBpedia URI); (iv) the enriched literal (word or expression) where the entity was identified. A total of about 360K enrichments were obtained for the 7 different tools or tool settings. Figures 3 and 4 show respectively the number of enrichments and the coverage of the evaluation dataset's records for each tool. More statistics and tool information can be found in our extended document and archive [12].

5 Creation of the Reference Annotated Corpus

Building a complete "gold standard" of correct enrichments for every object, as done in related work, is not feasible for us: the amount of objects and the variety of tools and targets are just too large. We instead tried to build a reference dataset starting from the enrichments themselves, reflecting their diversity *and* their commonalities.

The variety of target datasets hides cases where tools agree on the semantic level, i.e., they point to semantically equivalent resources in different datasets. We have "normalized" the targets of enrichments into a reference target dataset using existing coreference links (i.e., owl:sameAs or skos:exactMatch) between original targets, so that original enrichments can be "reinterpreted" as linking to resources from the reference dataset. We selected GeoNames (for places) and DBpedia (for other resources)

---

[11] http://www.locloud.eu/.

[12] http://pelagios-project.blogspot.nl/.

[13] http://ontotext.com/.

as reference datasets, as they benefit from the highest overlap across the output of all tools. It was possible to normalize 62.16 % of the results this way.

To build the corpus to be manually annotated, we compared normalized enrichments sets to identify the overlap between tools (i.e., enrichments with the same source object, target resource and enriched property) and sets specific to one tool. This gave 26 different sets that reflect the agreement combinations between tools (for more details see [12]). For each set, we randomly selected at most 100 enrichments (if the set contained less than 100, all enrichments were selected) resulting in a total of 1757 distinct enrichments.

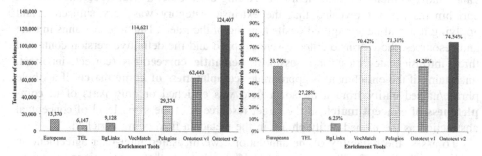

Fig. 3. Number of enrichments by tool.   **Fig. 4.** Dataset records enriched, by tool.

Table 1. Overview of the tools evaluated.

Part.	Tool	Entity types	Target datasets	Methods
EF	Europeana semantic enrichment framework[a]	Places, agents, time spans, concepts	DBPedia[b] (agents, concepts), GeoNames[c] (places), Semium Time (time spans)	Rule based tool, string normalization and matching
TEL	In-house dev.	Places, agents	GeoNames, GemeinsamenNormdatei	NERD, heuristic-based (places); coref. information [13, 14] (agents)
LoCloud[d]	Background link (BgLink), DBpedia Spotlight 0.6[e]	Wide range of entities and concepts	DBpedia	NERD; supervised statistical methods (English)
	VocMatch service and TemaTres[f]	Concepts	Several thesauri and taxonomies[g]	SKOS vocs, automatic term assignment
Pelagios	Recogito[h]	Places	Pleiades, Digital Atlas of the Roman Empire[i], Archaeological Atlas of Antiquity[j]	NERD; user verification and correction
Ontotext	v1: Ontotext Sem. Platform, GATE[k] v2: same v1	Concepts (English only), persons, places	MediaGraph (a custom KB including DBpedia and Wikidata)	NERD, rule based and machine learning

[a]http://pro.europeana.eu/page/europeana-semantic-enrichment
[b]http://wiki.dbpedia.org/
[c]http://www.geonames.org/
[d]http://support.locloud.eu/Metadata%20enrichment%20API%20technical%20documentation
[e]http://spotlight.dbpedia.org/
[f]http://www.vocabularyserver.com/
[g]http://vocabulary.locloud.eu/?p=36
[h]http://pelagios.org/recogito
[i]http://darmc.harvard.edu/icb/icb.do
[j]http://www.vici.org/
[k]https://gate.ac.uk/

This approach helps us to identify when a similar logic is shared across tools (such as using the same rule or same part of the target dataset). However it negatively affects the evaluation of some tools. VocMatch for instance used target datasets that have no co-reference links the reference datasets. Therefore it didn't share enrichment with any other tool, and its results were underrepresented in our corpus.

Next, 16 members of our Task Force manually assessed the correctness of the 1757 enrichments annotating the corpus accordingly. We prepared a first version of annotation guidelines looking at the extrinsic (user-focused "usefulness") and intrinsic (system-focus) value of an enrichment. The extrinsic criteria assessed the informational value and specificity of an enrichment. Testing these criteria with three raters on six enrichments per set revealed that the extrinsic category was very subjective and applying it would have required onsite training of the raters. Due to constraints in time and resources, the extrinsic criteria were dropped and the definitive version contained three intrinsic categories for assessment: **semantic correctness** (correct, incorrect, uncertain), if the enrichment is appropriate; **completeness of name match**, if a whole phrase/named entity from a metadata field was enriched or only parts of it; **completeness of concept match**, if the target resource is at the same level of conceptual abstraction as the named entity/phrase in the metadata field being enriched.

To assess the reliability of the annotations, we measured inter-rater agreement on the "semantically correct" assessment for 46 enrichments that were assigned to all 16 raters – resulting in 736 annotations. We selected the enrichments manually to make sure the low sampling rate would not result in missing interesting cases and losing variety of enrichments. Agreement was measured using the Fleiss Kappa [15], calculated with parameters $N = 46$, $n = 16$, $k = 3$. Inter-rater agreement is 0.329, i.e., "fair agreement" under the typical Kappa value interpretation, although the observed percentage agreement was high (79.9 %). One reason for this is that the ratings were not evenly distributed between the different categories as most of the enrichments were considered to be correct, so the prevalence of correct ratings was very high. We therefore also report on the free-marginal multirater Kappa [16], which is 0.698, an agreement we considered satisfactory.

6   Analysis of Enrichment Results

The results of enrichment tools were compared against the manually annotated corpus, adapting Information Retrieval's common precision (fraction of enrichments that were judged to be correct over all the enrichments found by a tool) and recall (correct enrichments found by a tool against all the correct enrichments that could have been found) measures. We chose to compute our measures for two 'aggregates' of the three correctness criteria above: **relaxed**, where all enrichments annotated as semantically correct are considered "true" regardless of their completeness; and **strict**, considering as "true" the semantically correct enrichments with a full name and concept completeness. Enrichments for which raters were unsure were ignored in the calculations.

The fact that we could not identify all possible enrichments for the evaluation set lead us to apply **pooled recall** [17], in the total amount of correct enrichments is replaced by the union of all correct enrichments identified by all tools. As mentioned in

Sect. 5, however, some tools' results are under-represented in the corpus. This especially impacts the pooled recall. To take this into account in our analysis, we computed the **maximum pooled recall**, i.e. pooled recall assuming that all enrichments from a tool are correct and applying the 'strict' approach as it gives an upper bound for this measure (NB: for precision and pooled recall, "true" depends on the choice of 'strict' or 'relaxed', while for max pooled recall we use only 'strict'):

$$\text{Pooled Recall} = \frac{\{\# \text{ "true" enrichments of a tool}\}}{\{\# \text{ "true" enrichments of all tools}\}} \tag{1}$$

$$\text{Max Pooled Recall} = \frac{\{\# \text{ enrichments of a tool}\}}{/\{\# \text{ "true" enrichments of all tools}\}} \tag{2}$$

In general one must keep in mind the coverage of enrichments (Figs. 2 and 3) when analysing the results of our evaluation. For example, from Ontotext v2's relaxed precision (92.4 %) and its total amount of enrichments (124,407), we can extrapolate that this tool probably produces over 100K correct enrichments, which is a good indicator of its performance in the absence of recall based on a complete gold standard.

Results of the evaluation are presented in Table 2. A first look at these, in particular the strict precision, shows a divide between two groups: EF and TEL (group A), and BgLink, Pelagios, VocMatch and Ontotext (group B). Tools in group A enrich records based only on metadata fields which typically contain (semi-) structured information (e.g., dc:creator) while tools in group B enrich using fields with any sort of textual description (e.g., dc: description). In semi-structured metadata fields, the difficulty of identifying the right named reference is lower since these fields tend to: (a) contain only one named reference, or several entities with clear delimiters (author names within a dc:creator field are often delimited by a semicolon); (b) often obey a normalized format or cataloguing practice (e.g., dates with a standardized representation); (c) contain references to entities whose type is known in advance (e.g., dcterms:spatial should refer to places and not persons).

Group A. The tools from **EF** and **TEL** rank first and second on relaxed and strict precision. Besides the fact that they focused enrichments mainly to semi-structured fields, they benefit from enriching only against a specific selection of the target vocabularies (made prior to enrichment), which reduces the chance of picking incorrect enrichments because of ambiguous labels (cf. Section on techniques and tools in the main Task Force report [4]). EF results drop to second place for strict precision since in case of ambiguity, the tool cannot select the right entity. A typical example is place references that may correspond to different levels of administrative division with the same name. TEL features a disambiguation mechanism to pick the entity most likely to be the one being referred, based on its description. In particular, for places it uses classification (e.g., 'feature type' in GeoNames) or demographic information as indicators for the relevance of an entity. Both tools do not take into account the historical dimensions of the object when selecting a geographical entity. For example, some objects from the 18th century with the named reference "Germania" are enriched with "Federal Republic of Germany" in EF. This can be seen as an avoidable side effect of using GeoNames, which mostly contains contemporary places. Finally, the results confirm previous findings that some incorrect enrichments could be avoided if the language of the metadata was taken into account [2].

Table 2. Precision, pooled recall and F-measure results.

Tools	Annotated enrichments (% of full corpus)	Precision		Max pooled recall	Estimated recall		Estimated F-measure	
		Relax.	Strict		Relax.	Strict	Relax.	Strict
EF	550 (31.3 %)	0.985	0.965	0.458	0.355	0.432	0.522	0.597
TEL	391 (22.3 %)	0.982	0.982	0.325	0.254	0.315	0.404	0.477
BgLinks	427 (24.3 %)	0.888	0.574	0.355	0.249	0.200	0.389	0.296
Pelagios	502 (28.6 %)	0.854	0.820	0.418	0.286	0.340	0.428	0.481
VocMatch	100 (05.7 %)	0.774	0.312	0.083	0.048	0.024	0.091	0.045
Ontotext v1	489 (27.8 %)	0.842	0.505	0.407	0.272	0.202	0.411	0.289
Ontotext v2	682 (38.8 %)	0.924	0.632	0.567	0.418	0.354	0.576	0.454

Group B. Pelagios has the best strict precision in group B, and its relaxed precision is slightly below BgLinks'. The fact that Pelagios is specialized for place name enrichments certainly helped achieving this. Its target vocabulary is smaller and more specialized than the datasets used by other tools, which makes it able to apply place-specific heuristics. This can explain why in terms of deviation between relaxed and strict precision it performs similarly to TEL and EF, which apply rules and target datasets that depend on the type of the entity expected to be found in certain fields. The most common reasons for incorrect or partial enrichments in Pelagios are related to issues with disambiguating between target entities. It does disambiguation, but the Wikidata target vocabulary that it exploits does not yet provide the necessary demographic information that it (as TEL) uses as indicator for the relevance of an entity. For example, "Siberia" is enriched with a place in California[14]. Pelagios also applies fuzzy matching between the named reference and the labels of the target entity, which leads to enrichments across different types of nouns, such as "people" with Peoples[15], a place in the U.S. Additionally, even though Pelagios aims at enriching old place names, it had issues determining whether an entity actually corresponds to the time frame of the description. The disambiguation problems did not significantly impact the overall performance since only a small amount of the enrichments evaluated were referring to text fields (about 20 % of the total number of enrichments, to be compared with an average of 50 % for other tools in group B[16]).

The two **Ontotext** versions perform differently, due to the fact that v1 applied enrichment only to objects with dc:language "en" and uses NLP methods for English as an attempt to increase precision. As a matter of fact nearly 100 % of v1 enrichments

---

[14] http://sws.geonames.org/5395524/.

[15] http://sws.geonames.org/4303909/.

[16] See Appendix A of [12] for the complete distribution of enrichments per property.

were also detected by v2 but v1 discarded about half of the ones detected in v2. Yet performance was reduced overall since dc:language gives the language of the object not that of metadata. A great amount of enrichments were identified for non-named references like verbs (e.g. think), adverbs (e.g. viz.), adjectives (e.g. valid), abbreviations (Mrs), simple nouns (e.g. purpose), etc., which do not really contribute to improving the description of objects and sometimes lead to wrong enrichments. For the remainder of the enrichments, Ontotext shows a good performance.

BgLinks appears just below Ontotext v2 and above Ontotext v1. These tools are the ones that share the biggest number of enrichments, which partly explains the proximity in their performance. A closer look shows that enriching acronyms is a particular challenge for BgLinks. Very few of these were correct. BgLinks performs significantly better in determining the right references within the text to enrich, compared to Ontotext and is also successful at enriching more complex named references. This comes from applying more relaxed approaches to name matching. An aspect that explains in part the difference between the results for relaxed and strict is that many partial enrichments are produced for terms that denote entities without an exact semantic equivalent in the target dataset. This is an issue for all tools, but is particularly found in group B, as references to such entities are more common in long text descriptions than in normalized or structured fields.

VocMatch had the lowest performance. The fact that it was exceptionally difficult for raters to identify the actual portion of the text that served as clue for the enrichment made it hard to assess its correctness. As already hinted, VocMatch's pooled recall is impacted by its use of specialized vocabularies not used by the others, and for which no coreference links were available to reconcile them with other enrichments. A closer look shows that some incorrect enrichments come from matching against all terms available in the target vocabularies, without disambiguation. An example is the word "still" as part of the term "still image". This approach is much more effective when applied to semi-structured fields like dc:subject or dc:type; this is quite visible when comparing VocMatch and EF, which applies the same methods as VocMatch but only to semi-structured fields. Subsequent investigations have shown that using only semi-structured fields, VocMatch reaches 86.7 % relaxed precision.

7 Conclusion

Our experiment is the result of an effort to gather representatives of several projects over a couple of months. While we have stumbled over issues in the evaluation procedure that in retrospect seemed obvious, the exercise has proven to be fruitful for exchanging practitioners' perspectives on the assessment of enrichment tools. Our work is important for users from the CH communities and/or owner of digital library applications using generic frameworks like GERBIL, which offers many options but little domain guidance and may lack test datasets that fit specific cases. As a matter of fact, we find it useful to articulate and share the following recommendations regarding the evaluation process:

1. **Select a dataset for your evaluation that represents the diversity of your (meta)data:** Covering language diversity, spatial dispersion, subjects and domains.
2. **Building a gold standard is ideal but not always possible:** Build a reference set of correct alignments manually if you have sufficient time and human resources, otherwise annotate the enrichments identified by the tool being evaluated or other enrichment tools. The trade-off is that the latter option does not allow one to obtain absolute recall figures
3. **Consider using the semantics of target datasets:** When datasets are connected by coreference links, these may be used in a process that "normalizes" enrichments to get a more precise view on how they compare across tools, or to reuse a gold standard from another evaluation.
4. **Try to keep balance between evaluated tools:** Some of the corpus creation strategies can result in a bias against some tools. Make sure bias is recognized and properly related to your evaluation strategy's motivations.
5. **Give clear guidelines on how to annotate the corpus:** Guidelines should be simple but still complete enough for raters to deliver appropriate judgements. Consider having examples for the cases that may raise the most doubt. Try testing the guidelines with raters early in the process.
6. **Use the right tool for annotating the corpus:** Choose or develop a tool that displays all the necessary information; respects your guidelines and guides raters to efficiently and effectively perform their task.

In addition, we want to share recommendations for enrichment tools:

1. **Consider applying different techniques depending on the values used with the enriched property**; e.g., semi-structured or textual descriptions, or values generally containing references for places, persons or time periods.
2. **Matches on parts of a field's textual content may result in too general or even meaningless enrichments** if they fail to recognize compound expressions. This especially hurts when the target datasets include very general resources that are less relevant for the application needs.
3. **Apply strong disambiguation mechanism** that considers the accuracy of a name reference together with the relevance of the entity in general (looking as its data properties) and in particular, i.e., within the context of reference and use. For example, we observed that most tools would benefit from identifying and comparing the temporal scope of both records and candidate target entities.
4. For most if not all cases in the Europeana context, **concepts so broad as "general period" do not bring any value as enrichment targets**. Additional logic should be added to the enrichment rules so that they are not used to enrich objects.
5. Our evaluation has confirmed that **quality issues originating in the metadata mapping process are a great obstacle to get good enrichments** [2]. Enrichment rules designed to work on specific metadata fields (e.g., spatial coverage of an object) should be applied carefully when these fields can be populated with values that result from wrong mappings (e.g., publication places).

We hope that future evaluations of enrichment tools can benefit from our experience, as they are essential to improve the design of the tools themselves, and help make the applications built on these, such as Europeana, deliver better performance.

References

1. Bunescu, R., Paşca, M.: Using encyclopedic knowledge for named entity disambiguation. In: Proceedings of the 11th Conference of the European Chapter of the Association for Computational Linguistics, pp. 9–16 (2006)
2. Stiller, J., Petras, V., Gäde, M., Isaac, A.: Automatic enrichments with controlled vocabularies in Europeana: challenges and consequences. In: Ioannides, M., Magnenat-Thalmann, N., Fink, E., Žarnić, R., Yen, A.-Y., Quak, E. (eds.) EuroMed 2014. LNCS, vol. 8740, pp. 238–247. Springer, Heidelberg (2014)
3. Agirre, E., Barrena, A., Lopez de Lacalle, O., Soroa A., Fernando S., Stevenson, M.: Matching cultural heritage items to Wikipedia. In: Proceedings of LREC 2012, Istanbul, Turkey (2012)
4. Isaac, A., Manguinhas, H., Stiller, J., Charles, V.: Report on Enrichment and Evaluation. The Hague, Netherlands (2015). http://pro.europeana.eu/taskforce/evaluation-and-enrichments
5. Usbeck, R., Röder, M., Ngonga Ngomo, A., et al.: GERBIL: general entity annotator benchmarking framework. In: Proceedings of 24th WWW Conference. ACM (2015). http://doi.acm.org/10.1145/2736277.2741626
6. Griffiths, J., Basset, S., Goodale, P., et al.: Evaluation of the second PATHS prototype. Technical report, Paths Project (2014)
7. Olensky, M., Stiller, J., Dröge, E.: Poisonous India or the importance of a semantic and multilingual enrichment strategy. In: Dodero, J.M., Palomo-Duarte, M., Karampiperis, P. (eds.) MTSR 2012. CCIS, vol. 343, pp. 252–263. Springer, Heidelberg (2012)
8. Monti, J., Monteleone, M., Buono, M., et al: Cross-lingual information retrieval and semantic interoperability for cultural heritage repositories. In: RANLP 2013. Hissar, Bulgaria (2013). http://aclweb.org/anthology/R/R13/R13-1063.pdf
9. Petras, V., et al.: Cultural heritage in CLEF (CHiC) 2013. In: Forner, P., Müller, H., Paredes, R., Rosso, P., Stein, B. (eds.) CLEF 2013. LNCS, vol. 8138, pp. 192–211. Springer, Heidelberg (2013)
10. Dobreva, M., Chowdhury, S.: A user-centric evaluation of the Europeana digital library. In: Chowdhury, G., Koo, C., Hunter, J. (eds.) ICADL 2010. LNCS, vol. 6102, pp. 148–157. Springer, Heidelberg (2010)
11. Tordai, A., van Ossenburg, J., Schreiber, G., et al.: Let's agree to disagree: on the evaluation of vocabulary alignment. In: Proceedings of 6th K-CAP. ACM (2011)
12. Isaac, A., Manguinhas, H., Charles, V., Stiller, J., et al.: Comparative evaluation of semantic enrichments. Technical report (2015). Report available at http://pro.europeana.eu/taskforce/evaluation-and-enrichments. Data archive available at: https://www.assembla.com/spaces/europeana-r-d/documents?folder=58725383
13. Freire, N., Borbinha, J., Calado, P., Martins, B.: A metadata geoparsing system for place name recognition and resolution in metadata records. In: ACM/IEEE Joint Conference on Digital Libraries (2011). http://dx.doi.org/10.1145/1998076.1998140
14. Charles, V., Freire, N., Antoine, I.: Links, languages and semantics: linked data approaches in the European Library and Europeana. In: Linked Data in Libraries: Let's make it happen! IFLA 2014, Satellite Meeting on Linked Data in Libraries (2014)

15. Fleiss, J.L.: Statistical Methods for Rates and Proportions, 2nd edn. Wiley Series in Probability and Mathematical Statistics, pp. 212–236. John Wiley & Sons Ltd (1981). (Chap. 13)
16. Randolph, J.: Free-marginal multirater Kappa (multirater κ free): an alternative to fleiss' fixed-marginal multirater Kappa. Joensuu University Learning and Instruction Symposium 2005, Joensuu, Finland, 14–15 October 2005. http://eric.ed.gov/?id=ED490661
17. Manning, C., Raghavan, P., Schütze, H.: Introduction to Information Retrieval. Cambridge University Press, Cambridge (2008). http://nlp.stanford.edu/IR-book/pdf/08eval.pdf

Digital Humanities

Data Integration for the Arts and Humanities: A Language Theoretical Concept

Tobias Gradl$^{(\boxtimes)}$ and Andreas Henrich

Media Informatics Group, University of Bamberg, Bamberg, Germany
{tobias.gradl,andreas.henrich}@uni-bamberg.de

Abstract. In the context of the arts and humanities, heterogeneity largely corresponds to the variety of disciplines, their research questions and communities. Resulting from the diversity of the application domain, the analysis of overall requirements and the subsequent derivation of appropriate unifying schemata is prevented by the complexity and size of the domain. The approach presented in this paper is based on the hypothesis that data integration problems in the arts and humanities can be solved on the theoretical foundation of formal languages. In applying a theoretically substantiated framework, integrative solutions on the formal basis of language specifications can be tailored to specific and individual research needs—abstracting from reoccurring technical difficulties and leading the focus of domain experts on semantic aspects.

1 Introduction

The research data landscape of the arts and humanities is characterized by a high degree of distribution, heterogeneity and autonomy, which often originate from the logical and geographical distribution of institutions and their data sources. Without the existence of authorities that focus on the consolidation and coordination of data structures and processing practices, the aspect of autonomy further promotes the development of heterogeneity on various abstraction levels [14].

With respect to traditional application domains of data integration, the term *heterogeneity* is often associated with its connotation as fundamental *data integration problem* [7]. In the particular context of the arts and humanities, however, heterogeneity—particularly on structural and semantic levels—reflects the diversity of the disciplines, their research questions and communities. With particular focus on metadata, heterogeneity typically exists in terms of applied schemata and the context- and discipline-specific knowledge that is required to correctly understand the schemata and contained data. Situated within a particular academic surrounding, the digitization, description and usage of research objects in digital collections is embedded within context-specific semantics, which we understand as *generative context*. In contrast to this original context, data can be interpreted and utilized with respect to research questions that are situated in other disciplines—representing the *interpretive or application context* of data. In this respect, our goal is to facilitate data integration for the diverse domains and

© Springer International Publishing Switzerland 2016
N. Fuhr et al. (Eds.): TPDL 2016, LNCS 9819, pp. 281–293, 2016.
DOI: 10.1007/978-3-319-43997-6_22

research questions of the arts and humanities. Traditional approaches, which often define a global, integrative view [8] lead to a unifying harmonization of data. For the overall case of the arts and humanities, this harmonization would, however, result in the loss of structural elements, which are not considered relevant for most disciplinary contexts in order to compose a view of reasonable complexity. Opposed to such harmonizing efforts, we assume the evolution of so called *semantic clusters* [5]. Such clusters are composed of data sources that are associated through a common application context such as a funded research project, which then explicates the relations between the sources. A cluster is modeled by domain experts in accordance with the specific information needs of researchers and can be reused for similar research or interconnected with other clusters.

In this paper we present an integration concept [4], which satisfies requirements that we consider critical for supporting the evolution of semantic clusters and hence the research-driven definition, transformation and integration of data: A fundamental requirement thereby consists in the *adaptivity* of the concept with respect to the structuredness of relevant data sources and the usage scenarios in the domains. To achieve such adaptivity, a high degree of *expressiveness* of data descriptions and transformations is required—allowing a detailed explication of the generative context and the conversions into application contexts. Despite this expressiveness, the concept *encapsulates technical aspects* of data integration, which can be solved generically—allowing domain experts to focus on the semantic aspects of integration. The remainder of this paper is structured as follows: After an overview of the context in Sect. 2 we will introduce the language theoretical foundation and the derived modeling concept in Sect. 3. In Sect. 4 we present the architecture of the implemented grammatical transformation framework as well as the web-based modeling tool. After presenting exemplary evaluation results, we conclude the paper with Sect. 5.

2   Context

In order to mediate between locally contextualized schemata, data integration involves an extensive requirements analysis to gain insight of the concepts that are contained in the local perspectives and need to be incorporated within a global schema [8]. The appropriateness of such a structure is traditionally expressed in terms of four requirements [2]:

- *Completeness*: The schema includes all concepts that are contained in the local schemata and that are relevant to the considered application domain.
- *Correctness*: Any local concept that is represented by the global schema must be reflected with equivalent semantics.
- *Minimality*: Concepts with identical semantics are represented only once.

– *Understandability*: Labels of elements are chosen in a fashion that prevents ambiguities and enhances overall understandability.

Approaches to data integration often follow the theoretical foundation expressed in [8] by validly employing the concept of a global view to provide integrative services (see e.g. ISIDORE [13], OAIster [6] and Europeana [12] in this context). Resulting from the consideration of the arts and humanities as application domain of data integration, the methodological execution of an extensive requirements analysis and the subsequent derivation of appropriate unifying schemata is prevented by the complexity and size of the domain: If the design of a global schema aims a the requirement of completeness, (1) a large amount of concepts have to be included within the schema that are irrelevant in specific use cases and (2) the understandability and correctness are potentially impacted by the complexity of the schema.

The approach presented in this paper is based on the hypothesis that the data integration problem in the particular context of the arts and humanities can be interpreted and solved individually by experts within the arts and humanities on the theoretical foundation of formal languages. The recognition of provided input, its transformation into an intermediate representation and the generation of desired output are all typical tasks of language applications. In applying a theoretically substantiated framework, integrative solutions on the formal basis of language specifications can be tailored to specific and individual research needs.

Traditional language applications are formed by compilers—i.e. computer programs, which parse source code specified in terms of a computer language and translate instructions to machine executable code. More generally, compilers can be understood as "computer programs that translate a program written in one language into a program written in another language" [3]. Compilation is performed against human readable source code and results in immediately executable binary code or an intermediate representation. In the latter case, a runtime environment is needed to interpret such representations at the execution time of the program. Despite the focus on the complex tasks of source code compilation or interpretation, the concept of *language application* can be interpreted from a wider perspective as "any program that processes, analyzes, or translates an input file" [10]. Although applications or components e.g. for processing configuration files or importing data from external files are typically not defined as language applications, they are based on the (often implicit) specification of a language, providing rules, which input should be considered as valid and how it should be further processed. Irrespective of being explicitly or implicitly defined, language applications implement the syntactical and semantic rules of a language by performing tasks according to sentences of the specified language.

3 Language Theoretical Foundation

Our concept is based on the formal description of schemata as regular tree grammars (RTG) [4,9,15], which results in the interpretation of a schema as a finite structure $\langle N, T, R, P \rangle$—an RTG with the finite sets of nonterminals (N) and terminals (T), the root symbol ($R \in N$) and the set of production rules (P). Due to the generalization from strings to trees, regular tree grammars allow production rules of the form $n \to te_c$, where

- $n \in N$,
- $t \in T$ and
- $e_c \subset N$ reflects the content model that is defined over the set of non-terminals.

Based on actual schemata and documents of the arts and humanities, the above definition for schemata is considered to represent the *parsing-oriented view*, which allows an initial validation and processing of external data, but does not necessarily reflect the full extent of the semantic structure and content that is encoded within a document. If the content of individual elements can be further described, decomposed or transformed into semantically enriched forms, a perspective different from the strictly parsing-oriented view is required.

To facilitate the representation of such substructures or alternative elements within a schema, we define a *semantic extension*—resulting in the definition of a schema as 6-tuple $S = \langle N, T, R, P, E, F \rangle$, where N, T, R and P form grammatical components and as such the parsing-oriented view as introduced above. The components of L and F provide the semantic extension of the original structure of the schema, where:

- L forms a set of labels and
- F is a set of labeling functions $x \to le_l$, where:
  - $x \in (N \cup L)$,
  - $l \subseteq L$ and
  - $e_l := \{I, op\}$ defining a function over a set of input values $I \subseteq N$ and an operation of the arity $|I|$.

For the creation of unifying views over heterogeneous semi-structured data, mappings between the relevant source schemata and the selected target schema need to be evaluated and applied. Data that is specified in terms of local schemata is transformed into the corresponding representation of the integrative view. Mappings can be defined as functions for relating source and target objects [4,14], which indicates that simple associations between individual elements are insufficient to represent related objects. We introduce the notion of *concept mappings* $cm = \langle E_{S_S}, E_{S_T}, f \rangle$ as correlation between a set of source elements E_{S_S} and a set of target elements E_{S_T} [4], where:

- $E_{S_S} = \{e_{S_S,i} \ldots e_{S_S,j}\} \mid e_{S_S} \in (N_{S_S} \cup L_{S_S})$
- $E_{S_T} = \{e_{S_T,k} \ldots e_{S_T,l}\} \mid e_{S_T} \in (N_{S_T} \cup L_{S_T})$
- $f : E_{S_S} \to E_{S_T}$

A schema mapping $M_{S_S \leadsto S_T}$ is then defined as a set of concept mappings $cm_1 \ldots cm_n$ defined over a source and target schema S_S and S_T. Aside from the labeling functions, the mapping functions f of concept mappings present the second use-case of our language specifications based framework.

Defining and Deriving Structure. Returning to the distinction between the generative and interpretive context of data, the approach presented in this paper identifies two separate tasks of data modeling:

- The *description task*, in which domain experts are enabled to describe data in terms of the language specifications—i.e. to express the information, which is required to describe data (the syntax and semantics) in terms of a technology-agnostic notation: context free grammars (CFG).
- The *transformation task* then allows researchers with a knowledge of particular research questions to express data transformation statements, which can be based on the enriched semantic representation that resulted from the description task.

The following figure shows the overall concept of the transformation rule framework, which forms a combination of the application of domain-specific languages (DSL) as the descriptive task and a transformation language for the formulation of transformation rules as the transformation task.

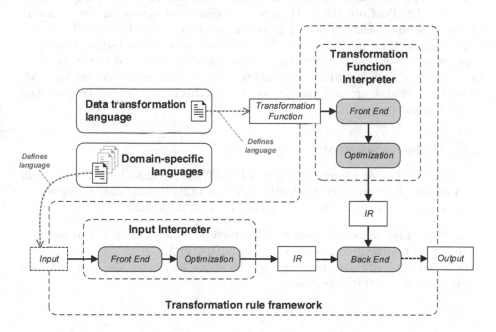

Fig. 1. Overview of the transformation rule framework

The building blocks within the framework are consolidated in two inter-preters, one for the validation and processing of input against a DSL and one for the validation and processing of a transformation function against the trans-formation language. Both are conceptualized as autonomous language applica-tion components accomplishing the tasks of input processing, the construction of intermediate representations, their traversal, optimization and transforma-tion [10].

In contrast to the typical structure of compilers, the back end of the rule framework needs to combine two parse trees, one with encoded transformation instructions and the other with a decomposed and structured version of the input. The back end finally concludes individual transformation tasks by:

- *interpreting input* executing the encoded transformation instructions and
- *generating the output* as output tree that forms a representation of the input in a target structure (mapping function) or that is included under the input element in the same structure (labeling function).

4   Data Description and Integration

In order to detail the principles behind our concept and the actual behavior of the developed framework, we present a exemplary metadata record that conforms to simple Dublin Core (DC) [1]. After modeling and extending the structure of the presented record in Sect. 4.1, we will model the correlation with another schema in Sect. 4.2 illustrate the idea of data integration in terms of our con-cept. Please note that the very simple example data presented in this paper is deliberately chosen in order to be able to focus on the theoretical background. Actual case studies are performed when writing this paper and will be published as applicable.

4.1   Modeling Schemata

The record in Listing 1 is available through the Open Archives Initiative Protocol for Metadata Harvesting (OAI-PMH)[1] interface of Pangaea, a data publisher for earth and environmental science[2] and serves us to illustrate

- problems that can arise, when simple, interoperability-oriented standards are chosen to represent complex data or metadata, and
- how the explication of language specifications facilitates the tasks of data description and transformation.

---

[1] http://www.openarchives.org/pmh/.
[2] http://ws.pangaea.de/oai/?verb=GetRecord&metadataPrefix=oai_dc&identifier= oai:pangaea.de:doi:10.1594/PANGAEA.50542.

```
<oai_dc:dc xmlns:dc="http://purl.org/dc/elements/1.1/" xmlns:xsi="http
    ://www.w3.org/2001/XMLSchema-instance" xmlns:oai_dc="http://www.
    openarchives.org/OAI/2.0/oai_dc/"     xsi:schemaLocation="http://
    www.openarchives.org/OAI/2.0/oai_dc/ http://www.openarchives.org/
    OAI/2.0/oai_dc.xsd">
    <dc:title>Ice rafted debris (&gt; 2 mm gravel) distribution in
        sediment core PS2646-5</dc:title>
    <dc:creator>Grobe, Hannes</dc:creator>
    <dc:source>Alfred Wegener Institute, Helmholtz Center for Polar
        and Marine Research, Bremerhaven</dc:source>
    <dc:publisher>PANGAEA</dc:publisher>
    <dc:date>1996-02-29</dc:date>
    <dc:type>Dataset</dc:type>
    <dc:format>text/tab-separated-values, 1148 data points</dc:format>
    <dc:identifier>http://doi.pangaea.de/10.1594/PANGAEA.50542
        </dc:identifier>
    <dc:identifier>doi:10.1594/PANGAEA.50542</dc:identifier>
    <dc:language>en</dc:language>
    <dc:rights>CC-BY: Creative Commons Attribution 3.0 Unported</dc:
        rights>
    <dc:rights>Access constraints: unrestricted</dc:rights>
    <dc:coverage>LATITUDE: 68.556667 * LONGITUDE: -21.210000 * DATE/
        TIME START: 1994-09-19T14:56:00 * DATE/TIME END: 1994-09-19
        T14:56:00 * MINIMUM DEPTH, sediment/rock: 0.0 m * MAXIMUM
        DEPTH, sediment/rock: 11.5 m</dc:coverage>
    <dc:subject>ARK-X/2; AWI_Paleo; Denmark Strait; Gravity corer (
        Kiel type); Ice rafted debris; IRD-Counting (Grobe, 1987);
        Paleoenvironmental Reconstructions from Marine Sediments @
        AWI; Polarstern; PS2646-5; PS31; PS31/162</dc:subject>
</oai_dc:dc>
```

Listing 1. Pangaea DC example

The document contains rather obvious weaknesses with respect to further data processing or integration. In the following, we focus on three particular issues that we resolve though modeling. Although an analysis of other records is required to specify data—for the purpose of simplicity, we assume the presented example as representative for the entire collection:

- The dc:creator element contains the full name of a person, which we separate into a last and first name.
- The dc:coverage element is composed of a substructure, whose explication results in a decomposition of the encapsulated data.
- The dc:subject element actually contains a list of subjects, which is split into individual elements.

As an initial step, both the structure required to parse data (the Extensible Markup Language (XML) schema) and the extensions to that schema are modeled. In order to provide a convenient interface for experts within the arts and humanities, we provide a web-based schema editor for this task as shown in Fig. 2. Within the displayed element model, blue nodes represent original elements of the XML schema, yellow nodes symbolize grammars (g: ...) and functions (f: ...), which produce the purple colored labels of the schema. On the left side of the screen, editor sessions can be managed and part of the result

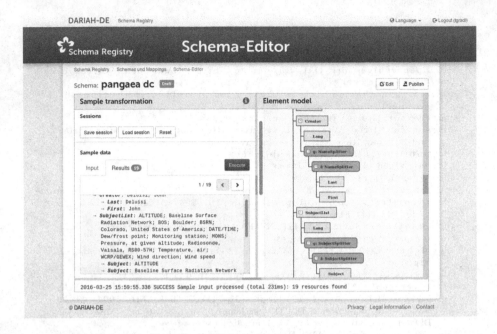

Fig. 2. User interface of the schema editor

of our example model is shown: the creator name has been decomposed as well as the subject list.

In order to fill subordinate labels, CFGs are formalized. In the cases of the `Creator` and `SubjectList`[3] elements, these grammars define a very simplistic language as shown in Listings 2 and 3. Both grammars define one rule for lexical analysis, which results in tokens separated by the specified character (',' and ';'). The original elements correspond to the `name` and `subjectlist` entry parser rules, which each produce the specified subsequent rules.

```
name: lname ', ' fname;

lname: STRING;
fname: STRING;

STRING: ~(',')+;
```

```
subjectlist: subject ('; ' subjecté)*;

subject: STRING;

STRING: ~(';')+;
```

Listing 2. Creator grammar   **Listing 3.** Subject grammar

One possible specification of the content within the *Coverage* element results in the grammar shown in Listing 4. Multiple lexer rules are defining types of

---

[3] The nonterminal element has been renamed to represent its content, the associated terminal references `dc:subject`—keeping the parsing-oriented structure intact.

tokens: IDs[4], DATE and SEPARATOR. The components of DATE are specified as lexical fragments, which is not strictly necessary in this case, but simplifies the parser grammar. The WS lexer rule specifies any tokens that are composed solely of whitespaces not to be passed to the parser. The parser rules then detail the logical composition of the content in Coverage. For further reference on such context free grammars we refer to guides on the Extended Backus-Naur Form (EBNF) or the ANTLR framework that we use to derive executable lexer and parser Java code from the grammatical specifications [10,11].

```
subelem      :    (longitude | latitude | start | end | minDepth | maxDepth |
                             otherElem) SEPARATORé;

longitude    :    'LONGITUDE' ':' value;
latitude     :    'LATITUDE' ':' value;
start        :    'DATE/TIME START' ':' value;
end          :    'DATE/TIME END' ':' value;
minDepth     :    'MINIMUM DEPTH, sediment/rock' ':' value;
maxDepth     :    'MAXIMUM DEPTH, sediment/rock' ':' value;

otherElem    :    key ':' value;

key          :    ID;
value        :    DATE
             |    ID;

ID           :    ~(' '|':'|'*') ~(':'|'*')+ ~(' '|':'|'*');
DATE         :    YEAR '-' MONTH '-' DAY 'T' HOUR ':' MIN ':' SEC;
SEPARATOR    :    ' 'é '*' ' é;

fragment     YEAR    :    [1-2][0-9][0-9][0-9];
fragment     MONTH   :    [0-1][0-9];
fragment     DAY     :    [0-3][0-9];
fragment     HOUR    :    [0-2][0-9];
fragment     MIN     :    [0-6][0-9];
fragment     SEC     :    [0-6][0-9];

WS  :  [ \t\r\n]+ -> skip ;
```

Listing 4. Grammar for the Coverage element

Part of the parse tree that results from the interpretation of the dc:coverage content in the example document is presented in Fig. 3—with intermediary nodes representing the applied *parser rules* and leaf nodes the *lexer rules*.

As discussed, grammars defines language constraints to which the content of a particular element (*nonterminal* or *label* with respect to the definition above) has to conform. The parser and lexer classes that ANTLR generates from provided grammars do not fail on any erroneous input, but for cases, where ambiguities are raised. Our modelling interface shows any data interpretation error or warning directly within produced parse trees.

In order to finally transform content and assign output to labels, transformation functions are specified, which relate to parser rules. In Listing 5 nodes of

---

[4] As pendant to STRING in the Creator and SubjectList grammars.

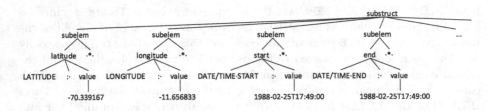

Fig. 3. Example parse tree of the `Coverage` grammar

the parse tree (`@latitude.value` and `@longitude.value`) are selected and (in lines 1 and 2) directly assigned to produced labels or used in commands (line 3). Such commands provide extension points of the framework and currently—based on actual use cases—include logical, string and arithmetic commands in the *Core* commands package. Othfronter packages wrap access to natural language processing (NLP) libraries, Wiki markup related functionality and geotemporal code. The specific command in line 3 of the listing concatenates the value of longitude, latitude and 0 to provide a label, which—continuing the example—is needed for mapping and thus data integration.

```
Lat = @latitude.value;
Long = @longitude.value;
LatLng = CORE::CONCAT(@longitude.value, "," , @latitude.value, ",0");
```

Listing 5. Transformation for the `Coverage` element

4.2   Modeling Mappings

The research around this paper is associated with the Digital Research Infrastructure for the Arts and Humanities (DARIAH-DE). A component that has been developed in DARIAH-DE consists in the Geo-Browser[5], a tool that visualizes spatial and/or temporal elements of data and allows the upload and analysis of data specified in terms of the Keyhole Markup Language (KML)[6]. Listing 6 shows an exemplary KML document.

```
<kml xmlns="http://www.opengis.net/kml/2.2">
    <Placemark>
        <name>Bamberg</name>
        <description>University of Bamberg</description>
        <Point>
            <coordinates>10.869461,49.902974,0</coordinates>
        </Point>
    </Placemark>
</kml>
```

Listing 6. KML placemark example

---

[5] http://geobrowser.de.dariah.eu/.
[6] https://developers.google.com/kml/.

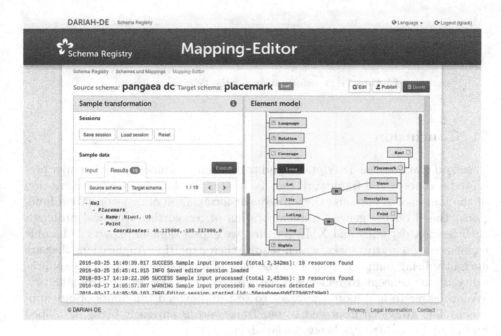

Fig. 4. User interface of the mapping editor

Based on the extracted latitude and longitude from the Pangaea record above—and especially the prepared LatLng label—the mapping between the two schemata can be modeled as shown in Fig. 4. In the screenshot, the element model for the assigned schemata is shown on the right side. On the left the user again finds access to the session management as well as sample transformation controls. Data that conforms to the modeled Pangaea schema has been uploaded and executed (1) against its schema definition to produce enriched content and (2) against the mapping specification—the result of the latter being shown in Fig. 4.

To fill the title of the KML document, we chose to utilize an example of the *GeoTemp* extension of our framework, which—in this particular case—is called to determine the nearest populated place based on the provided coordinates and GeoNames[7] data.

This reverse lookup deliberately uses an external, web-based interface and introduces latency. For this reason, the command is not executed in terms of the element mapping, but instead resulted in the definition of another element of the Pangaea structure. This results in the execution of the functionality when processing and indexing data, not when ad-hoc transforming data into another structure. Listing 7 shows the extended transformation function of the Coverage element.

---

[7] http://www.geonames.org/.

```
Lat = @latitude.value; Long = @longitude.value; LatLng =
CORE::CONCAT("[", @longitude.value, "," , @latitude.value, "]");
City = GEO::SIMPLEREVERSE(@latitude.value, @longitude.value);
```

Listing 7. Transformation for the **Coverage** element

5   Conclusion

Based on grammatical descriptions and transformation functions, this paper presented a concept and framework that facilitates the domain-specific description of the language of data and the subsequent formulation of transformation functions. Based on the generic implementation of reoccurring technical problems such as data decoding and XML parsing, we expect domain experts to be able to focus on integrative tasks, which require their expertise. Applications that are currently being built and tested on the base of our framework promise applicability of the concept to real-world data problems of the arts and humanities. We are eager to find more use cases to test and extend the limits of our approach. The presented front end to our framework is currently being developed and available at http://schereg.de.dariah.eu.

References

1. Dublin Core Metadata Element Set, Version 1.1 (2012). http://dublincore.org/documents/dces/
2. Batini, C., Lenzerini, M., Navathe, S.B.: A comparative analysis of methodologies for database schema integration. ACM Comput. Surv. **18**(4), 323–364 (1986)
3. Cooper, K.D., Torczon, L.: Engineering a Compiler, 2nd edn. Elsevier Morgan Kaufmann, Amsterdam (2012)
4. Gradl, T.: Concept and implementation of a rule framework to dynamically transform data and queries for heterogeneous collections. Master thesis, University of Bamberg, Bamberg (02-07-2014)
5. Gradl, T., Henrich, A.: A novel approach for a reusable federation of research data within the arts and humanities. In: Digital Humanities 2014 - Book of Abstracts, pp. 382–384, Lausanne, Switzerland (2014). http://dharchive.org/paper/DH2014/Paper-779.xml
6. Hagedorn, K.: OAister: a no dead ends OAI service provider. Libr. Hi Tech **21**(2), 170–181 (2003)
7. Hull, R.: Managing semantic heterogeneity in databases. In: Mendelzon, A., Özsoyoglu, Z.M. (eds.) Proceedings of the Sixteenth ACM SIGACT-SIGMOD-SIGART Symposium, PODS 1997, pp. 51–61 (1997)
8. Lenzerini, M.: Data integration: a theoretical perspective. In: Abiteboul, S. (ed.) Proceedings of the Twenty-First ACM SIGMOD-SIGACT-SIGART Symposium on Principles of Database Systems, p. 233. ACM, New York (2002)
9. Murata, M., Lee, D., Mani, M., Kawaguchi, K.: Taxonomy of XML schema languages using formal language theory. ACM Trans. Internet Technol. **5**(4), 660–704 (2005)

10. Parr, T.: Language implementation patterns: create your own domain-specific and general programming languages. The Pragmatic Programmers, The Pragmatic Bookshelf, Raleigh, NC, P3.0 Printing, Version: 2011-7-13 edn. (2011)
11. Parr, T.: The Definitive ANTLR 4 Reference. The Pragmatic Programmers, 2nd edn. Pragmatic Bookshelf, Dallas and Raleigh (2012)
12. Peroni, S., Tomasi, F., Vitali, F.: Reflecting on the Europeana data model. In: Agosti, M., Esposito, F., Ferilli, S., Ferro, N. (eds.) IRCDL 2012. CCIS, vol. 354, pp. 228–240. Springer, Heidelberg (2013)
13. Pouyllau, S.: ISIDORE: acces to open data of arts & humanities (2011). http://de.slideshare.net/stephanepouyllau/prsentation-gnraleisidore-generalenv
14. Sheth, A.P., Larson, J.A.: Federated database systems for managing distributed, heterogeneous, and autonomous databases. ACM Comput. Surv. **22**(3), 183–236 (1990)
15. Zhang, Z., Shi, P., Che, H., Gu, J.: An algebraic framework for schema matching. Informatica **19**(3), 421–446 (2008). http://dl.acm.org/citation.cfm?id=1454341.1454348

The Challenge of Creating Geo-Location Markup for Digital Books

Annika Hinze[✉], David Bainbridge, and Sally Jo Cunningham

Department of Computer Science, University of Waikato, Hamilton, New Zealand
{hinze,davidb,sallyjo}@waikato.ac.nz

Abstract. The story lines of many books occupy real world locations. We have previously explored the challenges of creating automatic location markup that might be used in location-based audio services for digital books. This paper explores the challenges of *manually* creating location annotations for digital books. We annotated three very different books, and report here on the insights gained and lessons learned. We draw conclusions for the design of software that might support this annotation process *in situ* and *ex situ*.

Keywords: Location · Location-based services · Semantic markup · Semantic annotation · Crowdsourcing · Annotation

1 Introduction

Location markup is the process of annotating elements of eBooks—typically, phrases or paragraphs but also pictures, audio or video—with references to geo-locations, e.g., as identified on a map. A common automated approach is to utilize a gazeteer to identify place names explicitly mentioned [10,15]. However, even in this simplified version of the problem, much effort is spent on trying to avoid false positives, with only moderate success.

To explore this difficulty of location annotation in depth, so that we may lay the foundations for how to create software to support the process, we examine here the challenge of location markup for digital books. We manually created the location-markup for three different books, and report here on the insights we gained during the process.

A strongly location-focussed book was used for initial case study; it was assumed to be the easiest to markup for locations as the very structure of the book mirrors a walk through a succession of locations and their descriptions. A second case study was done with a family history book, which was much less structured according to location, but rather according to the families timeline and memories. Marking-up the locations of this book required a considerable amount of online and map research. The third book was a science-fiction novel about aliens landing at our local university.

The remainder of this paper is structured as follows: Sect. 2 reports on our case study with a location-based book. Section 3 reports on our case study of

© Springer International Publishing Switzerland 2016
N. Fuhr et al. (Eds.): TPDL 2016, LNCS 9819, pp. 294–306, 2016.
DOI: 10.1007/978-3-319-43997-6_23

marking-up a family history book, and in Sect. 4 we report on marking-up a Science Fiction book. We then discuss the implications of our observations for the design of a location annotation service for crowd-souring location information for books in Sect. 5. Section 6 compares our observations and design decisions to related work. The paper concludes with a summary of what we learned and plans for future work.

2 Case Study 1: Location-Based Book

In our first case study, we created the location markup for a book about the Paradise Gardens in the Hamilton Gardens, New Zealand. We set eight pages of the text with wide margins to allow room for annotations, using 1.15 line spacing for easier reading. The same book had been previously used as an audio book in a user study that allowed participants to experience its content in situ [13]. The book describes the history of the gardens, how they were designed and the meaning of their layout and plants. In contrast to a dedicated exhibition guide, this book has a more loose structure referring to both specific items as well as larger areas. Texts for exhibition guides are purpose-built with a rigid structure on the assumption that each exhibited item can be clearly identified.

For this case study, we explored available topological maps, satellite images, and conceptual drawings as sources to support the markup process. The available detail of information varies greatly between maps and areas concerned. In performing the markup, we used a topographical map from Open Street Map and an image map.

The researcher used both maps for markup of the given print text, and was familiar with the Hamilton Gardens. Figure 1 shows the markup of a section of the book. The markup in this example, and the others that follow, has been

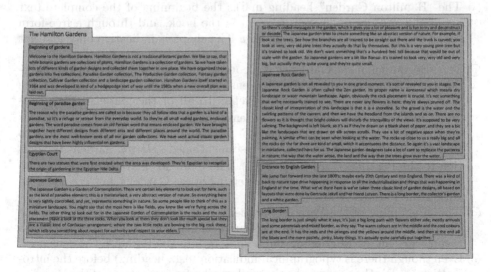

Fig. 1. Gardens text with location references markup (Color figure online)

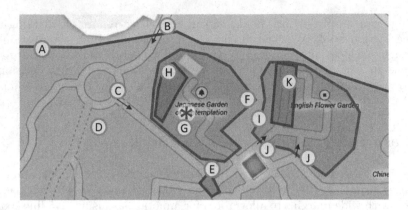

Fig. 2. Gardens map with location markup

transferred onto digital text and map for easier readability, after the markup was initially performed manually on paper. The colours have been introduced in the digital version for easier readability. Circled letters are used for cross-referencing to the map in Fig. 2; squared numbers with dashed arrows make identification of discussion points easier and are not part of the location markup. Figure 2 shows the location indication on the topological map corresponding to the location markup of the book text (cropped from original map). Four markup elements were used (area, path, point and gateway), which were developed during the markup phase in response to the text's location references. From initial inspection of the text, one can identify seven paragraphs that each have a heading that seems to refer to a location within the Hamilton Gardens. These seven elements are headed by the larger caption "The Hamilton Gardens". More specifically:

Ⓐ The "Hamilton Garden" heading marks the beginning of the complete text about the gardens, indicated in green in the book, and through a free-form shape on the map, indicating an *area*. The shape is indicative only and not precise.

Ⓑ The "Beginning of the Gardens" is a complete paragraph that provides introductory information about the Gardens. Even though it was marked on the map as a *gateway*, its information refers to all of the gardens. The paragraph mentions the "Hamilton Gardens" again (indicated by $\boxed{1}$), which is not marked separately as element Ⓑ is a sub-area of Ⓐ, which refers to Hamilton Gardens. The paragraph also mentioned *external* gardens in addition to the Paradise Gardens (see $\boxed{2}$). These are not marked-up as they are references to places within the wider setting of the Gardens, but are not sited within the Paradise Gardens Ⓑ.

Ⓒ The "Beginning of the Paradise Gardens" is a similar introduction into the Paradise Gardens as a subset of the Hamilton Gardens. Again, it is marked as a *gateway*—an arrow crossing the boundary into the Paradise Gardens area.

Ⓓ Even though there is typographical indication (e.g., heading) before the introduction (in D), the paragraphs that follow all refer to elements of the Paradise Garden and are therefore marked-up together (yellow highlight in text and yellow area on map). When analysing the shape on the map, one notices a second entrance (exit?) to this area; this will be discussed in Sect. 5.

Ⓔ The Egyptian court area is very small (about 5×5 m). The description also refers to two statues, which are not marked-up separately as they are essentially the only items in the court area.

ⒻⒼⒽ The Japanese Garden section contains two parts — one paragraph of general introduction (with specific reference to three rocks Ⓖ) and a separate paragraph about the Japanese Rock Garden Ⓗ. The text also contains a reference to the trees, though no specific tree is singled out (see 1). Note that the rocks are marked with a *cross* on the map (indicating a GPS location); the researcher considered using a very small area as a comparable alternative. The location of Ⓖ on the map was estimated; the correct location would require the researcher to be at location to allow direct input of GPS coordinates.

ⒾⒿⓀ The next two paragraphs refer to elements of the English Garden, which is not explicitly named but was marked Ⓘ in the text and as an area on the map. The "Entrance of the English Garden" provides an introduction to the English Gardens — it was marked at the two entrances (see Ⓙ on map); it could also be joined with Ⓘ and used as an introduction (similar to the Japanese Garden). The Long Border was marked-up as an *area*, but equally could have been marked as a *path*, depicted using a line. For identifying the location of the English Garden, the researcher had to use both the topographical and the image map.

We conducted an exploratory study with a group of participants to identify how difficult it would be for people to create meaningful markup. In parallel to the researcher creating the location markup on the book, six participants used the book text to mark up paper maps *in situ* at the Hamilton Gardens. As space limitations do not allows us to describe the study in detail, we briefly summarise our main findings. None of the participants found creating location markings to be easy. They used different markers with in the text (e.g., a number next to a paragraph or next to the paragraph heading, or they drew a frame around a word, phrase or paragraph). Very few annotations were both placed in the correct location on the map and indicated an adequate area. Many annotations were placed close to the correct location, but often merely as a number without any indication of area. Many annotations in the text did not identify exactly the text segment (e.g., sometimes only the heading was selected, sometimes nothing was selected but a number placed near the paragraph). Some annotations on the map were in a place very far from any place the text referred to (e.g., the text for the English Garden received an annotation in the Chinese Garden. This pilot study backs the idea that users will need strong support in order to create well-formed and usable location annotations.

3   Case Study 2: Family History

Next we explored the markup of a book about the history of a Waikato family.[1] We created location markup for five of the book's 300+ pages, using the original

---

[1] The author Tom Brough has kindly provided an electronic copy of the book [6].

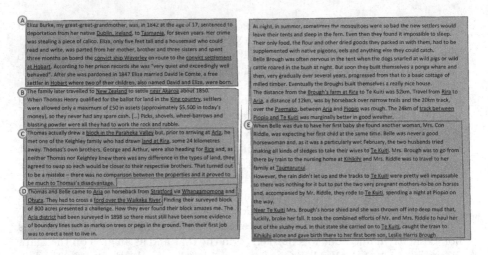

Fig. 3. Pages with location markup

print layout. To discuss the markup we re-formatted the text for easier reading with markup. It quickly became apparent that it was more difficult to decide on the location markup for this book. Many places names were mentioned, but the structure of the book followed a time-line rather than a defined physical path, with the narrative being driven by anecdotes relating to several family members rather than a single story line. In the Fig. 3, we indicate the place names and other location identifiers with underlining.

(A) The first paragraph briefly describes the deportation of Eliza Burke to and her life in Tasmania. Locations are "Dublin", "Tasmania" and the city "Hobart". It is not clear in the text if the "convict settlement at Hobart" is Port Arthur (60 km south of Hobart) or Maria Island (111 km north of Hobart). The "convict ship Waverley" is a *moving* location; its voyage is recorded to be to "Van Diemen's Land" (an older name for Tasmania 1803–1856).

(B) The following paragraphs are about events in New Zealand; we note but do not show the country-level and district-level annotations here. Thomas Henry Brough is the husband of Eliza who in turn is the daughter of Eliza and David. The narrative of the book often jumps back and forth between people and locations, relying on information given on earlier pages. This makes identification of the locations more difficult.

(C) This paragraph provides a short explanation about a place that is not part of the main story line (Paraheka Valley). The location of the block of land referred to cannot be easily determined but the wider area of Paraheka Valley area is known. We searched for a Paraheka Stream to identify the possible location of the valley; there are two in the Waikato/King Country region.[2] The whole paragraph could be marked-up with either or both locations of Aria and Paraheka Valley.

---

[2] Accessible by searching for "Paraheka" at Land Information New Zealand Gazetteer, www.linz.govt.nz.

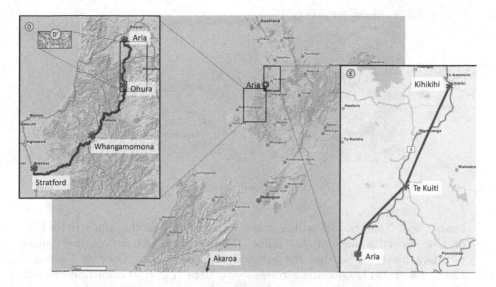

Fig. 4. Map with location markup for family history book

Ⓓ This short paragraph describes a journey and is marked-up as a *path*. The precise path is unknown so only a rough estimate can be provided by the path markers. The path they travelled has a direction (northward from Stratford to Aria). Determining the location of their crossing over the Waikaka river is somewhat challenging. A Google maps search only finds Waikaka (place) on the South Island of New Zealand; Wises[3] identifies the Waikaka Stream near Gore (South Island). Mapquest (using Open Street Map) knows about the Southland location and a location 15 km east of Aria, which may be the river-head of the Waikaka. Finally, the NZ Topo Map[4] provides rich information about rivers, hills, homesteads and places named Waikaka (18 entries). The LINZ database records that the name Waikaka Stream "is not official — this feature does not have an official name." Once the general area was identified, we acquired the relevant Topo50 (1:50,000) and Topo250 (1:250,000) maps to explore the possible paths between Ohura and Aria. At the time of this chapter in the book, New Zealand had not yet been officially mapped to any great detail, whereas there are now two roads, which may in previous times have been horse treks. Based on the story details of the Waikaka crossing, one is more likely than the other — as the road in question bears the name 'Matiere-Aria road', the researcher marked the path shown in Fig. 5. In order to record this reasoning about the river crossing, this markup example indicates that one may wish to add supporting documents to the location markup (e.g., maps or old newspaper articles) as part of the process.

Another location that is unclear is "Rira" – it is most likely the name of a place or homestead, which was developed from "a block of standing native

---

[3] www.wises.co.nz.

[4] www.topomap.co.nz.

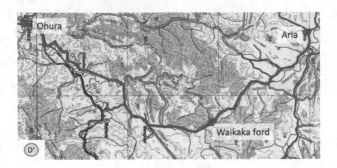

Fig. 5. Waikaka ford

bush" [6].[5] None of the online maps we checked nor the maps from 1841 or 1948 record a place of that name. However, it is not an imaginary place or anonymisation, because the regional newspaper King Country Chronicle (20 October 1917) contains a reference to Rira (Fig. 6). Given that the settlers were from Ireland, we speculate that Rira may refer to an Irish phrase (Rí Rá) meaning a ruckus or general mayhem. A few days after this first reference to Rira, we became aware of another one: the 1987 edition of the Wises New Zealand Guide (a gazetteer of New Zealand towns and places) contained an entry on Rira, see Fig. 7.

MR AND MRS T. H. BROUGH, of Rira, wish to express their deepest gratitude and thanks to the many friends for their expressions of sympathy and kind assistance rendered in their late bereavement.

RIRA Locality on the banks of the Huioteko Stream, a tributary of the Mokau River, Waitomo District, Waitomo. 8km south-east from Aria, 23km south-east of Piopio, 48km south-west of Te Kuiti.

Fig. 6. Reference to Rira   **Fig. 7.** A second Rira reference (1987)

ⓔ Belle's travel to Te Kuiti and Kihikihi is marked-up as another *path*; this time with three waypoints. The location markup is more indicative here, in comparison to the more detailed ones in ⓓ and ⓓ'. Furthermore, the location "near Te Kuiti" where the horse shied is another example of a location that cannot be precisely identified.

4   Case Study 3: Science Fiction Book

As a third option, we explored marking-up a science fiction book that describes the landing of a UFO at the grounds the University of Waikato.[6] We created

---

[5] The reading of "Rira" being a name introduced perhaps by the new settlers was suggested by the Waikato University Map Librarian John Robson, as he recognized that the word was not of English or Maori origin.

[6] This ebook by Owen Mooney, called "The visitors", is available at https://www.smashwords.com/books/view/449780.

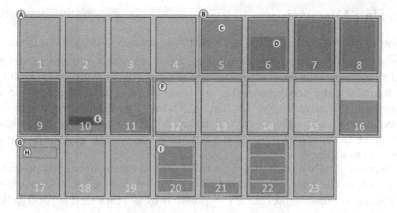

Fig. 8. Overview of SciFi text with location markup (Color figure online)

location markup for the prologue and the first three chapters (23 of the 93 pages of Volume 1). Not very many different places appear in these three chapters; in order to give an overview we summarise the pages visually as shown in Fig. 8, but omit providing a map.

Ⓐ The prologue (pg 1–4) is situated "deep in a mountain" in a US American space observation station. As a book set in the future, it was no surprise to determine that the place and organisation do not exist.

Ⓑ The first chapter (pg 5–11) describes the main protagonist remembering how he met his girlfriend Susan. All of the events happen in Hamilton, New Zealand.

Ⓒ+Ⓓ Still in Hamilton, the protagonist visits Susan's flat. He remembers how he met her in a supermarket and the supermarket carpark. The supermarket is not specified.

Ⓔ He remembers going with Susan to the Riverlea theatre in Hamilton.

Ⓕ In the second chapter (pg 12–16), a UFO is seen to land at the University of Waikato playing field ("the second from the road"). The protagonist goes to investigate. He writes a story about it for the newspaper (and seems to deliver it) and returns to the university.

Ⓖ+Ⓗ+Ⓖ Chapter 3 (pg 17–23) moves the action to Wellington, presumably the Prime Minister's office. The Prime Minister and his team meet, and watch a TV special report from Hamilton Ⓗ. Later in the chapter the Prime Minister has a phone conversation with the US President in Washington(?) Ⓖ.

We observe that a "university" was first mentioned on page 5 but only resolved to "Waikato University" at page 15. Also, page 16 contains an indirect reference to "arriving back", which also refers to the university grounds. Chapter 2 contains deceptions and memories, some of which are not related to any particular locations. It is unclear, whether the protagonist remained at the previous location (Susan's flat) while reminiscing.

5 Design Discussion

Based on our three case studies, we identify a number of design requirements.

1. As mentioned earlier, four location elements emerged and were introduced with respective individual markers for the maps (see case study 1): area (polygon boundary and area), path (uni- or bi-directional polyline graph), location (cross), and gateway (arrow). Within the book text, each of these is indicated by (coloured) consecutive text, which may span a paragraph or whole chapter that is set in that location, or a single word or phrase providing a passing reference to a place.

2. Gateways typically have a direction to be used as entrance or exit. Information presented to the user may need to be provided only when a user enters or exits an area—managing this is complicated by the fact that several gateways may exist for a place (see case study 1).

3. One needs a distinction between *here* and *there*, i.e., locations may be part of the current narrative of a protagonist, or a place about which they hear, talk, dream or remember (e.g., places in the family history). Piatti et al. developed interesting visual indicators for these modes of being 'in' locations [25].

4. Stories are not typically geographically linear; they split into alternatives, involve more than one place at the same time, describe alternative routes or merely list places (e.g., as in case study 1) and generally may freely jump through time (e.g., in the family history).

5. Not all places in a story are precisely specified: different markers may disagree about their placing. A place may not be specified sufficiently for mapping or might not be intended by the author that these map to specific existing spots (e.g., the supermarket and Susan's place; both in the SciFi book). One may allow for a number of alternatives to be coded as placeholders for 'locations that it might have been' or select one place (e.g., a supermarket close to the university; SciFi book). From the standpoint of literature scholars and literary geographers, these solutions may not be adequate [26], however from a literary tourism viewpoint these suggest workable alternatives. Acceptable solutions need to be explored through end-user involvement and analysis of targetted applications.

6. Not all places in a story may currently exist or have ever existed in the geographic current reality (e.g. the office in the mountain; SciFi book), or are reachable (most places in the *Hitchhiker's Guide*). That is, some places may only be mappable by intruding their own maps (e.g., such as the map provided in Stevenson's *Treasure Island*. Others might be handled by allowing "close-enough" matches (e.g., Platform 9 at King's cross standing in for the 'proper' Platform $9^3/4$ in *Harry Potter*).

7. 'Vanished' spaces (e.g., Rira) pose another problem, especially if the landscape has been significantly changed (e.g. Dickens' London). One may wish to identify the (prior) location *and* examples of elements similar to those that vanished (e.g., a typical rural farm in the King Country, as in the family history book).

8. Maps may not provide the necessary location markers for the person doing the markup to easily identify a location feature. Additional maps or other documents may be useful to support the annotation process or the decisions made; these may be kept within the Digital Library (see [21]) for ready access.

9. Location annotations may happen on several abstraction levels (e.g., University of Waikato, Hamilton, New Zealand, Earth; see SciFi book). Annotations capturing such information should generally be done on the lowest level possible for a given text element; aggregation for presentation can always be decided upon later.

6   Related Work

Inspiration for a location-annotation tool may be drawn from manual semantic annotation for named entity recognition [16]. Tools for automatic annotation (e.g., [27]) provide better results than simple location gazetteers, but are limited to known places and do not necessarily carry geo-location information. Native support for semantics in Digital libraries (e.g., JeromeDL [19] and Greenstone [14]) refer to the document metadata level and so do not provide the fine grained access needed for our task.

To the best of our knowledge, the geo-mapping of story lines is only supported in our own software prototype [17]. This, however, does not mean that it is not done outside of the digital domain—quite the opposite. The mapping of stories has a long tradition: Nabokov manually mapped Austen's Mansfield Park in the margins of his copy [3]; Janine Barchas maps and compares Austen's Bath and Joyce's Dublin [2]; Moretti even published a whole Atlas of the European novel [22]. Moretti highlights the transformative power of mapping stories, which allows the reader to identify circular patterns, linear trajectories, binary fields triangulations and multi-polar stories [23]. So far, the mapping of stories has predominantly been a tool of literature scholars or for private entertainment (published as sets of locations on private webpages [8]).

Mobile Geo-visualisation of literature is hitherto rare. In the Tipple project, we are concerned with location-based access of digital books while travelling and reading books [12]. Most existing approaches use static maps and are independent of the reader's location. Bollen and Holledge (manually) mapped production details of Ibsen's "A dollhouse." [5] Their maps reveal issues similar to those discussed in our work: travel across continents and travel involving parallel movements of several persons easily becomes confusing. Additionally, what can still be shown on a map (albeit a convoluted one) may create issues when deciding where to direct the reader/listener of the book. Berendt et al. [4] proposed a visualisation of story paths by exploring a story as a set of time-indexed streams of different text elements about a common topic. They refer to the underlying problem as *evolutionary theme patterns discovery, summary and exploration* They identified elements such as information being released, people going missing and a suspect being arrested, which they then visualised as 2D graphs ordered by time to show the interaction between the various topics. Their approach did not target locations, but one might speculate about a possible combination of the two approaches in which summaries of story elements are visualized by location.

Fluid Reader is a digital reader software that displays interactive hypertext narratives, whose alternative paths can be explored by the reader. The team also created a Fluid Writer system that uses a new tree-table visualization for constructing and managing alternative paths in a Fluid hypertext. The Fluid Writer interface may provide inspiration for travel path construction for complex narrative structures based on our location annotation (e.g., telephone conversations across continents or interleaving of story paths) [33]. Eccles et al. introduce a system that combines information and artefacts into a storyline, which is visualized using a map in order to identify geo-location patterns and support analytic sense-making of a situation in time and space [9]. None of these approaches deal with the problem of acquiring the location information.

Location tagging of landmarks in multimedia has been approached both manually and automatically [20]. Location tagging is a task that is related to the creation of location markup, but one that cannot simply be transferred onto the problem addressed here. Often the geolocation data embedded in the image metadata or even social media entries associated with an image can be mined. This may require integration of location data from different sources [31]. The location tagging approach is typically not feasible for digital books, as these rarely contain precise GPS location references to be mined.

A number of approaches exist for capturing and visualising travel data of actual human travellers [29,32] (i.e., not literary personas), devices and locations [7]. All of these approaches use automatically captured GPS travel records or existing map data. Stopher et al. [30] describe the process of using track points recorded to create maps and visual representations of the travel. They use the data to identify congestion, traffic delays, and periods and locations and acceleration and deceleration. The efficient (mobile) access to such location markup has been explored through location-based indexing and spatial access methods (e.g., [1,24]). Travel behaviour research is an area that is concerned with tracking of devices and people as well as human behaviour while travelling.

7   Conclusion

It is clear that the complex spatial dimensions present in fiction cannot be mapped through simplistic approaches. At the other end of the scale, the traditional approach of literary geography is similarly limiting: to employ "educated professional readers" [26] to analyse each work of literature and to create location mappings. The result will produce high-quality maps; but is not scalable.

Piatti et al. confirm that literary geography is based on individual readings, acknowledging its *double uncertainty* in both the primary material (uncertainty of location information in the text) and in the methodology (location annotation). We suggest to embrace his uncertainty and highly individually biased nature through our crowd-sourcing approach. Crowdsourcing, such as used for video geo-locating [11] or image annotation [28], has been successfully used to create large data collections or annotations. We do not aim to involve Amazon's Mechanical Turk [18] or any other online markets to outsource the task, but rather aim to support readers and book enthusiast in creating a valuable additional resource to accompany eBooks. We are therefore not concerned with cheat detection but assume these tasks to be a "labour of love" for enthusiasts.

We believe that, in contrast with video or image annotation tasks, some elements of location annotation may best be done *in situ*, i.e., at the location to which the book refers. The reasons are that one may not always be interested in the true or correct location, but the best fit (e.g., Harry Potter's Diagon Alley may not have a true geo-location but one or several "ideal" close matches). Clearly there are limitations to this approach, as the following the *Hitchhiker's Guide to the Galaxy* might lead to interesting travel plans! Video geo-locating was already recognized as being "difficult for both humans and machines," and the same applies in our case [11].

References

1. Al-Badarneh, A., Al-Alaj, A.: A spatial index structure using dynamic recursive space partitioning. In: Innovations in Information Technology (IIT), pp. 255–260. IEEE (2011)
2. Barchas, J.: Mapping Northanger Abbey: or, why Austen's Bath of 1803 resembles Joyce's Dublin of 1904. Rev. Engl. Stud. **60**(245), 431–459 (2009)
3. Barchas, J.: Matters of Fact in Jane Austen: History, Location, and Celebrity. JHU Press, Baltimore (2012)
4. Berendt, B., Subasic, I.: Identifying, measuring and visualising the evolution of a story: a web mining approach. In: Proceedings of WIS (2008)
5. Bollen, J., Holledge, J.: Hidden dramas: cartographic revelations in the world of theatre studies. Cartographic J. **48**(4), 226–236 (2011)
6. Brough, T.: The Way It Was: A Farming, Shearing, Hunting Life. Fraser Books, New Zealand (2011)
7. Burigat, S., Chittaro, L., Gabrielli, S.: Visualizing locations of off-screen objects on mobile devices: a comparative evaluation of three approaches. In: Human-Computer Interaction with Mobile Devices and Services, pp. 239–246. ACM (2006)
8. Cunningham, S.J., Hinze, A.: Supporting the reader in the wild: identifying design features for a literary tourism application. In: SIGNZ Human-Computer Interaction (CHINZ) (2013)
9. Eccles, R., Kapler, T., Harper, R., Wright, W.: Stories in geotime. Inf. Vis. **7**(1), 3–17 (2008)
10. Erle, S.: Project gutenkarte. www.gutenkarte.org
11. Gottlieb, L., Choi, J., Kelm, P., Sikora, T., Friedland, G.: Pushing the limits of mechanical turk: qualifying the crowd for video geo-location. In: WS on Crowdsourcing for Multimedia, pp. 23–28. ACM (2012)
12. Hinze, A., Bainbridge, D.: Listen to tipple: creating a mobile digital library with location-triggered audio books. In: Zaphiris, P., Buchanan, G., Rasmussen, E., Loizides, F. (eds.) TPDL 2012. LNCS, vol. 7489, pp. 51–56. Springer, Heidelberg (2012)
13. Hinze, A., Bainbridge, D.: Location-triggered mobile access to a digital library of audio books using tipple. Int. J. Digit. Libr. 1–27 (2015)
14. Hinze, A., Buchanan, G., Bainbridge, D., Witten, I.: Semantics in greenstone. In: Kruk, S.R., McDaniel, B. (eds.) Semantic Digital Libraries, pp. 163–176. Springer, Heidelberg (2009)
15. Hinze, A., Gao, X., Bainbridge, D.: The TIP/Greenstone bridge: a service for mobile location-based access to digital libraries. In: Gonzalo, J., Thanos, C., Verdejo, M.F., Carrasco, R.C. (eds.) ECDL 2006. LNCS, vol. 4172, pp. 99–110. Springer, Heidelberg (2006)

16. Hinze, A., Heese, R., Luczak-Rösch, M., Paschke, A.: Semantic enrichment by non-experts: usability of manual annotation tools. In: Cudré-Mauroux, P., Heflin, J., Sirin, E., Tudorache, T., Euzenat, J., Hauswirth, M., Parreira, J.X., Hendler, J., Schreiber, G., Bernstein, A., Blomqvist, E. (eds.) ISWC 2012, Part I. LNCS, vol. 7649, pp. 165–181. Springer, Heidelberg (2012)

17. Hinze, A., Littlewood, H., Bainbridge, D.: Mobile annotation of geo-locations in digital books. In: Kapidakis, S., et al. (eds.) TPDL 2015. LNCS, vol. 9316, pp. 338–342. Springer, Heidelberg (2015). doi:10.1007/978-3-319-24592-8_30

18. Ipeirotis, P.G.: Analyzing the Amazon Mechanical Turk marketplace. XRDS: Crossroads, ACM Mag. Students **17**(2), 16–21 (2010)

19. Kruk, S., Cygan, M., Gzella, A., Woroniecki, T., Dabrowski, M.: JeromeDL: the social semantic digital library. In: Kruk, S., McDaniel, B. (eds.) Semantic Digital Libraries, pp. 139–150. Springer, Heidelberg (2009)

20. Luo, J., Joshi, D., Yu, J., Gallagher, A.: Geotagging in multimedia and computer visiona survey. Multimedia Tools Appl. **51**(1), 187–211 (2011)

21. McIntosh, S., Bainbridge, D.: Integrating Greenstone with an interactive map visualizer. In: Joint Conference on Digital Libraries (JCDL), pp. 405–406. ACM (2010)

22. Moretti, F.: Atlas of the European Novel, 1800–1900. Verso, London (1999)

23. Moretti, F.: Graphs, Maps, Trees: Abstract Models for a Literary History. Verso, London (2005)

24. Osborn, W., Hinze, A.: Tip-tree: a spatial index for traversing locations in context-aware mobile access to digital libraries. Pervasive Mob. Comput. **15**, 26–47 (2014)

25. Piatti, B., Reuschel, A.-K., Hurni, L.: Dreams, longings, memories – visualising the dimension of projected spaces in fiction. In: WS on Cartography and Narratives, pp. 74–94

26. Piatti, B., Reuschel, A.-K., Hurni, L.: Literary geography – or how cartographers open up a new dimension for literary studies. In: International Cartography Conference (2009)

27. Ritter, A., Clark, S., Etzioni, O., et al.: Named entity recognition in tweets: an experimental study. In: Conference on Empirical Methods in NLP, pp. 1524–1534. Association for Computational Linguistics (2011)

28. Russell, B.C., Torralba, A., Murphy, K.P., Freeman, W.T.: LabelMe: a database and web-based tool for image annotation. Int. J. Comput. Vis. **77**(1–3), 157–173 (2008)

29. Shaw, S.-L., Yu, H., Bombom, L.S.: A space-time GIS approach to exploring large individual-based spatiotemporal datasets. Trans. GIS **12**(4), 425–441 (2008)

30. Stopher, P.R., Bullock, P., Jiang, Q.: Visualising trips and travel characteristics from GPS data. Road Transp. Res. **12**(2), 3–14 (2003)

31. Xu, Z., Chen, H.J., Wu, Z.: Applying semantic web technologies for geodata integration and visualization. In: Akoka, J., et al. (eds.) ER Workshops 2005. LNCS, vol. 3770, pp. 320–329. Springer, Heidelberg (2005)

32. Yu, H., Shaw, S.-L.: Representing and visualizing travel diary data: a spatio-temporal GIS approach. In: 2004 ESRI International User Conference, pp. 1–13 (2004)

33. Zellweger, P.T., Mangen, A., Newman, P.: Reading and writing fluid hypertext narratives. In: Hypertext and Hypermedia, pp. 45–54. ACM (2002)

Exploring Large Digital Libraries
by Multimodal Criteria

David Zellhöfer(✉)

Department of Information and Data Management,
Berlin State Library – Prussian Cultural Heritage, Berlin, Germany
david.zellhoefer@sbb.spk-berlin.de

Abstract. Digital library (DL) support for different information seeking strategies (ISS) has not evolved as fast as their amount of offered stock or presentation quality. However, several studies argue for the support of explorative ISS in conjunction to the directed query-response paradigm. Hence, this paper presents a primarily explorative research system prototype for metadata harvesting allowing multimodal access to DL stock for researchers during the research idea development phase, i.e., while the information need (IN) is vague. To address evolving INs, the prototype also allows ISS transitions, e.g., to OPACs, if accuracy is needed.

As its second contribution, the paper presents a curated data set for digital humanities researchers that is automatically enriched with metadata derived by different algorithms including content-based image features. The automatic enrichment of originally bibliographic metadata is needed to support the exploration of large metadata stock as traditional metadata does not always address vague INs.

The presented proof of concept clearly shows that use case-specific metadata facilitates the interaction with large metadata corpora.

1 Introduction

While digital library (DL) projects have been developed for more than two decades, their support for different information seeking strategies (ISS) has not evolved as fast as the amount of offered stock, presentation quality, and presentation variety.

Typical information seeking interactions with DLs are still predominantly text-centered although the DL stock often contains textual, visual, or audio material. Obviously, this phenomenon is linked to the fact that digital library metadata is often directly based on the *textual* catalogs of their "analog" counterparts, i.e., the physical library. Even though modern DLs such as Europeana or the Digital Public Library of America[1] feature *multimodal information access* to a certain extent utilizing faceted navigation, color-based browsing, or even geo-spatial presentations of their library stock, which eventually overcomes the usage scenarios of paper-based catalogs, the predominantly supported ISS is

---

[1] http://www.europeana.eu/; http://dp.la/.

© Springer International Publishing Switzerland 2016
N. Fuhr et al. (Eds.): TPDL 2016, LNCS 9819, pp. 307–319, 2016.
DOI: 10.1007/978-3-319-43997-6_24

still *directed search*, i.e., the submission of keyword-based queries or the usage of sophisticated query languages. Typically, directed searches lead to list-based result pages displaying both textual and visual content (frequently sorted by their relevance regarding the current query) or a display of digitized content within a dedicated viewer infrastructure. Roughly speaking, the first presentation approach mimics the catalog handling experience while the latter simulates the handling of printed media such as books or maps.

Complementing the aforementioned usage scenarios, DLs also target at researches from the Digital Humanities (DH) and other libraries/data aggregators by offering their metadata via standardized interfaces such as OAI-PMH[2] for further processing.

In any case, the described use cases have in common that they require clarity regarding the current information need (IN). While this is the case for libraries or data aggregators, it is well understood that this is not true for human users as various studies on information seeking behavior, e.g., [1,2], show.

For instance, the *cognitive viewpoint* on information retrieval (IR) argues that INs are not always well-defined and change during information seeking. Moreover, Kuhlthau's famous large scale study [2] observing 385 users at 21 different library sites shows that users apply multiple ISS to overcome different levels of clarity regarding their IN. Similar results are presented by [1,3,4] and many more. These ISS range from various browsing strategies used while the IN is vague, e.g., during the early development phase of a new research idea, to directed search if the IN is well-defined, e.g., if a specific medium has to be retrieved. Considering the age and reproducibility of these studies, it is surprising that they have gained little influence on practical DL applications or interfaces.

As indicated above, metadata regarding DL stock is multimodal. There is metadata describing geographic locations, entities such as natural persons, high-level textual features describing media content and automatically extractable low-level features such as color, shape, or audio properties. There is evidence on both experimental [5,6] and theoretic level [3] that a combination (or fusion) of multimodal features improves retrieval effectiveness. Furthermore, multimodal access to a document collection facilitates exploration as it allows its inspection from different "angles".

As a consequence, this paper proposes a multimodal approach for the exploration of large metadata repositories. Furthermore, the presented approach supports the transition between different explorative ISS and traditional directed search systems such as OPACs[3], discovery systems supporting full-text search, and digitized content viewers.

The main contribution of this publication is twofold. *First,* a primarily explorative research system prototype for metadata harvesting is proposed which allows multimodal access to DL stock for both DH researchers during the research idea development phase and librarians striving for data quality improvement.

---

[2] Open Archives Initiative Protocol for Metadata Harvesting; https://www.openarchives.org/pmh/.

[3] Online Public Access Catalog.

Second, this paper presents a curated data set for DH researchers that is automatically enriched with metadata derived by different algorithms including content-based image features. The data set is unique as it mostly represents the unique and rate stock of the Berlin State Library, one of the largest and oldest research libraries in the German-speaking world. The corpus is multi-lingual as the library's stock is not limited to European or German media alone. Present languages range from Latin over West European languages to Middle-Eastern and Asian languages.

To facilitate further research, the data set is prepared to be used with open source software common in data science such as the NumPy/SciPy/pandas stack[4] or graph visualization packages such as Gephi and D3[5]. For the sake of transparency, the data set is made available in company to the metadata processing tools used to derive it from raw metadata.

The paper is structured as follows. The next section outlines the theoretic motivation for explorative information seeking in the DL field. Section 3 discusses the prototypical system architecture serving as a proof of concept. The following Sect. 4 concentrates on its user experience (UI) and supportive visualizations. The last section concludes the paper and presents issues that need further research.

2   Theoretic Motivation and Background

The *cognitive viewpoint* on IR, commonly attributed to [7], takes a holistic perspective on information seeking. A representative of this approach is the principle of polyrepresentation (PoP) [8]. The principle's core hypothesis is that a document is best described by multiple representations such as textual content, low-level features or the user's context. The user's IN (or cognitive state) is also represented by different aspects such as the current work task or context. Each of these aspects is best addressed by different representations. Hence, a combination of multiple representations (the so-called cognitive overlap) is said to compensate the uncertainty about the current IN. Thus, the PoP can be interpreted as a theoretic justification for multimodal information access.

The cognitive viewpoint's core ideas are summarized by [8, p. 25] as follows:

1. Information processing takes place in senders *and* recipients of messages;
2. Processing takes place at *different levels*;
3. During communication of information any actor is *influenced* by its past and present experiences (*time*) and its social, organizational and cultural environment;
4. Individual actors *influence* the environment or domain;
5. Information is *situational* and *contextual*.

---

[4] http://www.numpy.org/; http://www.scipy.org/; http://pandas.pydata.org/.
[5] https://gephi.org/; https://d3js.org/.

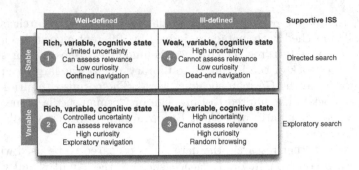

Fig. 1. Matrix of Ingwersen's intrinsic information need extreme cases [9, Fig. 6.4]

Reflecting the dynamic IN as recapitulated above, [3] describes the IN extreme cases users undergo during information seeking and how they can be supported best as illustrated in Fig. 1. The figure clearly suggests that the most supportive type of ISS changes as the state of the IN evolves. Similar findings are reported by [1, 2, 4, 10].

Consequently, a useful DL interface should foster the usage of different ISS and a transition between them. Addressing these ISS changes within the UI is also proposed, e.g., by [11] who suggests the usage of with different views supporting interactions based on varying ISS, or by the BRAQUE (*browsing* and *query*) system [12]. To sum up, [13] provides a comprehensible overview of alternatives to the direct search and query-response paradigm.

3   Data Curation and Prototypical System Architecture

As said before, the major strength of digital libraries is the support of directed ISS (see Fig. 1, #1) with the help of traditional OPACs and discovery systems. To complement other use cases driven by vague INs, we propose a prototypical research system, *SBBrowse*, focussing on the support of explorative ISS addressing IN extreme cases #2 and #3. Nevertheless, the prototype is not limited to explorative ISS alone as is tightly coupled to traditional retrieval systems as described in Sect. 4.

In our opinion, the support of vague INs has to be reflected *also* on the data side as typical bibliographic metadata aims at describing media most precisely. In fact, this level of preciseness might overstrain users developing their current IN while it at the same time lacks information helpful to explore a document corpus, such as the relation between different media objects. In other words, the data has to be *curated with respect to the targeted use case* – preferably without the continuous intervention of additional library staff in case of large libraries or collections. Hence, the presented system automatically simplifies the available metadata by removing typical cataloging information (see Sect. 3.1) and extends it with multimodal representations such as geographic coordinates, relationships (see Sect. 3.3), or content-based image features (see Sect. 3.2). We believe that

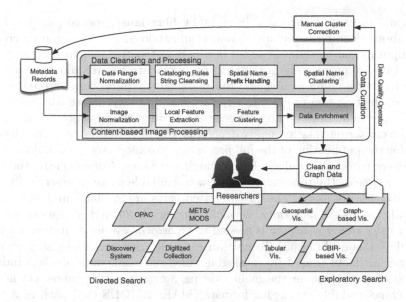

Fig. 2. Prototypical metadata processing and search architecture

a multimodal approach towards metadata exploration is crucial as a vague IN will be sharpened by inspecting a problem from different angles as suggested by the principle of polyrepresentation.

Figure 2 illustrates the main processing and visualization steps of SBBrowse including a data quality management feedback loop that is discussed in Sect. 4. The database of the prototype consists of 120,000 metadata records from the catalog of the Digitized Collections of the Berlin State Library publicly available via OAI-PMH. For the sake of simplicity, its Dublin Core[6] endpoint is used although the system is in principle not limited to a particular metadata standard.

3.1 Data Cleansing and Processing

The SBBrowse prototype focuses on the exploitation of visual similarity, temporal, and geographic information besides typical bibliographic data such as titles and publishers. To show the feasibility of the assertions made above, four exemplary processing steps illustrate the way from precise bibliographic metadata to metadata facilitating multimodal exploration based on computational methods without major human intervention. The steps have been chosen for illustrative purposes as the represent typical challenges of transforming metadata for different use cases, i.e., the transformation of bibliographic metadata that originally did not have DH scenarios in mind. In any case, the presented methods are not limited to the discussed fields.

As some of the processing steps utilize methods from machine-based learning, the derived metadata is not completely accurate with respect to the gold

---

[6] http://dublincore.org/.

standard, i.e., metadata manually input by librarians[7], special care has to be taken about cleaning metadata derived from catalogs. Hence, metadata created by computational methods is added in a non-destructive manner, i.e., this data is annotated appropriately in the resulting metadata records to ensure a maximum of *trustable data quality* and *transparency regarding data provenance*. For instance, the first two processing steps are mainly based on regular expressions yielding the same data quality for each processing run. In contrast, the results from the clustering vary in quality depending on the actual input data. Thus, we argue for the publication of the full processing workflow (see Sect. 3.3) to enable researchers to obtain resilient DH research outcomes. Consequently, the commented source of the processing workflow is available as an interactive Python-based Jupyter notebook. The CBIR-related parts are available in Java[8].

While the task of *date range normalization* needs no further explanation, *cataloging rules string cleansing* is needed to transform various strings into a form readable for non-librarians. To give an example, this processing step removes character artifacts caused by cataloging rules such as square brackets indicating texts not present on the main title page or separating characters needed by machine-readable cataloging formats (MARC, MODS etc.) such as $. This level of information neither does contribute to gaining an overview of a document corpus during exploration nor does it help DH researchers because the correct interpretation of bibliographic special characters often requires the study of extensive cataloging rules[9].

Typically, library metadata contains compound terms in various spelling variants such as "Saint Tropez" and "Saint-Tropez" in the case of spatial names. Such issues are usually handled by authority files which, at the same time, require the absence of spelling errors. These errors can be induced by faulty insertions, character encoding problems, or erroneous OCR. To tackle this problem group, the prototype relies on different clustering approaches to remove special characters and to handle the aforementioned kinds of "defective" strings. A particular challenge posed onto the clustering algorithm is the multi-linguality of the corpus. As a result, an appropriate algorithm has to be relatively robust against multi-lingual misspellings or phonetic transcriptions as well. Regarding the discussed corpus, the best clustering was obtained using the Jaro-Winkler distance [14] applied directly on Unicode strings of spatial names. Comparable distance functions such as Levenshtein and Winkler did not yield convincing results. In the first sub-processing step, the Jaro-Winkler distance was used to compute a distance matrix between spatial names which served as input for agglomerative (hierarchical) clustering using the single-linkage method. The resulting hierarchical clustering was then flattened so that the observations in each flat cluster

---

[7] Although one might argue whether humans do not err.

[8] For the sake of readability, we decided to publish the full source in form of a documented Jupyter (http://jupyter.org/) notebook as a supplement to this paper to limit the discussion of algorithmic parameters to a minimum.

[9] For instance, the *RAK* main manual, the German counterpart to Anglo-American Cataloguing Rules, is a 627 pages long document.

had no greater cophenetic distance t than 0.095. The dyadic clustering process has proven to be most usable because it allows the visual inspection of clusters with the help of dendrograms in order to choose parameters for the cluster flattening. Moreover, it does not require a-priori knowledge regarding the amount of clusters hidden in the data, as it would be the case with k-means clustering variants. The clustering reduced the number of spatial names from 7,273 to 3,855, a number that was controlled manually in order to assess their accuracy[10]. To further assess the quality of the built clusters, their representing spatial name was checked against Google Maps because its web service supports multi-liguality and is known to be robust against erroneous or ambiguous (spatial names contain cities, regions etc.) input. From all 3,855 clusters, Google rejected only four spatial names. The rejected names could all be traced back to character encoding issues, e.g., "KoiLn" for Köln (Cologne) could not be resolved. This data quality assurance step was also used to obtain geographic coordinates of the spatial names.

3.2 Content-Based Image Processing

The content-based image processing steps borrow methods from content-based image retrieval (CBIR). As said before, some DLs already support so-called *global* color or texture-based features for information access [15]. These features are describing an image globally, i.e., the image as a whole. Roughly speaking, this way of representing an images has the drawback that specific key points in an image are often underrepresented. For instance, the background of a historical book's page is often dominating the information-bearing content (e.g., its text) when two documents are compared by their color alone. In other words, the use of color-based information has limited discriminative power in DLs that primarily feature digitized paper-born content. In contrast, the SBBrowse prototype relies on *local* features that break down images into local information-bearing patches that are then used to determine the similarity between two images.

Before the local features could be extracted, all images were normalized to a maximum height of 512 pixels (with varying width respecting the original aspect ratio) and converted to the sRGB IEC61966-2.1 color space. All normalized images are based on manually quality controlled master TIFF images with at least 300 dpi resolution, which are also referenced in the corresponding metadata records[11].

The local features were extracted using LIRE [16] and OpenCV[12] using CEDD/SIFT descriptors. The resulting features were then clustered to obtain

---

[10] Because of the unavailability of ground truths similar to our corpus, the limited amount of data in the prototype, and the non-destructive extension of the metadata records, we decided against a full automation of the evaluation. However, the resulting clusters are stable enough to be checked against common authority files.

[11] The in principle optional normalization was carried out primarily to offer researchers a homogeneous image data set.

[12] http://opencv.org/.

bags of visual words with a codebook size of 512 and 1000 clusters. The core idea of the visual word approach is to convert local feature vectors to artificial "words" that are elements of the codebook (similar to the index vocabulary in textual information retrieval). The advantage of this approach is that such features can later be treated as normal "textual" documents also represented in the vector space model. In other words, DH researchers can apply their present data mining knowledge to this kind of features as well. For instance, the visual words are used for a similarity-based clustering of the images and the calculation of "centroid" images representing the resulting clusters that are later used to visualize the collection (see Sect. 4).

3.3 Data Curation, Data Publication, and Reproducibility

The final processing or data curation step consists of the metadata enrichment. In this phase, the original catalog-derived metadata is extended with geographic coordinates, spatial cluster names, and content-based local image features. Based on this additional metadata, the prototypical workflow creates different output formats in tabular and static graph form, e.g., graphs illustrating relations between creators and publishing location such as shown in Fig. 3. These static graphs are complemented by interactive, dynamic graphs addressed in Sect. 4.

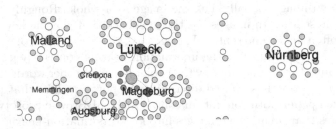

Fig. 3. Detail of a GML graph created in Gephi with ForceAtlas 2 layout of printer Lucas Brandis' publishing locations and available incunables (red highlight) (Color figure online)

Following a library's general objective to facilitate access to knowledge, the data publication is focussing on the support of platform-agnostic open source technologies allowing DH researchers further processing with their utilities of choice. For instance, all cleaned and enriched data is made available as Unicode CSV, pandas data frames, and JSON or GML in case of graphs.

The resulting data set consists of 120,000 enriched metadata records and 102,478 normalized color JPEG images of which 101,877 images are accompanied by local CEDD/SIFT features. The number of images is smaller than the total amount of records as periodicals were only represented by symbolic images which were therefore removed from the corpus. Table 1 gives a full overview of the data set.

Table 1. Characteristics of the data set

Description	Formats	No. of elements/avg. nodes & edges	Size
Enriched metadata records	CSV, pandas	120,000	24.7 MB
Graphs (publisher-creator-spatial-media)	GML	12,806.67/12,444.50	74.5 MB
Graphs (full network, per century)	GML	46,199.17/109,748.33	179,6 MB
Visual similarity cluster graphs	JSON, GML	1,622 clusters	163.7 MB
Normalized images	JPEG	102,478	6.9 GB
Local features	LIRE (ASCII)	101,877	1.66 GB

4   User Interface and Visualization

The main challenge regarding the UI is the sheer amount of displayed information in the case of a data set with more than 100,000 records. Hence, the user interface is developed around Shneiderman's ubiquitous visualization mantra "overview first, zoom and filter, then details on demand" [17]. At the *overview phase*, users are confronted with an interactive visualization of characteristic presentation images of various media (title pages, bindings, photographs etc.). At first sight, this visualization illustrates the variance of presentation images summarizing the nature of the collection.

To further facilitate orientation, the images are grouped by century and clustered by visual similarity (see Fig. 4). The amount of displayed images

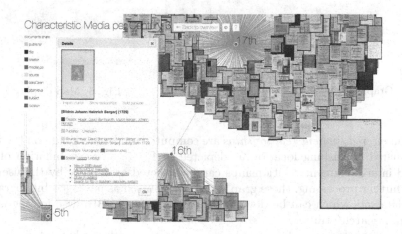

Fig. 4. Visual similarity-based overview with detail view and magnified document

(or clusters k) depends on the result of a mini-batch k-means [18] clustering whose k value is relative to the amount of available media per century. This visualization is helpful at overview level because it allows visual orientation while displaying the amount of available media per century as well in case users want to inspect the data deeper. The displayed images ("centroids" or most average image of a certain cluster) are representing further cluster levels that can be reached by clicking a centroid. The next filter level follows the same UI mechanisms.

To facilitate detailed inspections of the displayed media objects, the visualization can be zoomed and panned. Moreover, each image is magnified if touched with the cursor and each image offers a *detail view dialog* which shows its central multimodal representations in a tabular fashion (see Fig. 4; left) and allow access to additional functions. The detail view dialog is also central for the transition between different ISS. For instance, it links the inspected document to other IR systems such as the OPAC or enables users to display the document in a dedicated viewer infrastructure. The detail view is context-sensitive, i.e., in overview mode, it allows the inspection of the current cluster and the submission of queries relevant to the current document as its title or creator. In the map-based visualization, the dialog offers queries for spatial locations found in the examined metadata (see Fig. 6). Furthermore, the dialog allows the download of various metadata formats (e.g., METS/MODS) for the currently inspected or related documents.

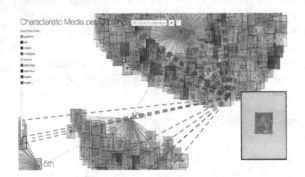 

Fig. 5. Graph displaying spatial relationships   **Fig. 6.** Map display

The *inter-document relationships* are computed on demand, e.g., on the basis of a shared publishing location as depicted in Fig. 5. The combination of displayed inter-document relationships can be selected arbitrarily by the user. To allow further processing, these graphs can be exported as packaged bibliographic metadata sets which can be downloaded directly from SBBrowse for the use in custom research utilities.

The supported steps from exploring a collection with the help of visual similarities or inter-document relationships to a directed search in library catalogs and the eventual download of metadata records is fully aligned with the cognitive viewpoint's assertions about different information processing levels (see Sect. 2). From a different point of view, the transition between different ISS can also be considered a reflection of choosing the appropriate toolset or data representation for overview, filter, and detail inspection tasks. This transition is also represented by metadata supporting different use cases as described above.

We have deliberately decided not to include facetted navigation in our proof of concept because of its lack of novelty. The utility of facetted navigation has been shown over the last decades and can be derived automatically from common bibliographic metadata as well as it can created on demand depending on the current user interactions.

Although the developed SBBrowse prototype primarily aims at supporting DH researchers, we have discovered that it can also be used as valuable utility for *metadata quality control*. By combing different visualizations and the background knowledge of librarians, the discovery process of problematic metadata records is simplified. For instance, by displaying media objects on a map, problematic metadata records are emphasized because they often appear at locations that are not typical for the Berlin State Library's collection profile. By utilizing features from CBIR, it becomes possible to find problems with structural bibliographic data such as the presentation of a digitized binding although a book contains a proper title page. Discovering such problems by hand would mean a systematic inspection of all metadata records and, in some cases, autoptic work in the closed stacks. The amount of human effort can be minimized by a machine-assisted reduction of the candidate set with the help of CBIR algorithms and the interactive and iterative use of visualizations which are connected with a feedback loop to the relevant processing steps as illustrated in Fig. 2 (right). This important feedback does not have to be limited to librarians. We believe that crowd or researcher-driven feedback can also be used as it might be potentially closer to the needed use cases of the published data.

The prototype's UI is completely Web-based and relies on modern browser's functionalities such as JavaScript, DOM, and SVG alone[13].

5   Conclusion

This paper presents a proof of concept of the explorative multimodal metadata harvesting utility *SBBrowse* and a data set for DH researchers. SBBrowse is based on research results from the cognitive viewpoint on IR motivating the dynamic support of different ISS. Consequently, the system is not limited to explorative ISS. Instead, it facilitates the transition between different ISS by combining a novel UI with well-established DL services such as OPACs or METS/MODS metadata endpoints. The exploration is based on an automated enrichment of traditional metadata records in order to adapt it to use cases occurring during early research, i.e., when the current IN is evolving. The resulting data set is made available to DH researchers in various forms in conjunction with

---

[13] http://sbb.berlin/sbbrowse.

the automated metadata enrichment toolchain. The utility of the prototypical system and metadata processing workflow is promising.

In the next release of the data set, OCR-derived full-texts will be added. The Berlin State Library is currently processing its full copyright-free stock in order to provide full-text to the DH community. We plan to release raw OCR results even if they contain errors (which we expect due to the large amount of text of the library's stock set in blackletter/fraktur). We believe that the presented multimodal approach will compensate weaknesses in the OCR performance during the interaction with the corpus. In any case, the proper handling and incorporation of OCR full-texts needs further research.

Regarding the prototype's functionality, we plan to add support for faceted navigation to extend the amount of available means for exploration. The research will be complemented by a user acceptance study.

References

1. Reiterer, H., Mußler, G., Mann, T., Handschuh, S.: INSYDER. In: Proceedings of the 23rd SIGIR 2000, pp. 112–119. ACM (2000)
2. Kuhlthau, C.C.: Inside the search process: information seeking from the user's perspective. J. Am. Soc. Inf. Sci. **42**(5), 361–371 (1991)
3. Ingwersen, P.: Cognitive perspectives of information retrieval interaction: elements of a cognitive IR theory. J. Doc. **52**, 3–50 (1996)
4. Ellis, D., Haugan, M.: Modelling the information seeking patterns of engineers and research scientists in an industrial environment. J. Doc. **53**(4), 384–403 (1997)
5. Thomee, B., Popescu, A.: Overview of the ImageCLEF 2012 flickr photo annotation and retrieval task. In: Forner, P., Karlgren, J., Womser-Hacker, C. (eds.) CLEF 2012, Online Working Notes, Rome, Italy, 17–20 September 2012 (2012)
6. Caputo, B., Muller, H., Thomee, B., Villegas, M., Paredes, R., Zellhofer, D., Goeau, H., Joly, A., Bonnet, P., Martinez Gomez, J., Varea, I.G., Cazorla, M.: ImageCLEF 2013: the vision, the data and the open challenges. In: Forner, P., Müller, H., Paredes, R., Rosso, P., Stein, B. (eds.) CLEF 2013. LNCS, vol. 8138, pp. 250–268. Springer, Heidelberg (2013)
7. deMey, M.: The cognitive viewpoint: its development and its scope. In: International Workshop on the Cognitive Viewpoint, CC 1977, Ghent, Belgium, pp. xvi–xxxii (1977)
8. Ingwersen, P., Järvelin, K.: The Turn: Integration of Information Seeking and Retrieval in Context. Springer, Dordrecht (2005)
9. Zellhöfer, D.: A preference-based relevance feedback approach for polyrepresentative multimedia retrieval. Ph.D. thesis, Brandenburg Technical University (2015)
10. Bates, M.: The design of browsing and berrypicking techniques for the online search interface. Online Rev. **13**(5), 407–424 (1989)
11. Marchionini, G., Geisler, G., Brunk, B.: Agileviews: a human-centered framework for interfaces to information spaces. In: Proceedings of the Annual Conference of the American Society for Information Science, pp. 271–280 (2000)
12. Belkin, N., Marchetti, P., Cool, C.: BRAQUE: design of an interface to support user interaction in information retrieval. Inf. Process. Manag. **29**(3), 325–344 (1993)
13. White, R., Roth, R.: Exploratory Search: Beyond the Query-Response Paradigm. Synthesis Lectures on Information Concepts, Retrieval, and Services. Morgan & Claypool Publishers, San Rafael (2009)

14. Winkler, W.E.: String comparator metrics and enhanced decision rules in the Fellegi-Sunter model of record linkage. In: Proceedings of the Section on Survey Research, pp. 354–359 (1990)
15. Fox, E.A., Leidig, J.: Digital Libraries Applications: CBIR, Education, Social Networks, eScience/Simulation, and GIS. Synthesis Lectures on Information Concepts, Retrieval, and Services. Morgan & Claypool Publishers, San Rafael (2014)
16. Lux, M., Chatzichristofis, S.: Lire: Lucene image retrieval: an extensible Java CBIR library. In: Proceedings of the 16th ACM MM 2008, pp. 1085–1088. ACM (2008)
17. Shneiderman, B.: The eyes have it: a task by data type taxonomy for information visualizations. In: Proceedings of the 1996 IEEE Symposium on Visual Languages, VL 1996, pp. 336–343. IEEE Computer Society (1996)
18. Sculley, D.: Web-scale k-means clustering. In: Proceedings of the 19th International Conference on WWW 2010, pp. 1177–1178. ACM, New York (2010)

Person-Centric Mining of Historical Newspaper Collections

Mariona Coll Ardanuy[1]([⊠]), Jürgen Knauth[1], Andrei Beliankou[2],
Maarten van den Bos[3], and Caroline Sporleder[1]([⊠])

[1] Göttingen Centre for Digital Humanities, Göttingen University,
Göttingen, Germany
{maria.coll-ardanuy,juergen.knauth,
caroline.sporleder}@cs.uni-goettingen.de
[2] Department of Computational Linguistics and Digital Humanities,
Trier University, Trier, Germany
a.beliankou@uni-trier.de
[3] Department of History and Art History,
Utrecht University, Utrecht, The Netherlands
M.J.A.vandenBos@uu.nl

Abstract. We present a text mining environment that supports entity-centric mining of terascale historical newspaper collections. Information about entities and their relation to each other is often crucial for historical research. However, most text mining tools provide only very basic support for dealing with entities, typically at most including facilities for entity tagging. Historians, on the other hand, are typically interested in the *relations* between entities and the *contexts* in which these are mentioned. In this paper, we focus on person entities. We provide an overview of the tool and describe how person-centric mining can be integrated in a general-purpose text mining environment. We also discuss our approach for automatically extracting person networks from newspaper archives, which includes a novel method for person name disambiguation, which is particularly suited for the newspaper domain and obtains state-of-the-art disambiguation results.

Keywords: Multilingual text mining · Historical text mining · Person name disambiguation · Semantic search

1 Introduction

From the outset, newspapers and other forms of printed mass media have played a prominent role in the formation of public opinion. For many years, they were the main sources of information. Thus they are not only records of the important events of a given time (or rather, of the events that were chosen to be reported by the newspaper editor) but they are also valuable indicators of public opinion and debate. In fact, due to their prominent role as an information source, newspapers played a significant role in transmitting opinions, ideologies and values, conditioned on the interests they served, thereby shaping public opinion and

© Springer International Publishing Switzerland 2016
N. Fuhr et al. (Eds.): TPDL 2016, LNCS 9819, pp. 320–331, 2016.
DOI: 10.1007/978-3-319-43997-6_25

stimulating public debate. Even so, printed mass media have not yet enjoyed the popularity among historians that would be expected for such a precious resource. This is probably due to two factors: First, historical newspapers have not always been easily accessible in the past, with different publications being dispersed and scattered in various libraries and archives. Fortunately, this is changing quickly with the mass digitization of newspapers, gazettes, magazines, pamphlets, and other kinds of materials which ideally puts these resources within easier reach for researchers from all over the world, making it likely that the practice of using them as sources in historical research becomes more common.

A second reason why printed mass media have not yet featured prominently as sources of historical research is that —in a Humanities context— newspaper archives constitute 'big data' and are therefore difficult to analyze with traditional Humanities methods, i.e., close reading. Historians are often not so interested in the detailed content of individual articles but in more general trends of how topics are discussed over time and how this differs from newspaper to newspaper or country to country. This is a distant reading application that requires dedicated text mining tools. The digital humanities community has developed a plethora of (typically) general purpose text mining tools over the past years, but most of these only support relatively shallow analyses. Needless to say, these can already be extremely valuable to researchers; a simple keyword search, for example, can already provide a historian with a sense of whether a collection contains material relevant for their research, thus saving many hours of visiting archives and skimming through pages. However, such approaches often fall short of fully supporting the specific needs of historical research.

In this paper, we present ASINO, a software infrastructure that can tackle large (i.e. terascale) textual collections of digitized texts. The tool has been designed with the help and advice of historians to tailor it to their needs. Apart from providing the basic functionalities of any digital humanities system, it facilitates a deeper examination of the texts, in order to reveal information and connections that might be hidden or difficult to find in large collections. One core aspect of ASINO is that it provides support for entity-centric text mining. Information about entities such as persons or locations is often crucial for historical research. However, most tools provide at best shallow named entity analysis, in which expressions in the text referring to a named entity are detected, classified according to the entity type (e.g., PERSON or LOCATION) and tagged. Users can then search for entities of a given type or restrict their analyses (e.g., computing co-occurrence statistics) to the set of all named entities or a given named entity type. However, historians are typically also interested in the *relations* between entities and the *contexts* in which they occur. For example, they might be interested in which contexts two people are mentioned together or during which time periods. Most existing text mining tools do not provide support for this.

Note that the extraction of named entity networks involves more than combining named entity recognition and tagging with computing co-occurrence information due to the fact that there is no 1-to-1 mapping between a linguistic expression referring to a named entity and its referent in the real world. Like all referring expressions, named entity expressions (also called *named entity mentions*) such as *Clinton* or *Michael Jackson* can be ambiguous and potentially have more than one referent.[1] Automatically extracting named entity networks thus requires automatic *named entity disambiguation* (i.e., distinguishing between different referents referred to by the same named entity expression) in addition to named entity recognition and tagging. While methods for named entity disambiguation exist, they are typically not included in text mining tools for the Humanities. We discuss how this can be done. We also provide a novel method for entity disambiguation which is particularly suited for the newspaper domain and obtains state-of-the-art disambiguation results. While ASINO will eventually provide support for person, location and other entities, in this paper we exemplary focus on the former.

The rest of this paper is organized as follows. In Sect. 2 we present the related work. Section 3 describes the system, with its possible inputs and outputs. The novel method for creating networks automatically and their impact in historical research is described in Sect. 4. We conclude the paper in Sect. 5.

2 Related Work

Text Mining Environments for Historical Newspaper Archives. The idea of creating an integrated text exploration and mining environment for digital humanists is not new. Many systems exist, each with different scopes and facilities. ASINO stands in the tradition of xTAS (*Text Analysis in a Timely Manner*) and Texcavator though it was developed from scratch and thus is not a direct descendent of these. xTAS is a system for historical text indexing that was proposed by de Rooij et al. [12]. It features a preprocessing pipeline for normalizing optical character recognition errors, spelling variants and stemming, and provides an interface for fast full text indexing. Texcavator [24] is a comprehensive system based on xTAS. It reuses its core implementation and adds rich facilities for faceted search, such as timeline visualization, word cloud cluster representation, and data export for additional processing. Torget et al. [32] pursued similar goals and added the geographic dimension in their project Mapping Texts,[2] producing an outstanding example of geospatial visualization and semantic modeling based on the concept of topic maps.

Entity Networks in Historical Research. The utility of social networks in historical research has been corroborated by several studies. Padgett and Ansell [23] investigated in detail the action and rise of the Medici during the period between 1400 and 1434 by means of social networks; Jackson [18] used social networks of the medieval Scottish elites from the period between 1093 and 1286 to find hidden relationships; Rochat et al. [29] focused on the Venetian maritime

---

[1] The converse also applies: a given referent can be referred to by various expressions. This is dealt with by automatic coreference resolution. We do not address this aspect in the current paper.

[2] http://mappingtexts.org.

empire from the end of the 13th century to 1453, and networks of ports and places were used to model maritime connections. Note that all these examples concern pre-modern history, where the source material is much more limited and thus networks can still be constructed by hand. In fact, most approaches to historical network analysis to date create the networks either manually or from structured data, in which cases person name disambiguation does not pose a problem. The potential of automatic approaches and the importance of performing name disambiguation is discussed by Stratford and Browne [31] in the context of analyzing connections and communities in Assyria. One of the few fully automatic approaches to network construction for historical research is provided by Coll Ardanuy et al. [9], who highlight the utility of automatically created social networks from newspaper archives in a case study on European integration. Automatically constructed networks have also been used in other Humanities areas, for example in computational approaches to literature studies [14]. Works of fiction, however, arguably pose less of a challenge for this task than newspaper collections as the number of distinct entities and thus the expected amount of ambiguity is significantly lower.

Automatic Person Name Disambiguation. Resolving and disambiguating person names across documents is an open problem of unquestionable importance in natural language processing. Its difficulty lies on the high ambiguity which is often associated with person names. The first work on person name disambiguation goes back to Bagga and Baldwin [2]. In it, a standard vector space model was used to cluster together documents referring to the same entity, thus exploiting the fact that the contexts in which a given referent occurs are typically similar. Several subsequent works [15,28] adapted, extended, and improved this approach. Yoshida et al. [33] proposed a distinction between weak features (context words of the document) [20,30] and strong features (named entities, biographical information, or temporal expressions) [1,6,21,22]. Among the latter, named entity information has been the most used feature [5,8,19,26]. Over the last years, the trend has moved towards adopting approaches that are more strongly knowledge- and resource-based, for instance exploiting knowledge bases (KB) [13] or Wikipedia. Furthermore, the person name disambiguation task has been in most cases subsumed by entity linking [7,11,17], where entity mentions in documents are linked to a reference knowledge base or Wikipedia articles. The main drawback of entity linking approaches, in particular when applied to network extraction from historical texts, is that it is often infeasible to create a knowledge base which is large enough to cover all entities mentioned in the corpus.

3   System Architecture

In designing ASINO, our aim was to build a specialized multilingual software that would assist historians exploring large collections of digitized texts. The software offers the basic functionalities for text mining and at the moment indexes a mixed collection of newspaper texts ranging from mid-19th Century to mid-20th

Century. The texts are mostly in German (from Germany and German-speaking Belgium) and Dutch (from the National Library in the Netherlands). Further newspaper collections from Luxembourg and Belgium are currently in the process of being preprocessed and indexed. The system can be expanded to work with sources in any language, possibly after mapping deviating metadata schemata.

ASINO makes use of a relational database[3] to index texts and metadata, enabling quick access and search of the indexed texts. Pre-processing is applied to texts and to the metadata. The latter involves converting metadata in non-standard XML or plaintext into a standardized format. The system includes an indexing server,[4] capable to cope with terascale data, which supports different languages using out-of-the-box tokenization, stemming, and named entity recognition, besides providing faceted search methods to narrow user queries by different data parameters. Finally, our system runs on an application server[5] which opens a Representational State Transfer (REST) interface for external queries and a web-based access for every modern browser.

ASINO supports different types of queries: atomic queries, quoted queries (which return exact matches), queries combined with Boolean operators, and queries employing regular expressions. An automatic query translation component is currently under development, which will enable the user to simultaneously search all language collections by providing a query in one language. The tool enables the user to restrict the query results according to the following criteria, present in the metadata: place of publication, type of publication, newspaper source collection, document language, and date range (see Fig. 1). ASINO offers various export options and supports typical output facilities such as:

- **List of Documents:** The user can view the full text of a retrieved document (hit), download the full text and the associated metadata as a CSV file, or select a number of hits and download their full text and associated metadata as a CSV file.
- **Wordcloud:** The user can preview a hit in the form of a word cloud (normalized and not normalized) and save the graphical representation.
- **Timeline:** The user can preview a hit or list of hits in the form of a timeline (normalized and not normalized) with given time granularity, and save the graphical representation.

In addition to this basic functionality, users will also be able to visualize the results of a given query as an enriched entity network.[6] We discuss this in the following section.

---

[3] We use PostgreSQL: http://www.postgresql.org/.

[4] Based on SOLR/Lucene: http://lucene.apache.org/solr/.

[5] Developed using Ruby/Rails/Angular.

[6] At the time of writing, the person network module has already been implemented, tested, and evaluated, but is not yet fully integrated in the main tool. Consequently, this visualization option is missing from the screenshot in Fig. 1.

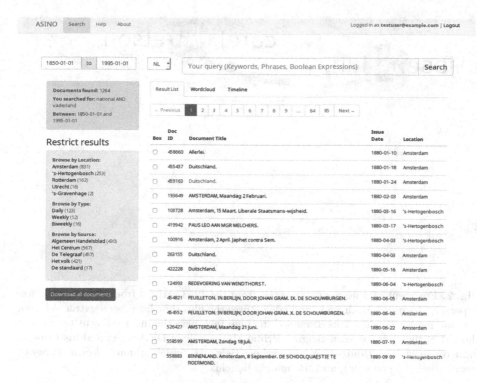

Fig. 1. Screenshot of the ASINO tool.

4 Person-Centric Text Mining

Displaying the retrieved documents as a social network representation draws the focus of the news articles to their social perspective. A social network captures the relations (edges) between different actors (nodes). We note that the terms of these relations, i.e., the context in which two people are mentioned together, might be as relevant for the historian as the relations themselves. Therefore, far from being empty structures, our networks store certain information for each node and edge. For each node, two attributes are automatically created: one with the list of documents (source files) in which the entity appears, and one with the list of the most relevant (tf.idf weighted) terms which are contained in the documents where the entity appears. Likewise, two attributes are created for each edge in the network: one with the list of documents in which the two people in question appear, and one with the list or relevant terms discussed in documents where the two entities co-occur (see Fig. 2).

The combination of (multilingual) query-based search and entity network representation provides the user with a very intuitive and flexible way to explore and visualize the collection. For example, the user might be interested in the key players in the European integration process. He or she can compile a list of relevant queries in one language, translate it (semi-)automatically into a specified

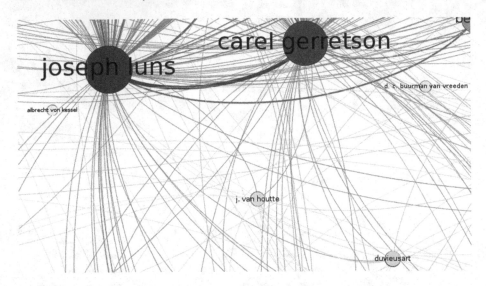

Fig. 2. Fragment of a network extracted from the 1953 articles from the Dutch newspaper *De Telegraaf*. Each node and edge stores the files they have been extracted from and the most relevant terms discussed in them (e.g. here the most relevant terms (not shown) of the edge between Belgian Prime Minister Jean Duvieusart and his Finance minister Jean Van Houtte are 'Belgische' (Belgian), 'unie' (union), 'lonen' (wages), 'economische' (economic), and 'Benelux' (Benelux)).

set of additional languages, find matching documents in the whole collection or a subset (for example a given time period). To quickly decide whether the subcorpus is usable, the result list can be browsed or the matching documents can be visualized as a word cloud. If the subcorpus seems relevant, the user can then display the entity network for the matching documents. This will provide information about (i) the important actors for the topic, (ii) which actors were commonly mentioned together and (iii) in which contexts individual actors or pairs of actors were mentioned. The network thus is a means to delve deeper into the source material and analyze it in ways that go beyond simply visualizing entity co-occurrences and that combine instead an entity and a topic-based exploration of the data. Eventually, ASINO will provide support for other entity types as well (locations, trademarks, etc.).

4.1   Named Entity Disambiguation

As we discussed in the introduction, it is important to note that 'person entity' is not a synonym for 'person name mention': the nodes in our networks correspond to entities, i.e., real-world referents, not to occurrences of a named entity expression in a text. This means that if two persons share the same name, they should have one node each, whereas if two mentions refer to the same person, they should be contained in the same node. The task of resolving which entity is referred to by each person name mention is known as *named entity disambiguation*, and is one of the biggest challenges of automatic network creation. In this section, we describe how we identify the nodes for the network, by means of a novel graph-based named entity disambiguation method.

Person name disambiguation is probably the greatest challenge automatic network creation faces: the same person name can refer to several entities whereas the same entity can be referred to by many several names. In order to illustrate the problem, we narrow our focus to a particular entity, the German politician Carlo Schmid, who was greatly involved in the reconciliation process between France and Germany after World War II. The combination of first and last name ('Carlo' and 'Schmid') is relatively uncommon, and thus the full name 'Carlo Schmid' is not very ambiguous. However, it is very often the case that names are not presented in their full form in newspaper articles. A good disambiguation solution should be able to decide that 'Schmid' in sentences 1 and 2 refers to the [Carlo] Schmid we are interested in, whereas in sentences 3 and 4 does not.[7]

> **Sentence 1.** *Erler and _Schmid_ stated that Khrushchev showed interest in the proposals.* [March 19th, 1959]
>
> **Sentence 2.** *Led by Bundestag Vice-President _Schmid_, the assembly pointed out that new hopes had arisen for the German problem after the visit of Khrushchev had been announced in Bonn.* [December 8th, 1964]
>
> **Sentence 3.** *Detective inspector _Schmid_ has made available all the files about Georg Vogel to be found at the police headquarters available.* [February 25th, 1960]
>
> **Sentence 4.** *Secondary actor _Schmid_ sold a pound note.* [November 14th, 1961]

An artificially-created network which does not deal with name disambiguation is likely to suffer from high recall loss (if every person name mention has its own node, then we loose the information that the 'Schmid' in sentences 1 and 2 refer to the same person), or high precision loss (if all matching person names are represented by one node, then the node 'Schmid' will also contain the information from sentences 3 and 4).

Most successful approaches to named entity disambiguation rely on a knowledge base of entities to which names can be linked (i.e., *named entity linking*). However, for a historical newspaper collection, no database exists containing all entities contained in it, and thus an alternative method had to be found which exploited only the information which the texts themselves provide. Our disambiguation method is based on textual context and network overlap, where ambiguity of a name plays a central role. Not all person names are equally ambiguous. Some person names are highly ambiguous, whereas others are fairly unambiguous. To date, very few approaches [4,25,34] incorporated name ambiguity. We include the degree of ambiguity of a person name in our disambiguation model in order to be able to estimate the probability that two networks containing the

---

[7] The examples are taken from the *St. Vither Volkzeitung*, a German newspaper based in St. Vith (Belgium). We present here the sentences in their translation into English for ease of understanding. We work with the original data.

same mention of a person name actually refer to the same entity rather than to two different entities with the same name.

Ambiguity of a name is directly proportional to its commonness: the more common a name is, and thus the more people sharing it, the more ambiguous this name will be. A name such as 'Edward Schillebeeckx' is very uncommon, and thus its ambiguity will be very low. Conversely, a name such as 'John Smith' is extremely common, and thus its ambiguity is very high, since it can possibly refer to many different entities. The ambiguity of a person name plays a big role in deciding whether two mentions of the same name refer to the same entity: to continue with our example, it is much more likely that two mentions of 'Edward Schillebeeckx' refer to the same Edward Schillebeeckx than that two mentions of 'John Smith' refer to the same John Smith. In designing our method for network creation, we took into account this observation.

Estimating Name Ambiguity. Person names are usually combinations of a first name, a last name, and occasionally one or more middle names. Only with the list of all the people in the world would it be possible to assess the true ambiguity of each person name. This is an unavailable resource, and alternative ways of approximating the ambiguity of person names need to be found. Zanoli et al. [34] use an Italian specific resource, the phonebook *Pagine Bianche*. It has wide coverage, but it could be argued that the use of *Pagine Bianche* leads to a gender-biased calculation of name ambiguity, since often only one person per household is included in its pages, which is often still the male head. Instead, we extract person names from a large corpus of raw text using a named entity recognizer. To optimize precision, we consider only names consisting of at least two tokens, since single tokens are often misidentified or misclassified by the recognizer. The recognized person names are then used to build three lists — corresponding to first, middle, and last names — in which each distinct name is associated with its occurrence frequency in the corpus.[8]

Disambiguation Strategy. As mentioned in the system architecture description (Sect. 3), all documents in the collection have been named entity tagged. Since we assume that the same person name within an article refers always to the same entity,[9] the first step is to group all matching surface forms of person names together.[10] We then turn every news article into a social network: the person names mentioned in it become the nodes of the network, and they are linked together through an edge whenever they co-occur in the same window of

---

[8] Note that this approach is optimized for the four languages in which we work: English, German, Dutch, and Italian. Dealing with languages with different naming conventions (such as Spanish and Chinese) would mean having to modify slightly the approach.

[9] This is a relatively safe simplifying assumption, reminiscent of the "one-sense-per-discourse" principle often adopted in word sense disambiguation.

[10] We implemented a set of heuristics to detect matching names. For example, a first name-surname combination matches with an identical surname string which does not contain a first name.

text (set to be a paragraph). In order to create cross-document networks, we compute for each pair of documents represented as person networks how many nodes they have in common. Our method assumes that the more names coincide in two documents, the higher the probability that these names refer to the same entities in both documents. However, the amount of overlap between two networks is not the only information source we take into account; we also factor in the inherent ambiguity of an entity mention. The confidence that we are talking about the very same person is noticeably higher if the entity mention has a low ambiguity (e.g. 'Edward Cornelis Florentius Alfonsus Schillebeeckx') than if it has a high ambiguity (e.g. 'John').

In order to decide whether the overlapping nodes of two networks correspond to the same entity, and thus should be combined into one node, we learn clustering probabilities from a development set. We use the Italian corpus CRIPCO [3] for this. For each node in common between two networks, we need to know how ambiguous this node is, how many other common nodes share the pair of networks, the inherent ambiguity of these common nodes, and the context similarity of the text behind the two networks (since an approach relying only on social networks alone produces high precision at the cost of recall, we also use context similarity between documents). Thus, for each pair of overlapping nodes, we learn the probabilities based on these criteria from the networks created from the texts of the development set.

Method's Performance. In the natural language processing field, the task of person name entity disambiguation is usually modeled as a clustering problem. Lacking a gold standard of the network creation task, this is the closest way to assess the performance of our method. We tested the performance of our method against three benchmarks in two languages: the CRIPCO Corpus in Italian [3], and the NYTAC Pseudo-name Corpus [27] and the John Smith Corpus [2] in English. Our method's performance is on par with the state-of-the-art reported for the CRIPCO dataset and proved to be competitive in different languages and throughout very different collections without need to retrain it.[11]

5 Conclusion

With the recent explosion of digitization projects, an increasing effort is being devoted to address the problem of information extraction from unstructured text. Several studies have hinted that deep text mining may offer great possibilities to broaden the boundaries of historical research through digitized texts. This paper introduced ASINO, an integrated working environment for digital humanists. The system, designed with the assistance and advise of historians, allows running queries over large collections of digitized texts and returns the results in a form which suits the needs of the researchers. One of the novelties of the system is that it supports person-centric data visualization, which allows a deep exploration of the original sources from a social perspective.

---

[11] For a more detailed description of the method, experiments, and results, see [10].

The visualization of results as lists of documents, wordclouds, or timelines is typically fairly shallow as it is based on the raw data matching of the query. This provides the historian with a good overview of potentially interesting data but does not provide a deeper exploration of the content. However, by visualizing the results as a social network, the historian obtains a person-centered overview of a collection, which can further be focussed in terms of dates, ideology, source, or topic (by means of query restriction). The resulting structure is likely to reproduce some expected results (expected people appearing in prominent positions in the network, expected relations between actors, expected relevant terms between one actor and another, etc.), which underlines the reliability of the approach; and, more interestingly, some unexpected results, which may potentially become the bases for new hypotheses that challenge the dominant narratives of history, which can be further explored by more traditional research methods.

References

1. Al-Kamha, R., Embley, D.W.: Grouping search-engine returned citations for person-name queries. In: Proceedings of the 6th ACM WIDM Workshop, pp. 96–103 (2004)
2. Bagga, A., Baldwin, B.: Entity-based cross-document coreferencing using the vector space model. In: Proceedings of Coling, pp. 79–85 (1998)
3. Bentivogli, L., Marchetti, A., Pianta, E.: Creating a gold standard for person cross-document coreference resolution in Italian news. In: Proceedings of LREC Workshop on Resources and Evaluation for Identity Matching, Entity Resolution and Entity Management, pp. 19–26 (2008)
4. Bentivogli, L., Marchetti, A., Pianta, E.: The news people search task at EVALITA 2011: evaluating cross-document coreference resolution of named person entities in Italian news. In: Sprugnoli, R. (ed.) EVALITA 2012. LNCS, vol. 7689, pp. 126–134. Springer, Heidelberg (2012)
5. Blume, M.: Automatic entity disambiguation: benefits to NER, relation extraction, link analysis, and inference. In: Proceedings of the International Conference on Intelligence Analysis (2005)
6. Bollegala, D., Matsuo, Y., Ishizuka, M.: Extracting key phrases to disambiguate personal name queries in web search. In: Proceedings of the ACL Workshop on How Can Computational Linguistics Improve Information Retrieval? (2006)
7. Bunescu, R., Pasca, M.: Using encyclopedic knowledge for named entity disambiguation. In: Proceedings of EACL, pp. 9–16 (2006)
8. Chen, Y., Martin, J.: Towards robust unsupervised personal name disambiguation. In: Proceedings of EMNLP-CoNLL, pp. 190–198 (2007)
9. Coll Ardanuy, M., van den Bos, M., Sporleder, C.: Laboratories of community: how digital humanities can further new European integration history. In: Aiello, L.M., McFarland, D. (eds.) SocInfo 2014 Workshops. LNCS, vol. 8852, pp. 284–293. Springer, Heidelberg (2015)
10. Coll Ardanuy, M., Sporleder, C.: You shall know people by the company they keep: person name disambiguation for social network construction. In: Proceedings of LaTeCH 2016 (forthcoming)
11. Cucerzan, S.: Large-scale named entity disambiguation based on Wikipedia data. In: Proceedings of EMNLP-CoNLL, pp. 708–716 (2007)
12. de Rooij, O., Vishneuski, A., de Rijke, M.: xTAS: text analysis in a timely manner. In: 12th Dutch-Belgian Information Retrieval Workshop (2012)

13. Dutta, S., Weikum, G.: Cross-document co-reference resolution using sample-based clustering with knowledge enrichment. TACL **3**, 15–28 (2015)
14. Elson, D.K., Dames, N., McKeown, K.R.: Extracting social networks from literary fiction. In: Proceedings of ACL, pp. 138–147 (2010)
15. Gooi, C.H., Allan, J.: Cross-document coreference on a large scale corpus. In: Proceedings of HLT-NAACL, pp. 9–16 (2004)
16. Han, X., Sun, L.: An entity-topic model for entity linking. In: Proceedings of EMNLP-CoNLL 2012, pp. 105–115 (2012)
17. Han, X., Zhao, J.: Named entity disambiguation by leveraging Wikipedia semantic knowledge. In: Proceedings of CIKM, pp. 215–224 (2009)
18. Jackson, C.A.: Using Social Network Analysis to Reveal Unseen Relationships in Medieval Scotland. In: Digital Humanities Conference, Lausanne (2014)
19. Kalashnikov, D.V., Chen, S., Nuray, R., Mehrotra, S., Ashish, N.: Disambiguation algorithm for people search on the web. In: Proceedings of IEEE International Conference on Data Engineering, pp. 1258–1260 (2007)
20. Kozareva, Z., Ravi, R.: Unsupervised name ambiguity resolution using a generative model. In: Proceedings of EMNLP, pp. 105–112 (2011)
21. Mann, G.S., Yarowsky, D.: Unsupervised personal name disambiguation. In: Proceedings of the 7th Conference on Natural Language Learning at HLT-NAACL, pp. 33–40 (2003)
22. Niu, C., Li, W., Srihari, R.K.: Weakly supervised learning for cross-document person name disambiguation supported by information extraction. In: Proceedings of ACL, pp. 598–605 (2004)
23. Padgett, J.F., Ansell, C.K.: Robust action and the rise of the Medici, 1400–1434. Am. J. Sociol. **98**(6), 1259–1319 (1993)
24. Pieters, T., Verheul, J.: Cultural text mining: using text mining to map the emergence of transnational reference cultures in public media repositories. In: Digital Humanities 2014 Book of Abstracts, pp. 299–301 (2014)
25. Popescu, O.: Person cross document coreference with name perplexity estimates. In: Proceedings of EMNLP, pp. 997–1006 (2009)
26. Popescu, O., Magnini, B.: IRST-BP: web people search using name entities. In: Proceedings of SemEval, pp. 195–198 (2007)
27. Rao, D., McNamee, P., Dredze, M.: Streaming cross document entity coreference resolution. In: Proceedings of Coling, pp. 1050–1058 (2010)
28. Ravin, Y., Kazi, Z.: Is Hillary Rodham Clinton the president? disambiguating name across documents. In: Proceedings of the Workshop on Coreference and its Applications, pp. 9–16 (1999)
29. Rochat, Y., Fournier, M., Mazzei, A., Kaplan, F.: A network analysis approach of the venetian incanto system. In: Digital Humanities Conference, Lausanne (2014)
30. Song, Y., Huang, J., Councill, I.G., Li, J., Lee Giles, C.: Efficient topic-based unsupervised name disambiguation. In: Proceedings of JCDL, pp. 342–351 (2007)
31. Stratford, E., Browne, J.: LinkedIn circa 2000 BCE: Towards a Network Model of Pušu-ken's Commercial Relationships in Old Assyria. Digital Humanities Conference, Sydney (2015)
32. Torget, A.J., Mihalcea, R., Christensen, J., McGhee, G.: Mapping texts: combining text mining and geo-visualization to unlock the research potential of historical newspapers. In: National Endowment for the Humanities (2011)
33. Yoshida, M., Ikeda, M., Ono, S., Sato, I., Nakagawa, H.: Person name disambiguation by bootstrapping. In: Proceedings of SIGIR, pp. 10–17 (2010)
34. Zanoli, R., Corcoglioniti, F., Girardi, C.: Exploiting background knowledge for clustering person names. In: Sprugnoli, R. (ed.) EVALITA 2012. LNCS, vol. 7689, pp. 135–145. Springer, Heidelberg (2012)

e-Infrastructures

Usability in Digital Humanities - Evaluating User Interfaces, Infrastructural Components and the Use of Mobile Devices During Research Process

Natasa Bulatovic[1](✉), Timo Gnadt[2], Matteo Romanello[3],
Juliane Stiller[4], and Klaus Thoden[5]

[1] Max Planck Digital Library, Amalienstrasse 33, 80799 Munich, Germany
bulatovic@mpdl.mpg.de
[2] Göttingen State and University Library, Platz der Göttinger Sieben 1,
37073 Göttingen, Germany
gnadt@sub.uni-goettingen.de
[3] Deutsches Archäologisches Institut, Podbielskiallee 69-71,
14195 Berlin, Germany
matteo.romanello@dainst.de
[4] Berlin School of Library and Information Science,
Humboldt-Universität zu Berlin, Dorotheenstr. 26, 10117 Berlin, Germany
juliane.stiller@ibi.hu-berlin.de
[5] Max Planck Institute for the History of Science, Boltzmannstr. 22,
14195 Berlin, Germany
kthoden@mpiwg-berlin.mpg.de

Abstract. The usability of tools and services that form a digital research infrastructure is a key asset for their acceptance among researchers. When applied to infrastructures, the concept of usability needs to be extended to other aspects such as the interoperability between several infrastructure components. In this paper, we present the results of several usability studies. Our aim was not only to test the usability of single tools but also to assess the extent to which different tools and devices can be seamlessly integrated into a single digital research workflow. Our findings suggest that more resources need be spent on testing of digital tools and infrastructure components and that it is especially important to conduct user tests covering the whole knowledge process.

Keywords: Usability · Evaluation · Digital humanities · Digital tools · Research process · Research infrastructures

1 Introduction

Usability describes the utility and operability of a product as seen from the perspective of its consumers. When considering the usability of research infrastructures, it is necessary to take into account not only the user interfaces, but also the degree of interoperability between the infrastructure components (e.g. tools, services) when they are used together in a research workflow. The more usable a given tool is, the higher

© Springer International Publishing Switzerland 2016
N. Fuhr et al. (Eds.): TPDL 2016, LNCS 9819, pp. 335–346, 2016.
DOI: 10.1007/978-3-319-43997-6_26

are the chances for this tool to be accepted among researchers. Therefore, usability plays a key role in the dissemination and impact of digital tools, consequently affecting the research output of their users.

The studies presented in this article were conducted in the project DARIAH-DE[1], the German contribution to the digital research infrastructures of the arts and humanities (DARIAH-EU[2]). It dedicates one of its clusters to evaluation research. The research studied the user behavior with regard to digital tools in the humanities[3]. Furthermore, the usability of several existing tools and their graphical user interfaces was evaluated. Another outcome was a model of the research process in the arts and humanities that included several digital tools and services and end user devices. The goal here was to examine the transition between tools and users' behaviour when performing device-specific tasks. This paper summarizes and presents the work and focuses on three different strands: (1) common usability problems for Digital Humanities (DH) tools and services, (2) problems arising from a purely digital workflow taking into account different end user devices and tools, and (3) usability factors beyond user interfaces.

2   Usability in the Digital Humanities[4]

The usability of tools and components of digital infrastructures is a paramount asset to increasing the acceptance of digital tools among researchers. The user-friendly design of workflows and user interactions can facilitate effective work and minimize operating errors. Usability tests nowadays are a standard element in most software development processes. The heuristics published by Nielsen [9] and often also the "Eight Golden Rules of Interface Design" by Shneiderman and Plaisant [12][5] form the basis of such testing methodology. Testing of human-machine interactions is currently carried out by means of qualitative methods ranging from interviews to observation as well as quantitative methods such as log file analysis. In order to evaluate the usability in information systems, interactive tests with users or experts are conducted, or software for automated interface testing is used [8].

In the Humanities, usability testing during and after the development of digital tools is not yet widely established. Burghardt [2] finds a gap between developers and scientific users, reflected for example in the lack of usability engineering for linguistic annotation tools. He employs a Heuristic Walkthrough[6] to generate design recommendations, differentiating between generic and domain specific usability problems.

---

[1] http://de.dariah.eu/.

[2] http://dariah.eu/.

[3] This paper is a translated, shortened and adapted version of a DARIAH-DE deliverable [10].

[4] Parts of this section are translated and adapted from [14].

[5] See an updated list at: https://www.cs.umd.edu/users/ben/goldenrules.html and in [12].

[6] A technique developed by Sears [11] combining heuristic evaluations, cognitive walkthroughs und usability walkthroughs and belonging to test with experts.

According to Gibbs et al. [4], missing investment in user interfaces is the reason for rejection of tools by frustrated users.

While usability of digital tools has to date not received much attention in Digital Humanities, it remains to be seen whether special requirements for this field exist. Nevertheless, the contribution of usability to an increased acceptance and usage of tools and services is undisputed and was confirmed also by surveys conducted in the second phase of DARIAH-DE (see [5, 13]).

The criteria from Nielsen [9] should therefore be applied to the development of DH tools and services. This could not only increase the user satisfaction, it will furthermore affect the validity and reproducibility of results produced with the help of these tools and ideally strengthen the confidence in DH tools and services development. Nevertheless, the tradeoff between additional costs of implementing usability criteria and the specificity of a tool i.e. the size of its target audience needs to be considered.

In the context of The William Blake Archive (see [7]), Kirschenbaum identifies a typical problem with interface design in the common tendency to deal with usability issues at a later stage of the development process (often when it is too late). In spite of this, user behavior and expectations are being increasingly considered in interface design as demonstrated e.g. by Boukhelifa et al. [1] and Heuwing and Womser-Hacker [6], who both use Participatory Design approaches to involve users during the software design phase.

3   Methodology

To get an overview of the different components of usability in the digital humanities, we tested (1) stand-alone applications and (2) complete workflows which were not bound to specific applications, but rather to the research questions researchers want to answer.

Usability tests of tools and components of the DARIAH-infrastructure were oriented towards classical usability tests using the methods of heuristic evaluation by experts and tests by users following special tasks (Think-aloud test). Additionally, we were interested in the interplay of different tools and services within one research process. Table 1 lists the number of testers and the methodology used for each usability test we carried out.

Table 1. Usability studies conducted within DARIAH-DE

Tested services	Methodology	Number of testers/evaluators
DARIAH Publish GUI	Heuristic evaluation	4
LAUDATIO[7] repository	Heuristic evaluation	2
LAUDATIO repository	User study	1
UX demonstrator	Field study	2–10 dependent on task

---

[7] LAUDATIO project, http://www.laudatio-repository.org/laudatio/

With regard to the tools offered by DARIAH-DE, we tested the functionality of the first prototype of the Publish-GUI[8], a web interface that allows users to archive digital objects and collections in the DARIAH-DE repository [3]. The results of the evaluation were reported back to the developers who subsequently translated them into implementation tasks. For external tools and services, we evaluated the "LAUDATIO" (Long-term Access and Usage of Deeply Annotated Information) repository[9], which provides access to historical language corpora. In this case, two experts performed a heuristic evaluation based on the guidelines defined by Nielsen, followed by a Think-aloud test. For this test, specific tasks determined by the repository's developers and researchers were executed by the tester [14].

Apart from usability testing, we designed a demonstrator workflow (User Experience demonstrator workflow, in short: "UX demonstrator") to examine the researchers' user experience when using various types of devices (smartphone, tablet, laptop, desktop computer) with different infrastructure components (Fig. 1). It comprised the following tasks from research processes of humanists working with images of real physical objects:

- integration of existing collections of images and their metadata;
- search for relevant locations to gather new material;
- collection of new material by means of mobile devices;
- metadata enrichment of collected material;
- visualization of material.

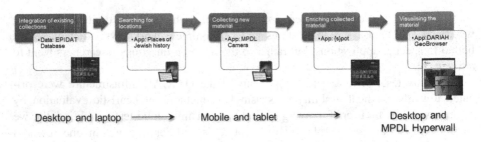

Fig. 1. UX demonstrator and the tools used for each step

The UX demonstrator was tested virtually and on-site at two locations with different participants. Table 2 shows the tasks completed by participants and the used device. The feedback from participants was collected with the Think-aloud method and in face-to-face meetings. Furthermore, the software development team directly provided feedback on technical efforts and functionality.

---

[8] The DARIAH-DE Publish-GUI will be accessible via https://de.dariah.eu/repository. Currently it is still in a beta status and not available publicly.

[9] http://www.laudatio-repository.org/repository/.

Table 2. Tasks within the UX demonstrator

Task	Participants (role)	Number of participants	Location	Device
Integration and aggregation of existing collections of images	Junior researcher, senior researcher, data infrastructure expert, data expert	5	Virtual	Desktop
Search for relevant locations to gather new material	Junior researcher, senior researcher	3	Munich	Mobile, tablet
Collecting new material with mobile devices	Junior researcher, senior researcher	4	Essen	Android camera, tablet
Enriching material with metadata	Senior researcher, junior researcher	2	Virtual	Desktop
Visualizing material	Data infrastructure expert, senior researcher, junior researcher	10	Munich	Desktop, hyperwall

The steps of the UX demonstrator which were assessed are detailed below:

A. **Integration and Aggregation of Existing Collections of Images.** In the first step of the workflow, we ingested a small part of the EPIDAT[10] database of Jewish epigraphy (ca. 1500 images and their metadata) into the {s}pot service of the Max Planck Digital Library (MPDL)[11]. Images were selected by participating researchers and were aggregated in two datasets, dating from 1980 and 2015. Metadata were provided in a tabular format (MS Excel file). Both images and metadata were uploaded to the {s}pot service via its REST-API using an ad-hoc MATLAB script[12].

B. **Search for Relevant Locations to Gather New Material.** To gather further material, the mobile application "Places of German-Jewish History"[13] was used. It offers local information about Jewish cultural history. The application identifies and shows places of interest near the device's geolocation via Google Maps and links them with contextual information gathered from several data sources such as Wikipedia, EPIDAT, Getty Thesaurus of Geographic Names[14] and others[15].

C. **Collecting New Material with Mobile Devices.** Together with three researchers from the Steinheim-Institute, new material for the research collection was accumulated on the Segesroth cemetery in Essen. For this purpose, an early version of

---

[10] http://steinheim-institut.de/cgi-bin/epidat.

[11] https://www.mpdl.mpg.de/.

[12] https://github.com/imeji-community/imejiLab.

[13] http://app-juedische-orte.de.dariah.eu/?ilanguage=EN.

[14] http://www.getty.edu/research/tools/vocabularies/tgn/.

[15] See: http://app-juedische-orte.de.dariah.eu/q?vpage=datenquellen.

the LabCam mobile application (LabCam App)[16] was used to take pictures of gravestones and upload them to the {s}pot service. The application was installed on an Android Camera and on a tablet device. The tablet was connected to the internet and pictures were automatically transferred to {s}pot. Pictures taken with the Android Camera were uploaded to {s}pot once the device was connected to the internet.

D. **Enriching Material with Metadata.** After the upload to {s}pot, users were able to add new metadata and enrich the existing metadata with further details, for each image separately or in a batch mode.

E. **Visualizing Material.** To illustrate the usability of different end user devices, we defined the following visualization scenarios:

1. Visualization of metadata with the DARIAH-DE GeoBrowser[17] in a browser, on a laptop and a desktop device. The visualization presents all tombs in Germany and the dates of their construction.

2. Visualization of data and display of pictures on the Hyperwall[18]. Hyperwall tests were performed with several controlling devices: XBox 360 Controller, Keyboard and Mouse and a tablet device.

3. Visualization of several Web of Science Citation Maps in a browser, on a laptop, desktop device and on the Hyperwall.

The UX demonstrator was focused on identifying aspects of usability and the potential of using devices such as smartphones, tablets or specialized display installations as the Hyperwall (Fig. 2). In parallel, we analyzed the feasibility of the workflow and addressed the impact on the implementation efforts.

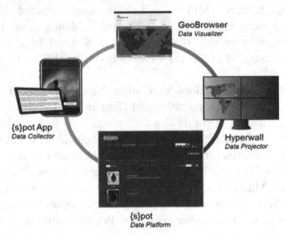

Fig. 2. The UX demonstrator on different devices

---

[16] For more information see http://labcam.mpdl.mpg.de/ - the LabCam App captures images or videos and transfers them to a repository based on the imeji software (http://imeji.org).

[17] https://de.dariah.eu/geobrowser.

[18] The Hyperwall is an installation of 4x4 K Monitors to show images and videos with high resolutions. It is located at the Max Planck Digital Library in Munich.

4 Findings and Results

In digital research workflows, the scope of usability is not limited to single interactions between the researcher and a device or to the design of the graphical user interface, but also extends to the seamless integration of different tools for the research process. Nevertheless, most of the usability problems which users encounter are related to graphical user interfaces (GUIs) and are not specific to the Digital Humanities. We found that the main problems can be summarized to the following list:

- Inconsistent use of vocabulary and terminology, unclear and ambiguous terms
- Violation of graphical conventions
- Insufficient or missing system status notifications
- Missing documentation and inadequate placing of information
- Error-prone conditions where the system does not prevent the users from making errors or recovering from errors fast
- Insufficient search and browsing functionalities

These issues can be considered as general software problems which users most often experience when using GUIs and are not specific to the tasks and functionalities of the tools.

Transitions Between Mobile and Desktop Devices are Feasible. Our initial hypothesis was that researchers will use the application "Places of Jewish history" to identify points of interest and the LabCam App to document their primary sources (gravestones), take pictures and transfer them to an online repository – the {s}pot service. This has been confirmed during the testing phase, which however, also pointed out the fact that mobile devices often do not have image quality necessary to digitize the epitaphs and inscriptions (of gravestones) in a resolution sufficient for studying the object. Researchers thus suggested an alternative workflow where the images taken with a mobile device are considered as a "draft" digitized object, enriched with some automatically generated metadata such as geo-coordinates. Participating researchers explicitly confirmed that they are willing to use mobile devices when performing their research activities.

Mobile Devices Need to be Practical for Use. UX demonstrator tasks for collecting new material were tested with an Android camera and a tablet device, both with an installed LabCam App. Due to its form and size, the Android camera was easier to work with than the tablet, although it produced images with much lower quality than a professional camera. The tablet device was ergonomically not appropriate to take pictures effectively, however, it was very convenient to add metadata to the pictures through the {s}pot service web application right on-site, after pictures were created and transferred to the {s}pot service.

Additional Devices Might be Necessary Depending on the Research Question. To produce high quality images, in most cases further equipment is required, e.g. professional lighting to emphasize the inscriptions on the tombstones or the magnitude of their damage, or a tachymeter to measure the distance, position and dimensions of the real-world object. Where high-resolution images are needed, "draft" images created with a mobile device will be replaced with an image produced by a professional camera at another (usually later) stage in the research.

Mobile Apps and Desktop Applications Differ by the Complexity of User Interactions. While desktop/web applications are frequently designed with the assumption that researchers will learn and perform even complex tasks and functionalities, for mobile applications this is rather not desirable. The LabCam App was designed and validated with researchers. It was concluded that it should offer substantially less functionalities than the {s}pot web application. Furthermore, it should provide a higher level of automation, i.e. several functions of the {s}pot web application should be integrated into a single application function. The LabCam App should provide different views of data (adjusted for mobile display) than those available for the {s}pot Web application.

User Expectations Towards Functionalities of a Desktop/Web Application and a Mobile Application are Different. We noticed several discrepancies between what users expected from the web application and the mobile application during testing. When uploading data through the web application, users demanded it to finish fast and to be complete. LabCam App expectations were slightly more sophisticated: the mobile application should automatically generate metadata such as the width/height of the tombstone, the distance to the next tomb, exact geo-coordinates of the tomb (not to be confused with the coordinates of the camera position).

The requirements for the data editing and metadata enrichment processes were almost opposite: manually adding/enriching image metadata was considered as an overload for the mobile app, an ability to add a comment or a "tag" was regarded as sufficient. This very same functionality was identified as a "must-have" feature for the {s}pot web application.

Such differences are bound to impact the implementation efforts. In scenarios where an application communicates with an online service, usually accessible through a browser-based web application, equivalent functions need to be implemented at least twice with different technical components and different interaction scenarios.

Large Displays with Higher Resolution are Not Always Better to Visualize Data. Displaying an image by using the operating system's default viewer on the Hyperwall, in our case the Windows Image viewer, was very effective, especially due to the high image resolution. Displaying the same image through the Digilib Image Viewer[19] on the Hyperwall was acceptable and scaled well according to the size of the physical display, while the representation was slightly blurred.

Display of data in the GeoBrowser web application was good both on standard monitors devices and on the Hyperwall. The visualization of a Web of Science Citation Map was good on standard monitor display, but not on the Hyperwall: some graphical elements did not scale according the size of the display and their content was practically not readable.

The Hyperwall tests were performed using a keyboard and mouse, an XBox 360 controller and a Tablet device. While the interaction with the tablet was easy for all

---

[19] See: http://digilib.sourceforge.net/.

users, the XBox controller usage required some upfront experience. The keyboard/mouse usage was hindered by the size of the mouse pointer: it was heavy to spot it on the Hyperwall display. For using the mouse on such a large screen, it was necessary to resize the mouse pointer in order to improve the interaction of the user.

These tests have shown that even when applications scale well to several different display sizes, they often fail to scale to the Hyperwall size. Optimizing the presentation for installations such as the Hyperwall should be taken into consideration already during implementation of the application, and should be tested thoroughly.

Optimized User Interfaces Often Depend on Research Questions, Which Change Continuously. The UX demonstrator was tested with researchers from various disciplines. Often the research question was not clearly specified or known in advance. Therefore, the goal of data collection and enrichment was primarily set to document the sources. The LabCam App enabled creation of images of primary objects (tombstones). Specific metadata such as the width or length of the tombstone, its position relative to other tombstones e.g. "row 1, left or right" or the identification of a group of tombstones could not be supported by the App. These are specific requirements, depending both on the nature of primary objects (tombstones) and the actual research question: same primary data could be used by linguists, or by archaeologists or historians. While for linguists, it would be relevant to link the metadata containing epitaph texts with digitized dictionaries, for archeologists forms and the distribution of the tombstones over time would be more important. Historians, finally, would be rather interested in the names and birthplaces of the buried persons and their personal relations.

In a further example, historians' requirements for the {s}pot service were focused on integration of controlled vocabularies – thus pursuing new developments on specific navigation and editing mechanisms. A generalization in the implementation may be achieved by the application of existing standards such as SKOS[20]. However, there is still no guarantee that the SKOS standardization will at the same time fulfill the requirements of linguists, especially since their dictionaries are seldom available in SKOS notation and have a different internal structure. To support such requirements, further changes in the application would have to be implemented, affecting existing navigation, editing and other functionality and user interfaces. New research questions bring new challenges for the application, especially for data visualization, data enrichment and formatting features. Optimization of the user interfaces thus has to be continuously addressed and implemented – as long as the application is in use.

A Data Expert is Beneficial for Input, Quality Assurance and Data Enrichment. Images and metadata of the ingested EPIDAT database collections and the information about "linking rules" between the image files and the metadata in the Excel table were acquired directly from participating researchers. Besides these, an additional Data Expert (DE) was involved during the process. DE contributed substantial experience with tools for data processing and relevant Python/MATLAB libraries and knew how to use the REST API of the {s}pot service to upload the data. DE performed validation of the data integrity and their completeness, fixed several quality and formatting

---

[20] https://www.w3.org/2004/02/skos/.

problems, and enriched the data with geo-coordinates, authoring and licensing information. The DE's feedback and proposals on improving the data quality were well accepted and adopted by the researchers.

An experienced DE acting as a mediator for the data entry, enrichment and quality assurance is often more valuable experience for researchers than a web application with good user interfaces and equivalent functionality, for the following reasons:

- researchers do not have learn rarely used features and technical details in advance
- researchers can focus on using tools they are most comfortable with when curating their data (e.g. MS Excel in many cases)
- the DE documents the process and therefore creates potentially reusable solutions
- the DE offers personal support for all questions related to processing of (highly valuable) researchers' data

Availability of Various Data Formats is Important Prerequisite for Better User Experience. An important question during establishing the UX demonstrator was if the uploaded data would be accessible from the {s}pot service, so that they could be presented and reused in different contexts. Data access via machine-readable interfaces and download of enriched data in formats such as JSON or CSV was regarded as essential by several researchers for reuse of data in other disciplines, making data publicly available and "branding" of data. Other features regarded as highly important were the provision of a familiar downloadable format (e.g. a ZIP archive) and a REST API to access the data and their metadata, so that these can be used for new types of analysis or visualization, and be easily linked from the web pages of the institution or a researcher.

5 Discussion

This paper presented the results from several usability studies in the realm of the Digital Humanities, performed on tools developed or used within the DARIAH-DE project landscape. The term "usability" was not only applied to the utility of tools and the design of graphical user interfaces, but was deliberately expanded to further aspects of the digital research process.

One of the main results of this study is that usability still only plays a marginal role in the tool development for Digital Humanities. Often, there are no resources for testing tools thoroughly and the schedule for tool and infrastructure development do not permit early iterative cycles of user testing and sufficient software adaptation to meet the user needs. Working with real users, as we have done in testing of the digital research process, requires resources for preparation and post-processing of the results. Finding subjects with the right expertise and the willingness to invest time into such a process is difficult, which is one reason why we could recruit only few participants. Heuristic walkthroughs usability tests can be completed faster and with fewer resources, but they do not reveal how researchers are actually working with the tools.

The studies showed that usability aspects should be considered early in the development process. Often it is enough to accompany the development cycle by iterative heuristic evaluations or think-aloud tests with a limited number of users. Standalone DH-Tools and -Services often fulfill very specific requirements and are highly specialized. They tend to be complex and are often not self-explanatory. Here, the question is how much time researchers are willing and able to invest in learning how to use a tool and utilize its functionalities. For this reason, addressing various usability perspectives of such tools directly with the researchers from the very beginning of the tools' development is pivotal. Wherever possible, usability testing should become an essential part of the development process.

Additionally, it is necessary to find the right level of granularity in tools specialization and interface design, which is not always the case in digital humanities. Our analysis has shown that profound human expertise in the research process can be more beneficial than an effective user interface which might require months of tool development and customization with an uncertain outcome. Well-documented REST APIs promote data reuse and lead to improvements, potentially beyond the original purpose of the creator tools.

References

1. Boukhelifa, N., Giannisakis, E., Dimara, E., Willett, W., Fekete, J.-D.: Supporting historical research through user-centered visual analytics. In: Bertini, E., Roberts, J.C. (eds.) EuroVis Workshop on Visual Analytics (EuroVA), pp. 1–5 (2015)
2. Burghardt, M.: Annotationsergonomie: Design-Empfehlungen für linguistische Annotationswerkzeuge. Inf. Wiss. Prax. **63**(5) (2012). doi:10.1515/iwp-2012-0067
3. Funk, S.E., Schmunk, S.: DARIAH-DE Repositorium – Prototyp. (M 4.3.2.1). DARIAH-DE. Report (2015). https://wiki.de.dariah.eu/download/attachments/14651583/M4.3.2.1-DARIAH-Repositorium-Prototyp-final.pdf
4. Gibbs, F., Owens, T.: Building better digital humanities tools: toward broader audiences and user-centered designs. Digit. Humanit. Q. **6**(2) (2012). http://www.digitalhumanities.org/dhq/vol/6/2/000136/000136.html
5. Gnadt, T., Stiller, J., Thoden, K., Schmitt, V.: Finale Version Erfolgskriterien. (R 1.3.3). DARIAH-DE. Report (2015). https://wiki.de.dariah.eu/download/attachments/14651583/R133_Erfolgskriterien_Konsortium.pdf
6. Heuwing, B., Womser-Hacker, C.: Zwischen Beobachtung Und Partizipation – Nutzerzentrierte Methoden Für Eine Bedarfsanalyse in Der Digitalen Geschichtswissenschaft. Inf. Wiss. Prax. **66**(5–6), 335–344 (2015). doi:10.1515/iwp-2015-0058
7. Kirschenbaum, M.G.: 'So the Colors Cover the Wires': interface, aesthetics, and usability. In: Schreibman, S., Siemens, R., Unsworth, J. (eds.) A Companion to Digital Humanities, pp. 523–542. Blackwell Publishing Ltd. (2004)
8. Lazar, J., Feng, J.H., Hochheiser, H.: Research Methods in Human-Computer Interaction. Wiley, Chichester (2010)
9. Nielsen, J.: 10 Usability Heuristics for User Interface Design (1995). http://www.nngroup.com/articles/ten-usability-heuristics/. Accessed 15 Jan 2016
10. Romanello, M., Bulatovic, N., Gnadt, T., Schmitt, V.E., Stiller, J., Thoden, K.: Usability von DH-Tools und -Services (R 1.2.3). DARIAH-DE. Report (2016)

11. Sears, A.: Heuristic walkthroughs: finding the problems without the noise. Int. J. Hum.-Comput. Interact. **9**(3), 213–234 (1997). doi:10.1207/s15327590ijhc0903_2
12. Shneiderman, B., Plaisant, C.: Designing the User Interface. Strategies for Effective Human-Computer Interaction, 5th edn. Addison-Wesley, Boston (2009)
13. Stiller, J., Thoden, K., Leganovic, O., Heise, C., Höckendorff, M., Gnadt, T.: Nutzungsverhalten in den Digital Humanities. (R 1.2.1/M 7.6) DARIAH-DE (2015). https://wiki.de.dariah.eu/download/attachments/14651583/Report1.2.1-final3.pdf
14. Stiller, J., Thoden, K., Zielke, D.: Usability in den Digital Humanities am Beispiel des LAUDATIO-Repositoriums. DHd 2016 Book of Abstracts - Vorträge. Leipzig (2016). http://www.dhd2016.de/abstracts/vorträge-058.html

CERN Analysis Preservation: A Novel Digital Library Service to Enable Reusable and Reproducible Research

Xiaoli Chen[1,5], Sünje Dallmeier-Tiessen[1(✉)], Anxhela Dani[1,2], Robin Dasler[1],
Javier Delgado Fernández[1,4], Pamfilos Fokianos[1], Patricia Herterich[1,3],
and Tibor Šimko[1]

[1] CERN, Geneva, Switzerland
{xiaoli.chen,sunje.dallmeier-tiessen,anxhela.dani,robin.dasler,
javier.delgado.fernandez,pamfilos.fokianos,
patricia.herterich,tibor.simko}@cern.ch
[2] Alexander Technological Educational Institute of Thessaloniki, Thessaloniki, Greece
[3] Humboldt-Universität zu Berlin, Berlin, Germany
[4] Universidad de Oviedo, Oviedo, Spain
[5] University of Sheffield, Sheffield, UK

Abstract. The latest policy developments require immediate action for data preservation, as well as reproducible and Open Science. To address this, an unprecedented digital library service is presented to enable the High-Energy Physics community to preserve and share their research objects (such as data, code, documentation, notes) throughout their research process. While facing the challenges of a "big data" community, the internal service builds on existing internal databases to make the process as easy and intrinsic as possible for researchers. Given the "work in progress" nature of the objects preserved, versioning is supported. It is expected that the service will not only facilitate better preservation techniques in the community, but will foremost make collaborative research easier as detailed metadata and novel retrieval functionality provide better access to ongoing works. This new type of e-infrastructure, fully integrated into the research workflow, could help in fostering Open Science practices across disciplines.

Keywords: Research data · Long-term preservation · Reproducible science · Digital repository · Research workflow

1 Introduction

In response to the pressing demand for scientific output to be reproducible and re-usable, as well as the call for science to generate wider societal impact, an increasing number of funding agencies and research organizations have started

---

The original version of this chapter was revised: The Fig. 2 was corrected. The erratum to this chapter is available at 10.1007/978-3-319-43997-6_45

© The Author(s) 2016
N. Fuhr et al. (Eds.): TPDL 2016, LNCS 9819, pp. 347–356, 2016.
DOI: 10.1007/978-3-319-43997-6_27

to mandate better data preservation practices and to call for the opening of data at varying levels and time intervals, with respect to disciplinary conventions [1].

Preservation and sharing of research results are the keys to scientific progress, and this fact has long been recognized within the high-energy physics (HEP) discipline as evidenced by its collaborative nature and well established preprint culture [2]. In order to accelerate communication, manuscripts have been shared among the HEP community prior to publication for many years, first by mail and more recently via open digital repositories after the invention of the World Wide Web at CERN.

Coordinated data preservation practices and research data sharing for HEP, on the other hand, have just gained momentum in recent years [3]. HEP experiments, which can involve from hundreds to thousands of collaborators, produce large datasets that undergo intricate quality assurance processes before entering analysis phase, leaving a trail of research outputs of different refinement levels and usage [4]. Data are available within the collaborations for analyses, and only the final results, usually appearing in the form of plots and tables, are included in the final publication. In 2014, after complex preparation, CERN launched its Open Data Portal to address the matter of data sharing [5]. The platform made preserved datasets and accompanying tools accessible to both the general public and the research community, yet through it emerged new challenges in research provenance preservation, which is essential for analyses reusability, reproducibilty and discoverability. Hence, the inception of an additional service, CERN Analysis Preservation.

CERN Analysis Preservation (CAP) is a digital library service designed with HEP's unique disciplinary research workflow in mind, aimed at capturing the research data analysis workflow steps and resulting digital objects. Emphasis is given to the contextual knowledge needed to reproduce an analysis. Hence, CAP can be seen as a concrete step towards better reuse of unique research materials and as a way to facilitate future reproducibility of results.

2   Concept

In contrary to traditional approaches, where documentation and preservation happen in separate formats after the completion of an analysis, CAP provides a centralized platform on which scientists can document their analysis as early as the start of a new project. In addition, scientists can keep track of any aspect or step of an analysis as well as related research objects within their collaboration.

Researchers can submit their content, datasets, code, intermediate documentation of processing steps, quality assurance processes or internal notes. The tool is connected to the collaborations' databases, which enables auto-completion of many parts of the analysis metadata. The service assists researchers in preserving implicit knowledge and facilitates sharing of and access to research materials for future use. Alongside the analysis details and processing steps, reusable information for the analysis and supplementary documents are also captured.

From the outset of its design and development, the CAP team at CERN has been working closely with the physics collaborations to deliver a tailour-made solution for the HEP community. Together with the community, the service providers in the Information Technology Department and the Scientific Information Service at CERN framed the use cases [6]. CAP aims to document and preserve tasks that traditionally are implicit to the research process - like personal log-book keeping for an ongoing analysis, exchange of preliminary results, quality assurance processes, links from an analysis to approval or publication processes - all in an integrated fashion. By supporting the physics analysis work flow electronically and centrally, CAP mitigates some long standing pains in the daily work, such as parallel versions, additional approval procedures, conflicted results resolution, and analyses combination or reproduction.

Overall this new digital library service helps researchers manage their research workflows better, by helping internal works and data become more accessible and findable despite being in a large collaboration. This will save time and effort on the researcher side, present collaborators with the chance to build on other peoples work more quickly, as well as offer new personnel a better chance to familiarise themselves with ongoing works in large scale collaborations.

From the service providers side, this new e-infrastructure service helps establishing enhanced data curation and standardization workflows. This approach is considered a milestone to facilitate reusable science, i.e. to use data or rerun an analysis in many years to come. With the unprecedented granularity, CAP provides a new lens for HEP scientists to look at on-going and past analyses. Further more, through its association with the CERN Open Data Portal, INSPIRE and CDS, CAP significantly lowers the barriers for HEP scientists to engage in data reuse and eventual reproducible analysis, and hence fosters Open Science practices.

3   Technology

The CAP service is built on top of the Invenio digital library framework [7]. Invenio provides an ecosystem of standalone independent packages that permit to build a custom digital repository solution oriented towards various use cases, such as an integrated library system, a digital document repository, a multimedia archive, or a data repository. In the context of CAP, the Invenio framework has been used and extended with several large-data oriented features.

The managed data are modelled in JSON format that is conforming to certain JSON Schema to ensure the compliance of captured JSON snippets with standard metadata requirements. The platform draws inspiration from the Open Archival Information System (OAIS) recommended practices to ensure long-term preservation of captured assets. The JSON snippets are stored in the Invenio digital repository database and sent to an Elasticsearch cluster that is being used for indexing and information retrieval needs.

CAP aims at preserving the complete environment. This necessitates to build connectors to various tools used in physics analyses in order to be able to take a consistent snapshot of the analysis process for knowledge capture and preservation.

The overall architecture (Fig. 1) of the CAP platform connects to various tools:

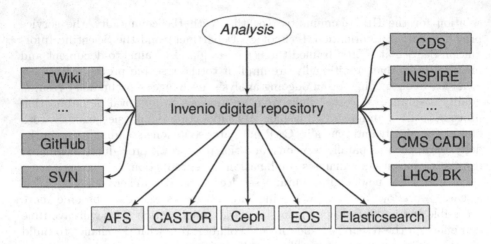

Fig. 1. The architecture overview of the CERN Analysis Preservation platform

The system notably includes:

- the "core" of the Invenio digital repository platform itself;
- the flexible file storage abstraction layer using the EOS storage backend;
- the connectors to harvest final datasets from various storage systems used by individual physicists, such as CASTOR;
- the connectors to harvest user code from Git and Subversion repositories used to develop the analysis code;
- the connectors to documentation systems such as TWiki are also being developed;
- the platform may automatically ingest information from internal collaboration systems (e.g. CMS CADI);
- for the approved open data, the platform may push parts of information to publishing platforms such as (CDS and INSPIRE).

In order to facilitate the reproduction of an analysis even many years after its initial publication, CAP aims at instantiating the original analysis environment on the CERN OpenStack cloud by means of technologies such as CernVM virtualisation or container-based solutions using Docker, Jupyter or Everware.

4   Functionality and UI

The service offers various entry points and functionalities to suit various users, so that the tool will more easily become an intrinsic part of the research process. Researchers can use a submission form (Figs. 2 and 3 show parts of an example) to submit their content, or they can select a more automated means to do so (via the REST API). Based on the flexible JSON schemata tailoured input formats can be offered for experiments and working groups. Auto-complete functionality

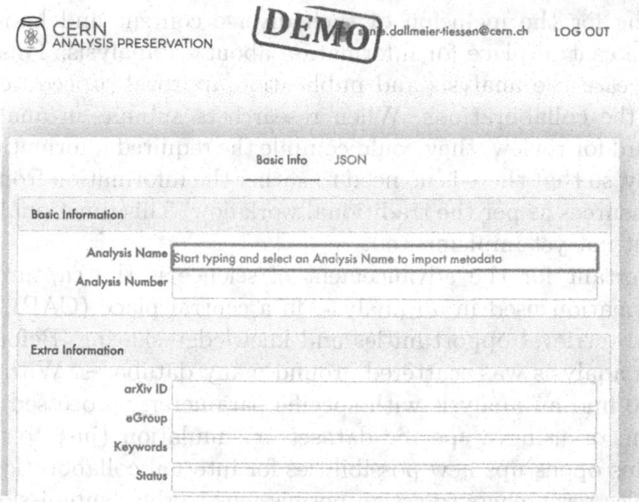
CERN
ANALYSIS PRESERVATION
DEMO
LOG OUT
Basic Info
JSON
Basic Information
Analysis Name
Start typing and select an Analysis Name to import metadata
Analysis Number
Extra Information
arXiv ID
eGroup
Keywords
Status

Fig. 2. First part of a submission form for one experiment. Auto-complete functionality (using information from a range of existing databases) allows researchers to easily fill in the form without much extra effort.

for the submission form (based on connections made to existing databases within the collaboration) should reduce the time researchers need to spend on using this e-infrastructure service for submitting their content.

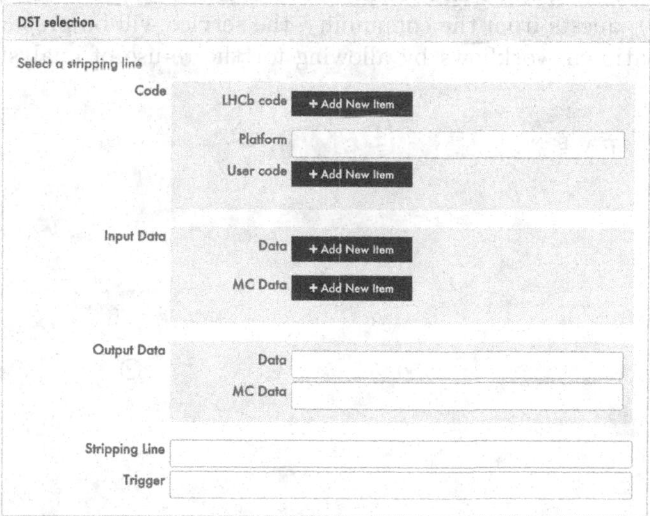
DST selection
Select a stripping line
Code
LHCb code
+ Add New Item
Platform
User code
+ Add New Item
Input Data
Data
+ Add New Item
MC Data
+ Add New Item
Output Data
Data
MC Data
Stripping Line
Trigger

Fig. 3. Second part of a submission form for one experiment. It contains links to code, data and physics information.

By allowing for the inclusion of such diverse content and knowledge, the service offers a central place for information about an analysis. This knowledge could help to ease the analysis and publication approval procedures that take place within the collaborations. When researchers submit an analysis to the dedicated board for review, they could compile the required information from the tool very easily, so that there is no need to gather the information from disparate, unconnected sources as per the traditional work flow. This functionality is under discussion, but not yet implemented.

Most important for the advancement of science is the capture of all the essential information used in an analysis in a central place (CAP), which permits advanced retrieval opportunities and knowledge sharing. Before, information about an analysis was scattered around many databases. With CAP users will be able to find an analysis with specific parameters, processed with a specific algorithm, or using a specific dataset or simulation (just to name a few examples). This opens ups new possibilities for internal collaboration.

CAP enables researchers to assign permissions to their submissions, so as to allow individuals control over the privacy of their submission while accounting for the collaborative nature of the discipline. The submitter can invite a group to view and/or edit the submission. A permissions tab allows the specification of each invitee's rights (Fig. 4). Given the work in progress nature of the content in CAP, this functionality has been considered crucial by the community.

While Analysis Preservation is considered an internal digital library service, it will offer open publishing options via partner portals. The researcher himself needs to trigger this process and content should be approved internally prior to publication. A DOI registration workflow for such content makes sure that reuse of materials can be tracked and credit can be attributed to the contributors.

Based on requests from the community, the service will be prepared to facilitate computational workflows by allowing for the re-use of analysis data and

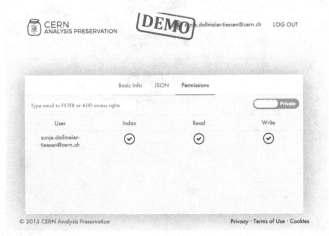

Fig. 4. Permissions setting for an analysis record on CAP. This enables researchers to share their analysis internally depending on their needs.

materials in an automated fashion. Additionally, it is under investigation how Jupyter notebooks might be integrated to accommodate and facilitate modern research practices in support of open science goals.

5 Content and Metadata

The service providers, together with preservation managers from the collaborations, identified the core components of a physics analysis and the metadata therein. This is very specific to the individual collaboration and, in many cases, even within the individual working groups of a collaboration. The tailoured JSON schemata have been developed with the community to make sure researchers can submit all content that is relevant for their analysis (the JSON schema then translates to the submission form; an example excerpt for the LHCb experiment can be seen in Fig. 5). This approach should make sure sufficient information is in place for better internal accessibility, future reuse and reproducibility.

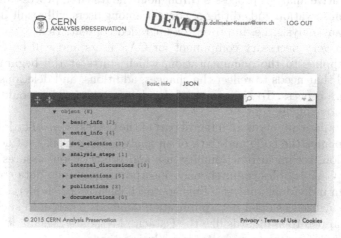

Fig. 5. An example of the high-level JSON components for a physics analysis of the LHCb experiment.

The service is connected to the core databases of the partnering collaborations. At the moment the service has access via APIs to CMS and LHCb databases. Work is in progress to integrate with other collaboration's tools as well. Harvesting databases facilitates auto-completion for most fields of the analysis submission form. Connecting to the existing databases within CMS, for example, provides basic information, such as authors and title, based on the unique ID of an analysis. Further important information about the ongoing analysis includes the operating system, the analysis software, the code, input data files and the output data. By following the ID in the CMS internal databases, one can retrieve information such as internal discussions, presentations and publications related to an analysis when they are available. Also, detailed physics information, which is needed for future reuse, can be available: e.g. details about the primary and simulated (Monte Carlo) datasets; physics information that refers to the final state particles, cuts and vetos.

Likewise in LHCb's databases the service harvests all the needed content and displays basic information like the distinctive analysis name and number. Additional information is provided about the responsible working groups and the status of an analysis. Furthermore, the tool gathers valuable components of an analysis, such as the code, the input and output data as well as their simulated (Monte Carlo) data. In the future, collaborations' Twiki pages should also be integrated (as preserved snapshots), as they are actively used within the collaborations for internal discussions. Lastly, whenever available, the tool incorporates information related to presentations, publications and any other documentations of an analysis. All these connections should help researchers in the submission process. They can open the tool and enter the unique internal ID for analysis so that much of the information is auto-completed. Then, they only need to add the remaining pieces of information.

This iterative analysis processes throughout the research process and the collaborative preservation and editing of content among users will result in different versions of an analysis, i.e. improved or upgraded content. Hence, versioning is considered a very necessary component for CAP users and will be facilitated. The same applies to the documentation, which starts at the beginning of an analysis and that needs to reflect the changes, additions, and deletions throughout the entire process. To keep track of the changes the service would maintain explicit versions of the content.

High metadata quality is an integral part of the CAP system. The information that accompanies the code and data of an analysis for purposes of description, administration, technical functionality, use and preservation, assures the long-term usage. CAP enables comprehensive incorporation of metadata components specific to an individual analysis. This approach ensures a rich metadata source that can be utilized as open data publication demands emerge. In such cases, it would be mapped to standards like the Datacite Metadata Schema [8] to enable the publishing of complete and citable analysis records.

High quality metadata makes the data more visible, accessible and usable in the long term and is considered a key benefit for researchers who make use of the tool's search function. Complete and interoperable metadata are considered crucial components for future reproducibility. At this point it should be noted that CAP is not an effort to enforce a standard across experiments, which is being done in other disciplines, but it instead works flexibly with each collaboration's requirements. The overall similarity of analysis components allows for mapping between them, rather than globally enforced requirements. To address the standardization challenge in the long term, work is underway, together with the DASPOS project, to establish an ontology for the main analysis building blocks.

6   Conclusions and Next Steps

CERN Analysis Preservation puts the concept of digital library services into new use. By being embedded into the actual research workflow, research outputs throughout the analysis process can be preserved and shared internally. This creates new challenge in terms of (frequent) versioning and handling of very high data volumes. An analysis can be up to 10TB in volume currently and is expected to grow further over the course of the next few years. Overall, the main goal of CERN Analysis Preservation is to facilitate the reusability and reproducibility of an analysis even many years after its initial start or publication. To do so, partnerships are created with other tool providers that enable researchers to easily reuse, validate and execute preserved content.

For the future it is needed to extend the first prototype that is available, and integrate it even more with existing databases and processes of the experiments. This will be a key factor to foster adoption. Furthermore, the second phase of the project will begin considering the possibility of connections to analysis environments, in order to enable researchers to embed the discovered content easily into Jupyter notebooks and other tools.

The Open Source developing approach enables other disciplines to reuse the concept and tool. In particular, service providers for data-intensive disciplines might be interested to build service components flexible enough to serve a fast paced research environment with diverse, dynamic and possibly large scale research objects and better documentation.

Acknowledgements. CERN Analysis Preservation builds on the collaborative work from various teams and experiments. We are thankful for all the hard work that the LHC experiments and the Invenio team have been putting into this service so far.

References

1. European Commission. Open data: an engine for innovation, growth and transparent governance, December 2011. http://eur-lex.europa.eu/LexUriServ/LexUriServ.do?uri=COM:2011:0882:FIN:EN:PDF
2. Gentil-Beccot, A., Mele, S., Brooks, T.C.: Citing and reading behaviours in high-energy physics. Scientometrics **84**(2), 345–355 (2009)

3. Kogler, R., South, D.M., Steder, M.: Data preservation in high energy physics. J. Phys.: Conf. Ser. **368**, 012026 (2012)
4. Herterich, P., Dallmeier-Tiessen, S.: Data citation services in the high-energy physics community. D-Lib Mag. **22**(1/2) (2016)
5. Cowton, J., Dallmeier-Tiessen, S., Fokianos, P., Rueda, L., Herterich, P., Kunčar, J., Šimko, T., Smith, T.: Open data and data analysis preservation services for LHC experiments. J. Phys.: Conf. Ser. **664**(3), 032030 (2015)
6. Dallmeier Tiessen, S., Herterich, P., Igo-Kemenes, P., Šimko, T., Smith, T.: CERN analysis preservation (CAP) - Use Cases (2015). http://dx.doi.org/10.5281/zenodo.33693
7. Kunčar, J., Nielsen, L., Šimko, T.: Invenio v2.0: a pythonic framework for large-scale digital libraries, June 2014. http://urn.fi/URN:NBN:fi-fe2014070432294
8. DataCite Metadata Schema for the publication and Citation of Research Data, DataCite Metadata Working Group (2014). https://doi.org/10.5438/0011

DataQ: A Data Flow Quality Monitoring System for Aggregative Data Infrastructures

Andrea Mannocci[1,2]($\boxtimes$) and Paolo Manghi[1]

[1] Consiglio Nazionale delle Ricerche,
Istituto di Scienza e Tecnologie dell'Informazione "A. Faedo", Pisa, Italy
{andrea.mannocci,paolo.manghi}@isti.cnr.it
[2] University of Pisa, Pisa, Italy
andrea.mannocci@for.unipi.it

Abstract. Aggregative Data Infrastructures (ADIs) are information systems offering services to integrate content collected from data sources so as to form uniform and richer information spaces and support communities of users with enhanced access services to such content. The resulting information spaces are an important asset for the target communities, whose services demand for guarantees on their "correctness" and "quality" over time, in terms of the expected content (structure and semantics) and of the processes generating such content. Application-level continuous monitoring of ADIs becomes therefore crucial to ensure validation of quality. However, ADIs are in most of the cases the result of patchworks of software components and services, in some cases developed independently, built over time to address evolving requirements. As such they are not generally equipped with embedded monitoring components and ADI admins must rely on third-party monitoring systems. In this paper we describe DataQ, a general-purpose system for flexible and cost-effective data flow quality monitoring in ADIs. DataQ supports ADI admins with a framework where they can (i) represent ADIs data flows and the relative monitoring specification, and (ii) be instructed on how to meet such specification on the ADI side to implement their monitoring functionality.

Keywords: Monitoring · Data flow · Data quality · Aggregative Data Infrastructures

1 Introduction

In recent years, the cross-discipline nature of science and the need of researchers to gain immediate access to research material often led to the realization of ADIs [5,12]. ADIs are information systems offering functionalities to integrate content collected from multiple data sources in order to form uniform and richer information spaces and support communities of users with enhanced access services to such content. In particular, ADIs offer functionalities for (i) the collection and processing of content (metadata descriptions or files), possibly availing of advanced tools and workflows that combine independent components, (ii) the population of uniform graph-like information spaces, and (iii) the provision of

© Springer International Publishing Switzerland 2016
N. Fuhr et al. (Eds.): TPDL 2016, LNCS 9819, pp. 357–369, 2016.
DOI: 10.1007/978-3-319-43997-6_28

these information spaces to consumers (humans or services) via web portals and/or standard APIs.

Examples of ADIs in the scholarly communication, hence tailoring scientific communication literature data sources, are Google Scholar, OpenAIRE [13], CORE[1]. Other examples from thematic-domain and/or supporting access to different research products are Europeana[2], the European Film Archive [2], the Heritage of the People's Europe [5], the Europeana network for Ancient Greek and Latin Epigraphy [14] and several others.

Aggregating a potentially steadily-increasing amount of data and processing it are some of the major challenges in delivering high-quality and accurate applications based on the resulting information space. To this aim, consumers of the information space typically demand for guarantees on its "quality" and "transparency" over time both in terms of structure and semantics and in terms of the processes generating it [20, 25, 27].

First of all, the perceived data quality at a data source may not be perceived equally at the ADI side as, in general, data quality falls under the "fitness for purpose" principle [7, 23, 25–27]. Determining whether data has quality or not mainly depends on the application they are intended for. Secondly, data sources may be unstable and unreliable and part of their data may "appear and disappear" entirely because of unpredictable reasons, be them physical (e.g. HW failures, I/O problems, network errors) or logical (e.g. records moved to another collection or deleted by accident). As original data may be unreliable in terms of both quality and presence over time, any knowledge inferred on top of them can be faulty too, thus leading to unwanted, and potentially harmful, end-user dissatisfaction [20]. Lastly, ADIs are often built incrementally in order to adapt to natural evolution of data sources and requirements of the ADI consumers. As a consequence, services, methods, algorithms used by an ADI to collect and process data are subject to changes and updates over time and may introduce unforeseen subtle errors.

Given such premises, in order to deliver a trustful and reliable service to their consumers, ADI administrators must deliver tools for monitoring the internal status, the ongoing processes, and the consistency of results in the running infrastructure. Unfortunately, this operation is often far from trivial. In many cases, ADIs are "systems of systems", i.e. patchworks of individual components possibly developed by distinct stakeholders with possibly distinct life-cycles, whose interactions are regulated by policies and mediation software developed at the infrastructural level. Accordingly, although there are exceptions, ADIs are generally not equipped with built-in application level monitoring systems and must introduce monitoring capabilities as *ex-post* software integration. Alas, this typically results in excruciating efforts and in a hardly sustainable monitoring software in the long run. Traditional monitoring tools can keep an eye on the system regarding lots of indicators valuable at "system administration level" (e.g. Nagios, Icinga, Ganglia), but they can do little when the focus is on the application level or, in particular, on the data

---

[1] CORE - The UK Open Access Aggregator, https://core.ac.uk.

[2] Europeana, http://www.europeana.eu.

content and its quality. Application monitoring frameworks such as Prometheus[3] and the Elastic stack[4] work great for exporting metrics from the application level, but again are mainly devised for platform operations and performance monitoring, while fail to capture time constraints between metrics generation and address typical problems of data quality.

In this paper, we describe DataQ, a general-purpose system for data flow quality monitoring in ADIs. DataQ supports ADIs administrators with the following features:

- *Workflow-sensitive monitoring*: models and monitors workflow entities of datum, data collection, and process;
- *Cross-platform compatibility*: its adoption is not bound to specific ADI technologies;
- *Cost-effective*: minimizes the amount of work required to integrate monitoring tools on top of ADIs.

DataQ models the *data flows* of a generic ADI taking inspiration from manufacturing processes, where "raw material" corresponds to data passing through sequential processing activities until final data products are produced. The ADI administrator is provided with a Web User Interface (WebUI) in which she can formally describe the data flows to be monitored in the infrastructure, intended as sequences of basic building blocks inspired by the Information Production Map (IP-MAP) model [4,24]. Once the ADI's data flows have been defined, DataQ allows administrators to assign *sensors* to the relative building blocks (or sets of them). Sensors specify a set of *metrics*, each corresponding to a number of observations collected over time.

Each defined sensor will require a corresponding *sensor hook* that is an ADI-specific software counterpart, capable of extracting observations for the identified metrics and send them to DataQ. Given the specification of the data flows and the relative sensors, DataQ supports ADI administrators at implementing the corresponding hooks in order to minimize the amount of code to be written and let them focus on the business logic required to compute the metrics observations.

Finally, the WebUI allows administrators to define which *controls* should be enforced over the metrics generated by sensors (e.g. rates, thresholds, upper bounds, moving averages, etc.), visualize trends of the monitored features, analyse comprehensive *reports* and receive *notifications* and *alerts* about potential troubles happening in the operation of the infrastructure.

Outline. Section 2 introduces a motivating example based on the experience gathered with the OpenAIRE project which builds an ADI that represents at best the cardinality of the real-scale problem. Section 3 describes the DataQ framework, its languages for data flows, sensors, and controls. Section 4 describes an implementation of the system as deployed in the production system of the OpenAIRE infrastructure. Section 5 discusses related work and similar methodologies applied in other fields. Finally, Sect. 6 concludes the paper and points out future developments.

---

[3] Prometheus, http://prometheus.io.
[4] The Elastic stack, https://www.elastic.co.

2   The OpenAIRE (Aggregative Data) Infrastructure

The OpenAIRE[5] infrastructure fosters Open Access (OA) to research products in Europe and beyond and provides an European data e-infrastructure enabling: (i) the generation of statistics measuring the impact of OA across Europe and individual countries joining the initiative, and (ii) the creation of a service for searching people, publications, projects, and datasets produced as scientific output within H2020 (as well as FP7) and browse their interconnections. The infrastructure collects metadata about millions of publications, scientific datasets, organizations, authors and projects from over 700 heterogeneous data sources and aggregates such content into an information graph in which nodes are interconnected by semantically-labeled relations. Infrastructure services then *harmonize* the structure and semantics of the information graph and *mine* full-text articles (4 million today) to infer extra metadata or links and *enrich* the information graph. Finally, the resulting information graph is exposed for search and stats (portals), bulk-download (OAI-PMH), and navigation (LOD). Certifying the quality of such "data product" and how much reliable is the information provided to the end-users is a non-trivial, yet vital, task for providing trustful and valuable search and statistical services.

However, data quality in OpenAIRE is dependent from a plethora of variables such as the variety and variability of data source, the refinement of algorithms currently in use for curation and mining, the cross-influence between the presence/absence of entire sets of information and the chosen mining algorithms, the presence of network and I/O errors, etc. Also, OpenAIRE persists the data collected and processed in several different back-ends with different purposes. More precisely, the so-called "provision workflow" materializes the information graph (which is stored in HBase[6] for mining and de-duplication purposes) into: (i) a full-text index, implemented with Apache Solr to support search and browse queries from the OpenAIRE Web portal, (ii) a PostgreSQL and a dedicated Redis key-value cache for statistics, (iii) a NoSQL document storage based on MongoDB in order to support OAI-PMH export of the aggregated records, and finally (iv) a triple store realized with OpenLink Virtuoso to expose OpenAIRE's information system as LOD via a SPARQL end-point.

OpenAIRE is indeed an ADI, and features serious monitoring challenges. Looking at the provision workflow above, we shall consider three representative examples of monitoring: (i) assessing (and keep assessing over time) whether the total number of "publications funded by Horizon2020 projects" indexed by Solr matches the same number of records delivered via the corresponding OAI-PMH Set; (ii) the number of publications harvested from a content provider should be monotonic increasing with a max percent variation between each harvesting over time; and, (iii) evaluating the "completeness" of records index by Solr evaluated as the ratio between number of mandatory attributes and number of not empty mandatory attributes.

---

[5] The OpenAIRE EU project, http://www.openaire.eu.
[6] Apache HBase, https://hbase.apache.org.

3 DataQ: A Data Flow Quality Monitoring System

This section introduces DataQ, a general-purpose, flexible, and cost-effective system for data flow quality monitoring. It first presents its data flow description language and data flow monitoring language, finally it gives an overview of its architecture.

3.1 Data Flow Description Language

Taking inspiration from the seminal work present in literature [3,4,10,24,28], a generic Aggregative Data Infrastructure (ADI), or a subpart of it, can be accurately modeled as an Information Manufacturing System (IMS): a series of processing steps that transforms raw input data into final data products. In this work, we adopted a subset of the construct blocks introduced in the Information Production Map (IP-MAP) model [24], reported in Fig. 1a. The *Data Vendor* block (VB) represents a generic source, either internal or external, providing data for downstream elaboration. Similarly, the Data Consumer block (CB) represents an entity downstream consuming final data information produced by the manufacturing process. The *Data Processing* block (PB) represents a generic processing step which processes a (stream of) *datum* from a status to another one. Finally, the *Data Storage* block (SB) stands for a back-end of choice persisting collections of data conforming to a common structure and semantics (e.g. E-R database, NoSQL, file system). According to this model, PBs and SBs handle data units of different granularity: *datum*, i.e. the observable unit of a PB, and *data collection* (a collection of data sharing a common property or semantic), i.e. the observable unit of a SB.

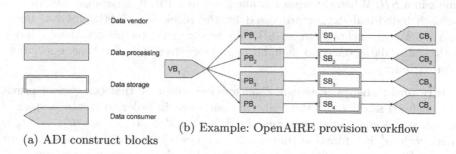

(b) Example: OpenAIRE provision workflow

(a) ADI construct blocks

Fig. 1. ADI modeling description language

ADIs data flows are modeled as sequences of instances of these blocks interconnected by directed *edges* representing time-order relations among them. The language does not model locality assumptions, i.e. ADI data flow blocks may refer to services running on different administrative domains. The example in Fig. 1b describes the OpenAIRE provision workflow (introduced in Sect. 2) with the formalism here introduced. The workflow ingests data from VB_1 (i.e. the OpenAIRE information graph) and transforms it through separated processing blocks PB_1, PB_2, PB_3 and PB_4. Then, the result of each PB_i is stored sep- arately in a dedicated backend SB_1, SB_2, SB_3 and SB_4. Finally, each storage block SB_i serves a specific use case, hence the customers CB_1, CB_2, CB_3 and CB_4 are, at least logically, distinct.

3.2 Data Flows Monitoring Language

Once the ADI *data flows* have been described, the ADI admin can start config- uring the relative monitoring scheme. To this aim, DataQ supports four main concepts: *sensors, metrics, controls* and *actuators*.

Sensors. A sensor s is a piece of software capable of generating an *observation* o about a data unit u (either a datum or a data collection) according to a metric m. A metric is a function intended to measure a specific quality feature of a target data unit u (e.g. the completeness of a XML record to be processed, the cardinality of a collection). A sensor is "anchored" to a block of reference b, be it a PB or an SB, and defines a set of metrics m_i. More specifically, a sensor s is described as a tuple of the form $s = \langle b, m_1, \ldots, m_I \rangle$, which lives in the ADI and is triggered by it in order to produce observations o_i, described as tuples of the form $o = \langle time, sensor, metric, value \rangle$. When a sensor s, anchored on block b, is triggered at time t on data unit $u_{b,t}$ it produces a set of observations, one for each of the metrics m_i it declares:

$$trigger(s,t) = \{o_1, \ldots, o_I\} = \left\{ \langle t, s, m_1, v_1 \rangle, \ldots, \langle t, s, m_I, v_I \rangle \right\} \qquad (1)$$

where $v_i = m_i(u_{b,t})$ and $u_{b,t}$ is the data unit relative to the block b at time t, e.g. the datum processed by a PB at time t or the *data collection* content at time t in a SB. When the sensor is anchored to a PB, it generates observations for each individual datum processed by the block (e.g. XML, JSONs, texts, images). When anchored to an SB, the sensor generates observations relative to the entire data collection at hand (e.g. query over SQL database, query over full-text index).

Compound Sensors. *Compound sensors* are sensors s' that can enclose previ- ously defined sensors and thus inherit their *scope* in order to create *compound metrics* m'_k. A compound sensor can be defined as $s' = \langle s_1, \ldots, s_J, m'_1, \ldots, m'_K \rangle$, where each m'_k is a function that produces a new v'_k by combining observations produced by a subset of the metrics anchored to the sensors in the scope of s'. While basic sensors are triggered by the ADI, compound sensors are triggered by DataQ according to a given admin-defined schedule. When a compound sen- sor is triggered at time t, it produces a set of observations according to the m'_k compound metrics defined, whose observation time is t:

$$trigger(s',t) = \{o'_1, \ldots, o'_K\} = \left\{ \langle t, s', m'_1, v'_1 \rangle, \ldots, \langle t, s, m'_K, v'_K \rangle \right\} \qquad (2)$$

To better understand the implications of this, Fig. 2a depicts two examples of compound sensors: an *in-sync* sensor s_6 wraps two storage sensors s_4 and s_5, which monitor two blocks SB_1 and SB_2; a *not-in-sync* sensor s_3 wraps s_1 and s_2, which monitor two sequential blocks PB_1 and PB_2 respectively. When an in-sync compound sensor wraps two SBs as s_6 in Fig. 2a it has the opportunity to synchronize the observations produced by them. If O_{m_i} is the time series of observations produced by the metrics m_i of s_4 and s_5, metrics m'_k in s_6 generate a new compound value $v'_k = m'_k(O_{m_{1_k}}(t), \ldots, O_{m_{I_k}}(t))$ – where $O_{m_i}(t)$ "reads" the time series O_{m_i} at time t. In this case, a compound sensor triggered at time t produces a new derived observation for every defined m'_k by combining into a function the values produced by the enclosed metrics of interest at time t.

A not-in-sync compound sensor cannot synchronize the enclosed metrics (e.g. s_3 in Fig. 2a). Hence, new compound value v'_k is obtained as a function of partial evaluations, one for each O_{m_i} of interest for m'_k, i.e. $v'_k = m'_k(O_{m_{1_k}}, \ldots, O_{m_{I_k}}) = g_{m'_k}(f_{m'_k,1}(O_{m_{1_k}}), \ldots, f_{m'_k,I_k}(O_{m_{I_k}}))$, where $\{f_{m'_k,1}, \ldots, f_{m'_k,I_k}\}$ is a set of functions dedicated to process one O_{m_i} each and produce partial evaluations, while $g_{m'_k}$ is the function combining such partial evaluations.

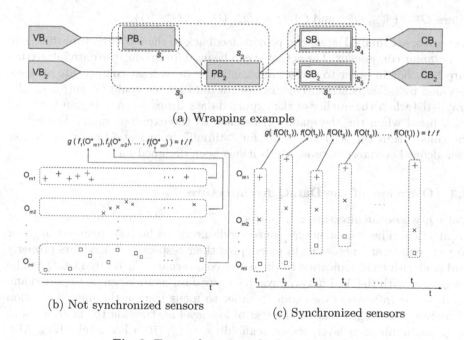

(a) Wrapping example

(b) Not synchronized sensors

(c) Synchronized sensors

Fig. 2. Types of sensor and relative controls

Controls. Once sensors and metrics are defined, the admin can configure the *controls* she wants to enforce over time against its sensors. Controls depend on the scope (possibly inherited) of the sensors to which they are anchored and their relative time ordering, hence DataQ can guide the user in their definition. More precisely, a control c is defined in terms of one or more *selectors* x_i and one *analyzer* a as a function $c(x_1, \ldots, x_K, a) = \{true, false\}$.

Given a set of observations of a metric O_m and two predicates p_t, p_v on observations' time and value, a selector x returns a subset O_m^* of O_m satisfying the predicates:

$$x(O_m, p_t, p_v) = \{o_1^*, \ldots, o_I^*\} = O_m^* \subseteq O_m \tag{3}$$

With *analyzer* we refer to a comparing function a that accepts one or more groups of selected observation $O_{m_j}^*$ as input and returns a binary result $\{true/false\}$ reflecting whether the controls has been passed or not. Two possible types of analyzers are provided, one working with not synchronized observations (Fig. 2b)

$$a(O_{m_1}^*, \ldots, O_{m_J}^*, f_1, \ldots, f_J, g) = g\big(f_1(O_{m_1}^*), \ldots, f_J(O_{m_J}^*)\big) = \{true, false\} \tag{4}$$

and one working with synchronized ones (Fig. 2c).

$$a_{sync}(O_{m_1}^*, \ldots, O_{m_J}^*, f, g) = g\Big(f(O^*(t_1)), \ldots, f(O^*(t_i))\Big) = \{true, false\} \tag{5}$$

where $O^* = \bigcup_{j=1}^{J} O_{m_j}^*$ and $O^*(t) = \{O_{m_1}^*(t), \ldots, O_{m_J}^*(t)\}$.

Actuators. Finally, DataQ can provide feedback to the infrastructure thanks to an *actuator* component. The actuator can be used for driving the data infrastructure behavior in order to correct automatically or at least compensate an issue revealed by failed controls. For example, a data source could be automatically graylisted when the quality of the exported data drops down a certain level and whitelisted when the the quality rises back to the expected value. The actuator runs within the ADI and waits for "stimuli" from DataQ in order to take user-defined countermeasures which it has been designed for.

3.3 Overview of the DataQ Architecture

DataQ, whose architecture is reported in Fig. 3, is designed as a client-server application. The infrastructure source code needs to be instrumented in order to put sensors and actuators in place, plug their respective behaviours properly and deal with communication to the server counterpart. Currently DataQ can be integrated with the ADI in two ways: (i) a core Java module can be imported and used by the infrastructure's code in order to focus just on the implementation of metrics' core logic, leaving the rest of low-level abstraction to the framework (e.g. communication layer, sensor scaffolding, etc.); (ii) a low-level REST API, which has to be called by the ADI components in order to provide observations to DataQ server component. For the sake of configurability and extendibility, when importing DataQ's Java libraries, a catalog of common metrics (e.g. for XML inspection) is provided off-the-shelf from the framework.

The DataQ *server* component is a stand-alone web application that enables the infrastructure manager to create, via a user friendly WebUI, one or more *monitoring scenarios*, each one designed to monitor a particular functional area or aspect of the ADI. Within a monitoring scenario, the infrastructure manager is able to model the data flows of the ADI as seen in Sect. 3.1, the deployed sensors

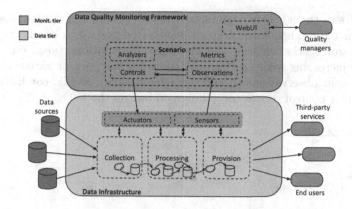

Fig. 3. DataQ architecture

and actuators, and the controls to be enforced over time against the generated metrics. Similarly to sensors, analyzers used by controls come with off-the-shelf implementations for most common comparison functions such as equality, upper and lower bound, peak/valley detection, threshold, percent variation, and so forth. In any case, analyzers can be customized and extended at need by the user requiring more specific behaviours. The DataQ server will separate the stream of incoming observations into different streams, each one related to a specific metric, and will store them as points of time series. In this way, observations can be queried and examined either as charts or tabular data via WebUI.

Furthermore, DataQ enables the generation of an exhaustive *report* about the defined metrics and controls providing insights, via the WebUI, in a quick glance about key features and potential issues present in the infrastructure. Given a set of controls, the monitoring service also takes care of raising *alerts* and *notifications* informing the infrastructure manager about the status of the infrastructure and its operation.

Finally, regarding actuators, the communication between DataQ and the actuators deployed in the ADI takes place through standard Web API; the developer using DataQ has the option (i) to use a Java ready-to-use actuator scaffolding which has only to be specialized with custom business logic, or (ii) take charge of low-level implementation in order to detect and react to such feedback.

4   The OpenAIRE Use Case

An implementation of DataQ has been realized and deployed in the OpenAIRE production infrastructure to monitor the "provision workflow" described in Sect. 2. For the sake of space, we selected a representative use case for demonstration (Fig. 4). Here, the metric "publications", has been extracted from two storage blocks (a Solr index and a Redis key-value DB instances) thanks to two sensors extracting the total number of publications collected by OpenAIRE.

As we can see, though in recent history, the values of the two time series are perfectly matching, hence the control passes (the green box on the right). The second control checks whether the last three observations taken on Redis are monotonic increasing with no "bumps" (meaning the percent variation between two subsequent observations should not exceed 15 %); as this condition is verified, the other control passes too (the green box on the left).

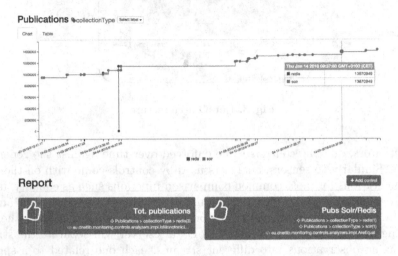

Fig. 4. Screenshot of the DataQ dashboard for the metric "Publications" (Color figure online)

5   Related Work

Data quality has been recently investigated in the field of eScience; in [21,22], the authors proposed a framework for the evaluation of the quality of information during Finite Element Method (FEM) simulations carried out on a machine availing of Workflow Management Systems (WfMSs), a situation in which poor data quality generally yields a waste of time, especially during long running experiments and simulations.

In [17,19], the authors developed a framework able to manage quality attributes in the context of eScience based on formal Information Quality ontology and proposed a case study in the proteomics field WfMS with experiments based on Taverna[7].

Other works have been presented in the field of (heterogeneous) data integration systems [1,6,8,15,16,18]. In [9,11], the authors described the QBox platform, a system capable of assessing quality in data integration systems and enabling mediation among several quality evaluation and data profiling tools.

None of these approaches has tackled monitoring with the idea of providing the tools to model data flows, sensors, metrics, and controls in a conceptual sense,

---

[7] Taverna, http://www.taverna.org.uk.

trying to push complexity of the monitoring framework away from ADI admins, who should only focus on the business logic of the metrics. To our knowledge, DataQ is the first framework of this kind, its first implementation being already in use in the production system of OpenAIRE.

6   Conclusions

In this paper we introduced DataQ a system for Data Flow Quality Monitoring of Aggregative Data Infrastructure (ADI). The proposed solution poses its foundation on the seminal work presented in [3,4,10,28] to model such data infrastructures as Information Manufacturing System leveraging a subset of the construct blocks available in IP-MAP [4,24].

DataQ enables the automatic extraction of quality features from data thanks to specially devised software sensors. Such observations, persisted as time series, can be inspected and automatically queried in order to check for used-defined controls about the expected behavior of the infrastructure.

The infrastructure manager is provided with alerts and notifications when the infrastructure as soon as the ADI encounters anomalies. We also foresee a mechanism for providing feedback to the monitored infrastructure in order to trigger a prompt reaction to anomalies and attempt, to some extent, an automatic correction of the misbehaviour.

An implementation of DataQ is currently deployed in the OpenAIRE infrastructure [13] and helps the infrastructure managers to spot anomalies and prevent the dissemination of erroneous data and statistic to the general public and to EU commission.

References

1. Akoka, J., Berti-Équille, L., Boucelma, O., Bouzeghoub, M., Comyn-Wattiau, I., Cosquer, M., Goasdoué-Thion, V., Kedad, Z., Nugier, S., Peralta, V., Sisaid-Cherfi, S.: A framework for quality evaluation in data integration systems. In: 9th International Conference on Enterprise Information Systems, ICEIS (2007)
2. Artini, M., Bardi, A., Biagini, F., Debole, F., Bruzzo, S.L., Manghi, P., Mikulicic, M., Savino, P., Zoppi, F.: The creation of the European film archive: achieving interoperability and data quality. In: 8th Italian Research Conference on Digital Libraries, IRCDL, pp. 1–12 (2012)
3. Ballou, D.P., Pazer, H.L.: Modeling data and process quality in multi-input, multi-output information systems. Manag. Sci. **31**(2), 150–162 (1985)
4. Ballou, D., Wang, R., Pazer, H., Tayi, G.K.: Modeling information manufacturing systems to determine information product quality. Manag. Sci. **44**(4), 462–484 (1998)
5. Bardi, A., Manghi, P., Zoppi, F.: Aggregative data infrastructures for the cultural heritage. In: Dodero, J.M., Palomo-Duarte, M., Karampiperis, P. (eds.) Metadata and Semantics Research. Communications in Computer and Information Science, vol. 343, pp. 239–251. Springer, Heidelberg (2012)
6. Batini, C., Barone, D., Cabitza, F., Grega, S.: A data quality methodology for heterogeneous data. Int. J. Database Manag. Syst. **3**(1), 60–79 (2011)

7. Batini, C., Cappiello, C., Francalanci, C., Maurino, A.: Methodologies for data quality assessment and improvement. ACM Comput. Surv. **41**(3), 16 (2009)
8. Boufares, F., Ben Salem, A.: Heterogeneous data-integration and data quality: overview of conflicts. In: 6th International Conference on Sciences of Electronics, Technologies of Information and Telecommunications, SETIT, pp. 867–874 (2012)
9. González, L., Peralta, V., Bouzeghoub, M., Ruggia, R.: Qbox-services: towards a service-oriented quality platform. In: Heuser, C.A., Pernul, G. (eds.) ER 2009. LNCS, vol. 5833, pp. 232–242. Springer, Heidelberg (2009)
10. Huh, Y., Keller, F., Redman, T., Watkins, A.: Data quality. Inf. Softw. Technol. **32**(8), 559–565 (1990)
11. Lemos, F., Bouadjenek, M.R., Bouzeghoub, M., Kedad, Z.: Using the QBox platform to assess quality in data integration systems. Ing. Syst. d'inf. **15**(6), 105–124 (2010)
12. Manghi, P., Artini, M., Atzori, C., Bardi, A., Mannocci, A., La Bruzzo, S., Candela, L., Castelli, D., Pagano, P.: The D-NET software toolkit. Program **48**(4), 322–354 (2014)
13. Manghi, P., Bolikowski, L., Manola, N., Schirrwagen, J., Smith, T.: OpenAIREplus: the European scholarly communication data infrastructure. D-Lib Mag. **18**(9–10), 1 (2012)
14. Mannocci, A., Casarosa, V., Manghi, P., Zoppi, F.: The Europeana network of ancient Greek and Latin epigraphy data infrastructure. In: Closs, S., Studer, R., Garoufallou, E., Sicilia, M.-A. (eds.) MTSR 2014. CCIS, vol. 478, pp. 286–300. Springer, Heidelberg (2014)
15. Marotta, A., Ruggia, R.: Quality management in multi-source information systems. In: Quality (2002)
16. Marotta, A., Ruggia, R.: Managing source quality changes in a data integration system. CEUR Workshop Proc. **263**, 1168–1176 (2006)
17. Missier, P., Preece, A.D., Embury, S.M., Jin, B., Greenwood, M., Stead, D., Brown, A.: Managing information quality in e-Science: a case study in proteomics. In: Akoka, J., et al. (eds.) ER Workshops 2005. LNCS, vol. 3770, pp. 423–432. Springer, Heidelberg (2005)
18. Peralta, V., Ruggia, R., Kedad, Z., Bouzeghoub, M.: A framework for data quality evaluation in a data integration system. In: SBBD, pp. 134–147 (2004)
19. Preece, A.D., Jin, B., Pignotti, E., Missier, P., Embury, S.M., Stead, D., Brown, A.: Managing information quality in e-Science using semantic web technology. In: Sure, Y., Domingue, J. (eds.) ESWC 2006. LNCS, vol. 4011, pp. 472–486. Springer, Heidelberg (2006)
20. Redman, T.C.: The impact of poor data quality on the typical enterprise. Commun. ACM **41**(2), 79–82 (1998)
21. Reiter, M., Breitenbücher, U., Dustdar, S., Karastoyanova, D., Leymann, F., Truong, H.L.: A novel framework for monitoring and analyzing quality of data in simulation workflows. In: IEEE 7th International Conference on E-Science, pp. 105–112 (2011)
22. Reiter, M., Breitenbücher, U., Kopp, O., Karastoyanova, D.: Quality of data driven simulation workflows. J. Syst. Integr. **5**(1), 3–29 (2014)
23. Scannapieco, M., Missier, P., Batini, C.: Data quality at a glance. Datenbank-Spektrum **14**(January), 6–14 (2005)
24. Shankaranarayanan, G., Wang, R.Y., Ziad, M.: IP-MAP: representing the manufacture of an information product. In: Proceedings of the 2000 Conference on Information Quality, pp. 1–16 (2000)

25. Strong, D.M., Lee, Y.W., Wang, R.Y.: Data quality in context. Commun. ACM **40**(5), 103–110 (1997)
26. Tani, A., Candela, L., Castelli, D.: Dealing with metadata quality: the legacy of digital library efforts. Inf. Process. Manag. **49**(6), 1194–1205 (2013)
27. Tayi, G.K., Ballou, D.P.: Examining data quality. Commun. ACM **41**(2), 54–57 (1998)
28. Wang, R., Storey, V., Firth, C.: A framework for analysis of data quality research. IEEE Trans. Knowl. Data Eng. **7**(4), 623–640 (1995)

Short Papers

Scientific Social Publications for Digital Libraries

Fidan Limani[1(✉)], Atif Latif[2], and Klaus Tochtermann[2]

[1] Faculty of Contemporary Sciences and Technologies,
South East European University, Tetovo, Macedonia
f.limani@seeu.edu.mk
[2] ZBW - German National Library of Economics Leibniz Information
Center for Economics, Kiel, Germany
{A.Latif,K.Tochtermann}@zbw.eu

Abstract. Social web content is an important development in the scientific workflow. In this context, scientific blogs are an important medium: they play a significant role in the timely dissemination of scientific developments, and provide useful grounds for discussion and development via the readers feedback. Blogs from the domain of economics are no exception to this practice. A possible extension to Digital Libraries (DL) services, content- and service-wise, is to enable its users access to these blogs. This paper demonstrates an approach for seamlessly integrating scientific blogs in DLs and, with the developed proof of concept application, showcases the resulting benefits for the users and DLs.

Keywords: Digital Libraries · Scientific social publications · Web 2.0

1 Introduction

Going beyond the daily news-, entertainment-, or commercially-related consumption, there is an emergence of a new breed of social web publications in the form of scientific blogs covering many domains and serving an ever-increasing audience, which is lately receiving more and more attention in the scholarly community (Latif et al. 2015). It is via this type of publication that scholars present their ongoing research, discuss latest scientific events, or follow and treat emerging developments (Powell et al. 2012).

Libraries are more and more emphasizing the social component, striving for a community-centered, "social DL" (Calhun 2013) that engages its users more with their collections (Miller 2006). The role and awareness of social web resources is already present in the research workflow supporting the whole research lifecycle as a complement to the "mainstream" scholarly publications (CIBER report 2010). Furthermore, integrating non-library resources with DL collections would allow for a seamless navigation across the resulting collection (W3C LLD Incubator Group Report 2011).

With over 150 million professional and amateur blogs in the blogosphere (Ragner and Bultitude 2014) collecting and disseminating science to larger audiences in a wide variety of domains, a significant and growing portion of which

© Springer International Publishing Switzerland 2016
N. Fuhr et al. (Eds.): TPDL 2016, LNCS 9819, pp. 373–378, 2016.
DOI: 10.1007/978-3-319-43997-6_29

scientific blogs, we are prompted to further explore their contribution to the DL ecosystem. For this there remains a need for an efficient approach to seamlessly integrate and make these resources accessible within a DL environment.

2   Related Work

We see different levels of social web resource integration with different DLs, starting from cases that enable user-provided contributions annotate DL resources for improved search or user engagement, to cases that bring together separate, heterogeneous collections for increased resource selection. Chenu-Abente et al. (2012) explore the opportunity of integrating several components and services of the research process lifecycle by proposing a platform that brings together research publications and researchers with the aim of integrating social and scientific services. In another undertaking, Mutschke and Thamm (2012), via the ScholarLib model architecture, treat the opportunity of relating social web resources with DL repositories in the domain of social sciences. In enabling Web 2.0 features within the DL ecosystem, García-Crespo et al. (2011) with their semantic DL implementation – CallimachusDL, demonstrate the inclusion of user-provided semantic annotations of social web multimedia resources for improved resources browsing and search. In this line of contribution, Gazan (2008) stipulates that the inclusion of Web 2.0 user-annotations into DL can result with "encouraging increased exploration and engagement" of users in the DL environment.

At another level, Wielemaker et al. (2008) exploit the availability of controlled vocabulary (CV) annotation and requirements for integrating heterogeneous sources to support users with search capabilities for the domain of cultural heritage. Another related contribution, that of Holgersen et al. (2012), demonstrates the possibility of enriching bibliographic records of DLs by including user-generated contribution in the form of ratings, comments, tags, etc.; in this case, a set of DL resources is associated with corresponding user feedback, thus enabling the DL to enlist social contributions alongside its resources.

3   Test Dataset Components

For our experimentation purposes, we needed domain-specific social web and DL collections of high quality, and a CV. We choose The Wall Street Journal's (WSJ) economic blog section[1], EconStor[2] and STW, respectively.

EconStor – The DL Dataset. The publishing/archiving platform EconStor includes Open Access publications from the areas of economics, business administration, and social sciences. The publications in this portal contain high metadata quality and users can rely on a set of feature-rich services to find publications of their interest.

---

[1] http://blogs.wsj.com.
[2] http://econstor.eu/.

The Wall Street Journal – The Scientific Blog Dataset. The WSJ as a domain-specific collection has its focused categories, such as: Bankruptcy beat, Digits, Private equity, Venture capital, etc. Publications from this collection contain high-quality content, making it an attractive proposal to the EconStor portal and a good candidate for our use cases. It is important to mention that blogs usually rely on their own set of terms (tags) when describing posts, different that any complex CV that a DL might use. "Data processing" has the details of how we deal with the "terminology gap".

Thesaurus for Economics (STW) – The CV. With about 6,000 descriptors, 20,000 non-descriptors, as well as rich set of "hierarchical" and other associations, STW represents a valuable asset used to index publications on Econ-Stor portal. Moreover, its (manual) alignments to other thesauri allow users to retrieve publications indexed with vocabularies other than STW terms. Any collection can use STW for indexing its resources as long as they are from the domain of economics and related sciences.

4   Approach

The proposed approach entails several activities that streamline blog posts to the DL environment. Following are the details of each activity.

Blog Post Collection Retrieval. Based on the URI pattern of the WSJ blogs categories and the Cascading Style Sheet (CSS) rules identifying blog post elements, we retrieved over 41K blog posts and their corresponding elements during the time period of July–August 2015.

Blog Post (Meta)Data Augmentation. Regardless of the blog post category, each WSJ post has the same set of elements: a URL address, title, tags (that describe the post), author, content, publication date, reader comments, share counts, etc. In addition to the original tags, we assign our own set of descriptors to each post using the same vocabulary used by EconStor (see "Data processing" next for more).

Data Processing. In addition to the available blog post elements, we conducted automatic indexing of posts based on the STW thesaurus[3]. For this task we used MAUI[4], a mature and open source solution for term assignment with CVs. We decided to assign no more than 5 STW terms to every blog post as the average number of terms (or, in blog's parlance, tags) per blog post dictated this limit. This effectively brings blog posts at the same vocabulary level as publications from EconStor.

---

[3] http://zbw.eu/stw/version/latest/about.
[4] https://code.google.com/p/maui-indexer/.

5 Proof of Concept Application

In this section we introduce the proof of concept prototype to support use case scenarios motivating our research. We first explain the typical use case scenario steps, and then conclude with an example demonstration.

5.1 System Prototype

The system prototype is given in Fig. 1; in completing users' requests, it interfaces with both the DL portal (EconStor) and the blog post collection.

The user submits a query; EconStor look-up service processes it and displays the results (Fig. 1, "EconStor Results"). The user selects an article from the result set and its thesaurus-related metadata are retrieved to further support her refine the results (Fig. 1, "Filtering options"). The prototype then uses the same metadata to query the blog post collection for related results (Fig. 1, "Blog post Results").

5.2 A Use Case Scenario: Description and Example

The proposed system architecture lends us different use case scenarios to explore. Let's follow with the search scenario illustrated in Fig. 1. A search with "ICT industry growth in EU" presents 272 results from EconStor; the STW terms used in this search are "ICT industry" and "economic growth". The user narrows down

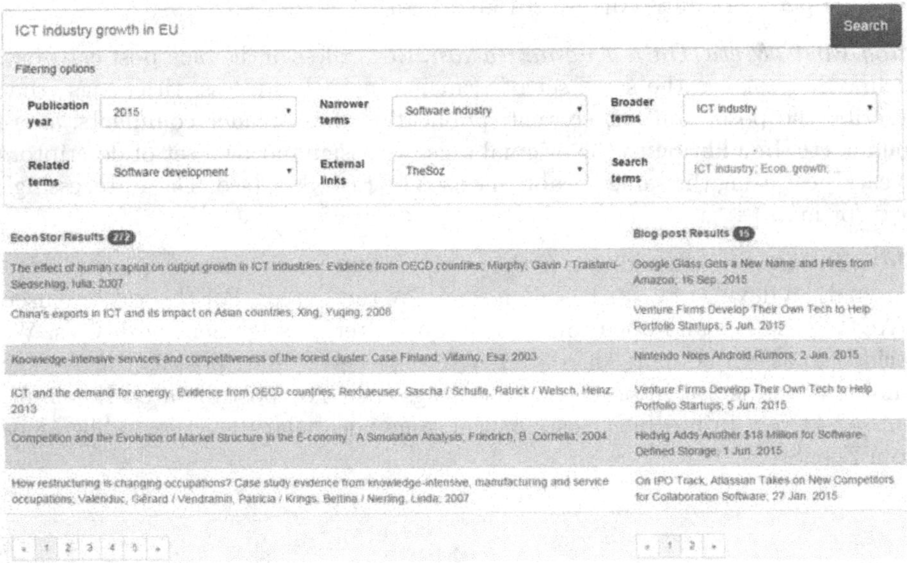

Fig. 1. System prototype

the search to "software industry"[5] which reduces the results to 246. In a similar way she can use STW structure and associations to further (re)define her search.

She selects the top-ranked article from the result set; its STW descriptors are "human capital", "ICT industries", and "Economic growth". The prototype queries the blog collection with the terms "ICT industry" and its narrower term "software industry" with no results. She chooses another related term, "telecommunications industry", which results in 36 matches; a related term, "telecommunications", results in 13 blog posts. She can further filter out blog posts by date, leaving only the most recent ones in the result set.

The selected EconStor article and its related blog posts show a meaningful relationship. Some of the top-ranked posts discuss the relationship of human capital and ICT-related developments. One post treats the initiative to revamp a Google project (Google Glass) for current and new market segments as well as information on new hires of engineering background to support the initiative, whereas another post discusses applications that could help venture capitals identify investment opportunities. The underlying key terms of these results are: "software development", "software", and "enterprise". There are other terms associated with these top-ranked posts that affect the results such as "team" and "engineers" for the first post, "self help" (used to mean empowerment in the context of social relations) for the second, and "short selling" (in the context of securities trading) for the third post.

6   Resulted Benefits

In this section, in perspective of our proof-of-concept application, we describe the resulted benefits for DL and users as a whole. Lastly we highlight the challenges and limitations faced in the experimental setup.

DL Benefits. (1) Indexed over 41K blog posts with DL repositorys CV of choice. This, in turn, enables a seamless search across the resulting collection; (2) Augmented the original blog post meta-data with STW term description, effectively bridging the terminology gap between the DL and blog post collections; (3) Applied a process flow that automates streamlining blog posts to the DL environment; (4) Enabled reuse of existing EconStor services on the newly-added blog posts. **User Benefits.** (1) Increased resource selection; (2) Social blog authors can rely on the presented process flow in describing their posts with CVs used by the target DL. **Challenges and Limitations.** (1) Short blog posts render MAUI unable to propose description terms. (2) Small blog post collection, which, although primarily designed to support our use cases, needs a continuous update in order to provide good basis for resources match-up. (3) Our process flow relies on CSS rules to identify blog post elements, which limits scaling.

7   Conclusion and Future Work

Scientific blogs are getting closer to the traditional research workflow with ever-increasing quality and audience. In this paper we have proposed a strategy to

---

[5] http://zbw.eu/stw/versions/latest/descriptor/24708-1/about.en.html.

streamline them to DL collections. Moreover, we have developed a proof-of-concept application that highlights the potential benefits of this approach. We are of a view that the integration of DL repository with scientific blogs introduces the issue of information overload for users and needs to be dealt with. In the future, we plan to experiment with various similarity measures to tackle this problem.

References

Latif, A., Scholz, W., Tochtermann, K.: Science 2.0 - mapping European perspectives. Report on the General Stance of Organizations on European Commissions Public Consultation on Science 2.0 (2015)

CIBER report: Social media and research workflow. University College London, Emerald Group Publishing Ltd. (2010)

Calhun, K.: Digital libraries and the social web: scholarship. In: Exploring Digital Libraries: Foundations, Practice, Prospects. Facet Publishing/ALA Neal-Schuman, London/Chicago (2013, 2014). ISBN: 9781856048200

Chenu-Abente, R., Menéndez, M., Giunchiglia, F., De Angeli, A.: An entity-based platform for the integration of social and scientific services. In: 8th International Conference on Collaborative Computing: Networking, Applications and Worksharing, Collaboratecom 2012, Pittsburgh, PA, United States, 4–17 October 2012

García-Crespo, Á., Gómez-Berbís, J., Colombo-Palacios, R., García-Sánchez, F.: Digital libraries and Web 3.0. The CallimachusDL approach. Comput. Hum. Behav. **27**(4), 1424–1430 (2011)

Gazan, R.: Social annotations in digital library collections. In: D-Lib Magazine, vol. 14, no. 11/12 (2008). http://www.dlib.org/dlib/november08/gazan/11gazan.html. Accessed 17 Jan 2016

Holgersen, R., Preminger, M., Massey, D. Using semantic web technologies to collaboratively collect and share user-generated content in order to enrich the presentation of bibliographic records development of a prototype based on RDF, D2RQ, Jena, SPARQL and WorldCats FRBRization web service. 4Lib J. (17) (2012). ISSN: 1940-5758. http://journal.code4lib.org/articles/6695

Library Linked Data Incubator Group Final Report. W3C Incubator Group Report (2012). http://www.w3.org/2005/Incubator/lld/XGR-lld-20111025. Accessed 15 January 2016

Miller, P.: Library 2.0: The challenge of innovation. Talis white paper (2006)

Mutschke, P., Thamm, M.: Linking social networking sites to scholarly information portals by ScholarLib. In: WebSci 2012, Evanston, Illinois, USA, 22–24 June 2012. ACM (2012). 978-1-4503-1228-8

Powell, D., Jacob, C., Chapman, B.: Using blogs and new media in academic practice: potential roles in research, teaching, learning, and extension. Innov. High. Educ. **37**(4), 271–282 (2012). Springer, Netherlands

Ragner, M., Bultitude, K.: The kind of mildly curious sort of science interested person like me: science bloggers' practices relating to audience recruitment. Public Underst. Sci. (2014). doi:10.1177/0963662514555054

Wielemaker, J., Hildebrand, M., van Ossenbruggen, J., Schreiber, G.: Thesaurus-based search in large heterogeneous collections. In: Sheth, A.P., Staab, S., Dean, M., Paolucci, M., Maynard, D., Finin, T., Thirunarayan, K. (eds.) ISWC 2008. LNCS, vol. 5318, pp. 695–708. Springer, Heidelberg (2008)

Formal Representation of Socio-Legal Roles and Functions for the Description of History

Yoonmi Chu[✉] and Robert B. Allen

Yonsei University, Seoul, Korea
yoonmichu@gmail.com, rballen@yonsei.ac.kr

Abstract. We propose a modeling approach for formal descriptions of historical material. In our previous work, we defined the formal structures of social entities such as roles, rights and obligations, activities, and processes which appear in the Roman Constitution, as an application of Basic Formal Ontology (BFO). In this paper, we extend that approach by incorporating aspects of the Information Artifact Ontology (IAO) and the emerging Document Acts Ontology (DAO). We use these to describe relationships among realizable entities (role and function), rights and obligations that are aligned to Socio-Legal Generically Dependent Continuants (SGDCs) of DAO, and activities as subtypes of directive information entity of IAO. Two examples are discussed: a passage from a digitized historical newspaper and a description of citizenship in ancient Rome.

Keywords: Community models · Direct representation · Legal and political ontologies · Rich semantics · Roman republic

1 Introduction

Many ontologies for history such as HEML [10] focus on high-level named historical events and eras. We have been exploring ways to provide more detailed coverage with rich semantics. For instance, in organizing and indexing the history of communities described in the millions of pages of newspapers which have now been digitized, we believe that the best approach is creating "community models" which are rich semantic descriptions of those communities [4]. Similarly, we have considered the possibility of rich semantics for organizing the history of "the decline and fall of the Roman Empire" along with commentary about that history from Gibbon [9]. Specifically, in [7], we modeled the formal structure of roles, activities and procedures for describing governmental structures of the Roman Empire. The models of communities and of historical analysis, along with other work on modeling scientific communications, are examples of an approach we call "direct presentation" [2].

In this paper, we extend the analysis of social roles in [7] to include not only the relationships between social roles and socio-legal entities (e.g., rights and obligations) but also the relationships among social functions of institutions that compose a society. We concretize it by using the structures of the Information Artifact Ontology (IAO) and by considering how the emerging Document Acts Ontology (DAO) might be applied. We consider two applications: first, descriptions of the contents of a digitized historical newspaper and second, descriptions of citizenship in ancient Rome.

© Springer International Publishing Switzerland 2016
N. Fuhr et al. (Eds.): TPDL 2016, LNCS 9819, pp. 379–385, 2016.
DOI: 10.1007/978-3-319-43997-6_30

2 Rich Semantics

2.1 The Basic Formal Ontology (BFO) and the Information Artifact Ontology (IAO)

Ontological realism views ontologies as representations of the reality that is described by science to ensure mutual consistency of ontologies over time and to ensure that ontologies are maintained in such a way as to keep pace with advances in empirical research [12]. Bioinformatics has found great success in building ontologies such as the GO (Gene Ontology) which is based on the realist Basic Formal Ontology (BFO). [7] was the first study to consider how the BFO could be applied to historical modeling.

BFO has been extended with the Information Artifact Ontology (IAO) [13]. The IAO considers the ways information artifacts relate to the *Independent Continuants* with which they are associated. IAO defines *Information Content Entities* as a subtype of a *Generically Dependent Continuant* that is dependent on independent continuants that can serve as its bearers, and that migrates from one bearer to another. Using concepts from the IAO we can update our previous model. We align activities to *Action Specification* and procedures to *Plan Specification* that are subtypes of *Information Content Entity* in the IAO.

2.2 Towards a Document Acts Ontology (DAO)

Social entities do not fit in a purely realist ontology. To model social entities such as rights and obligations, Smith and his colleagues have explored approaches which go beyond the realist BFO. Indeed, [11] proposes that some social entities are "quasi-abstract". Document Acts, which are formal and often legally binding, actions based on written agreements, are central to this approach. [6] took initial steps towards a Document Acts Ontology (DAO). DAO is focused on socio-legal entities such as claims (or more precisely, rights) and obligations. These entities presuppose the existence of a person (or other legal agent) whose rights or obligations they are, and can be reassigned from one person to another. Thus, the Document Acts Ontology defines these entities as *Socio-legal Generically Dependent Continuants* (SGDCs) which is a subtype of *Generically Dependent Continuants* (GDCs) that depend on *specifically dependent continuants* (SDCs). Although *Generically dependent continuants* are not realizable entities [5], IAO allows concretization of GDCs as realizable entities by not restricting *concretized* to hold only between GDCs and qualities. That means that socio-legal entities such as rights and obligations can be realized through their bearer's roles by allowing concretization of GDCs.

Using the concept of concretization, DAO identifies the relationships between realizable entities and SGDCs. We extend our previous work [7] by describing the way that social roles concretize SGDCs. Thus, we propose a formal structure describing the relationships among realizable entities (*role* and *function*), SGDCs, and *activity and procedure* as subtypes of *directive information entity* of IAO (see Fig. 1). Socio-Legal Roles are defined as the specifications of rights and obligations that the bearer of a role can (or should) exercise. In turn, rights and obligations can be expressed by the activities describing what one is allowed (or required to do).

In contrast with Socio-Legal Roles, Socio-Legal Functions "inhere in" the bearer by virtue of the bearer's make-up and support achievement of the bearer's goal. Social Functions are concretized as processes to achieve the bearer's goal and realized in some process in which their bearer participates. For instance, schools are a kind of social institution which has the goal of educating students. People employed to teach students have a Teacher Role and Rights. Their rights specify activities in accordance with the teaching plans of the schools that concretize the social function of education. Ultimately, the legitimacy of the Superintendent and Teacher Roles extends back to the government which underwrites the legitimacy of many Document Acts [3, 14]. We also recognize that people's activities may not follow formal specifications even if they are complying with the expectations of a role [8].

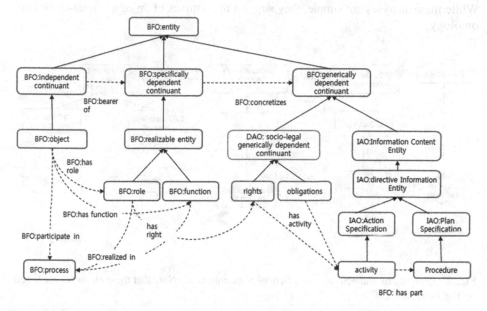

Fig. 1. Proposed representation of relationships among social roles, activities and procedures using constructs from the IAO and the evolving DAO.

3 Toward a Community Model for Norfolk, Nebraska

We now consider specific examples for applying the approaches derived from the IAO and DAO to historical descriptions. Some aspects of society and social activities such as school buildings and the fact that students attend schools can fit under a realist ontology, but a more complete analysis of social objects needs to include intangible aspects such as rights, roles, and obligations.

As mentioned above, we have been exploring the development of historical models to support interaction with the large number of digitized historical newspapers and other historical materials. To focus that work, we have been developing a testbed collection for the town of Norfolk, Nebraska [1]. In this section, we consider the activities of a specific person in that town. We picked an individual who was mentioned several times in the *Norfolk Weekly News Journal* from about 1895 to 1905. This was D.C. O'Connor who was the School Superintendent for the City of Norfolk, Nebraska[1]. While some social roles and functions are ad hoc and implicit, the roles of a School Superintendent are generally well delineated. We did not find documentation of the job requirements for the Superintendent of Schools of Norfolk Nebraska. But, we did find several accounts of O'Connor's activities based on that role. For instance, he made reports at School Board meetings and there is a description of a trip he took from Norfolk to Lincoln (capital of Nebraska) to hire a new teacher to replace one who had left. We have modeled this latter activity in Fig. 2. We also found descriptions of O'Connor chairing a session at an annual state-teachers association meeting. This could also be modeled as an activity which was associated with the Superintendent's role. While these analyses are simple, they suggest the outlines of an educational-institution ontology.

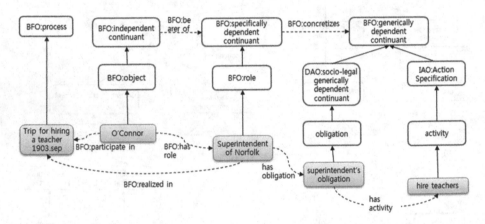

Fig. 2. Obligations and activities of a School Superintendent. Note that these elements are based on the DAO and IAO.

4   Rights and Roles of Citizens in the Roman Republic

We also modeled a description of citizenship for an adult male in the Roman Republic.[2] The legal status of Roman citizenship was a prerequisite for many important legal rights such as the right to trial and appeal, to marry, to hold office, and to enter binding

---

[1] *The Norfolk Weekly News Journal* (Norfolk, Neb.), 29 March 1900. *Chronicling America: Historic American Newspapers*. Library of Congress. http://chroniclingamerica.loc.gov/lccn/sn95070060/ 1900-03-29/ed-1/seq-7/ *The Norfolk Weekly News-Journal*. (Norfolk, Neb.), 04 Sept. 1903. *Chronicling America: Historic American Newspapers*. Library of Congress. http://chroniclingamerica.loc. gov/lccn/sn95070058/1903-09-04/ed-1/seq-6/.

[2] Roman Republic: Legislative Assemblies, https://en.wikipedia.org/wiki/Roman_Republic#Legislative_ Assemblies (accessed 18. Mar. 2016).

contracts, and to eligibility for special tax exemptions. A citizen has the political right to vote for assemblies. In turn, these assemblies elected magistrates, enacted legislation, and presided over trials of capital cases. As shown in Fig. 3, activities and procedures are modeled with elements of IAO while rights are modeled with the extended DAO. In Fig. 3, an adult male has a citizen role which is comprised of political rights and legal rights. These rights specify permitted activities such as to stand trial and to participate in electing assemblies. The assemblies elected by Roman citizens have the socio-legal function of governing the Roman Republic. This includes activities such as electing magistrates, enacting legislation, and presiding over trials. Note that we do not necessarily agree with some of the distinctions made between political rights and legal rights in the source article but we reflect those here as written. Ultimately, procedures will be needed for validating the content of the ontologies and the coordination of sets of interlocking ontologies.

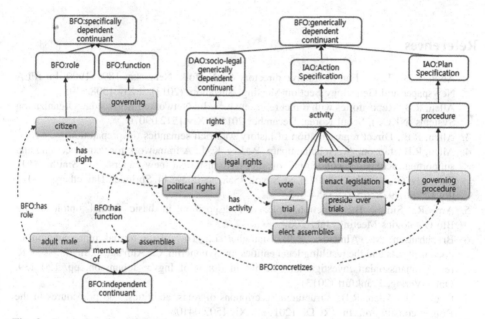

Fig. 3. Representation of rights, activities, and procedures associated with Roman citizenship and institutions.

5 Discussion and Conclusion

This proposal is a step in developing techniques for rich semantic modeling of historical scenarios. We illustrate this approach with examples from digitized historical newspapers and from a description of the rights and obligations of citizens of the Roman Republic. We can begin to consider a rich set of domain ontologies which build on the emerging Document Acts Ontology. A set of inter-related social and community ontologies could be coordinated just as the Open Biological Ontology (OBO) Foundry

[13] is developing a growing set of ontologies for the life sciences. This approach supports a relatively flexible modular approach to modeling complex environments. An initial set of ontologies based on IAO/DAO could be applied to systems of Social Rights such as those explicitly stated in the U.S. Bill of Rights. However, the line between political rights and legal rights is fuzzy and the sub-ontologies may have to be closely coordinated. Indeed, we view the coordination of multiple sub-ontologies, whether biological or social ontologies, as major future issue for information organization. In addition, while this approach can capture formally specified social entities, activities, functions, and roles, there will almost certainly be inconsistencies in such formal specifications. Even when there is a formal prescription, actual behavior may not reflect those expectations. Furthermore, in writing histories, there are many details which will never be known. In developing systems to describe history, we will have to go beyond the basic approaches to ontology by supporting alternative explanations, state-changes, probabilities, versioning, and argumentation.

References

1. Allen, R.B.: Toward an interactive directory for Norfolk, Nebraska: 1899–1900. In: IFLA Newspaper and Genealogy Section Meeting, Singapore (2013). arXiv:1308.5395
2. Allen, R.B.: Repositories with direct representation. In: Networked Knowledge Organization Systems (NKOS), Seoul Korea, December 2015, arXiv:1512.09070
3. Allen, R.B.: Direct representation of history with rich semantics (in preparation)
4. Allen, R.B., Japzon, A., Achananuparp, P., Lee, K.-J.: A framework for text processing and supporting access to collections of digitized historical newspapers. In: Smith, M.J., Salvendy, G. (eds.) HCII 2007. LNCS, vol. 4558, pp. 235–244. Springer, Heidelberg (2007). doi:10.1007/978-3-540-73354-6_26
5. Arp, R., Smith, B.: Function, role, and disposition in basic formal ontology. In: Bio-Ontologies Meeting, July 2008
6. Brochhausen, M., Almeida, M.B., Slaugther, L.: Towards a formal representation of document acts and the resulting legal entities. In: Svennerlin, C., Almäng, J., Ingthorsson, R. (eds.) Johanssonian Investigations: Essays in Honor of Ingvar Johansson, pp. 120–139. Ontos-Verlag, Frankfurt (2013)
7. Chu, Y.M., Allen. R.B.: Structured descriptions of roles, activities, and procedures in the Roman constitution. In: IRCDL (2015). arXiv:1502.04108
8. Clancey, W.J., Sachs, P., Sierhuis, M., van Hoof, R.: Brahms: simulating practice for work systems design. Int. J. Hum. Comput. Stud. **49**, 831–865 (1998)
9. Gibbon, E.: The History of the Decline And Fall of the Roman Empire (1845). http://www.gutenberg.org/files/731/731-h/731-h.htm
10. Robertson, B.: Exploring historical RDF with HEML. Digit. Humanit. Q. **3**(1) (2009). http://www.digitalhumanitites.org/dhq/vol/003/1/000026/000026.html
11. Smith, B.: Searle and DeSoto: The New Ontology of the Social World. In: Smith, B., Mark, D., Ehrlich, I. (eds.) The Mystery of Capital and the Construction of Social Reality, pp. 35–51. Open Court, Chicago (2008)

12. Smith, B., Ashburner, M., Rosse, C., Bard, J., Bug, W., Ceusters, W., Goldberg, L.J., Eilbeck, K., Ireland, A., Mungall, C.J., OBI Consortium, Leontis, N., Rocca-Serra, P., Ruttenberg, A., Sansone, S.A., Scheuermann, R.H., Shah, N., Whetzel, P.L., Lewis, S.: The OBO foundry: coordinated evolution of ontologies to support biomedical data integration. Nature Biotechnol. **25**(11), 1251–1255 (2007)
13. Smith, B., Ceusters, W.: Aboutness: towards foundations for the information artifact ontology. In: International Conference on Biomedical Ontology (ICBO) (2015)
14. Zaibert, L., Smith, B.: The varieties of normativity: an essay on social ontology. In: Tsohatzidis, S.L. (ed.) Intentional Acts and Institutional Facts, pp. 157–173. Springer, Netherlands (2007)

Evaluating Co-authorship Networks in Author Name Disambiguation for Common Names

Fakhri Momeni[(✉)] and Philipp Mayr

GESIS Leibniz-Institute for the Social Sciences, Cologne, Germany
{fakhri.momeni,philipp.mayr}@gesis.org

Abstract. With the increasing size of digital libraries it has become a challenge to identify author names correctly. The situation becomes more critical when different persons share the same name (homonym problem) or when the names of authors are presented in several different ways (synonym problem). This paper focuses on homonym names in the computer science bibliography DBLP. The goal of this study is to evaluate a method which uses co-authorship networks and analyze the effect of common names on it. For this purpose we clustered the publications of authors with the same name and measured the effectiveness of the method against a gold standard of manually assigned DBLP records. The results show that despite the good performance of implemented method for most names, we should optimize for common names. Hence community detection was employed to optimize the method. Results prove that the applied method improves the performance for these names.

Keywords: Author name homonyms · Co-authorship network · Community detection · Louvain method · Gold standard

1 Introduction

In scholarly digital libraries (DLs) authors are recognized via their publications. It is important for users to know about the author of a particular publication to access possible other publications of this author. For this purpose DLs provide search services using the publication information in their databases. However, when several authors share the same name or authors provide their works with different names DLs need more analysis on author's oeuvres. Many different approaches have been proposed in the field of author name disambiguation. Manual author identification in large DLs is very costly. The consequence is that large sets of ambiguous author names need to be analyzed automatically. In addition the demographic characteristics such as name origin and frequency of names used for authors influence the identification of authors. Therefore, all constrains of the underlying data should be considered to choose the appropriate method for author name disambiguation.

Author assignment method and author grouping method [3] are the two main methods for author name disambiguation. Author assignment method constructs a model that represents the author and assigns proper publications to the

© Springer International Publishing Switzerland 2016
N. Fuhr et al. (Eds.): TPDL 2016, LNCS 9819, pp. 386–391, 2016.
DOI: 10.1007/978-3-319-43997-6_31

model. It requires former knowledge about the authors. Nguyen and Cao [7] used this method and proposed to link the author names to the matching entities in Wikipedia. The author grouping method clusters the publications on the basis of their properties (co-authors, publication year, keywords, etc.) to assign a group of publications to a certain author. Following this framework, Caron and van Eck [2] applied rule-based scoring to clustered publications. In their approach the authors suppose that there is enough information about authors and their documents. Also, Gurney et al. [4] clustered publications with employing different data fields and integrated a community detection method. Some authors [5,8,9] used social networks (mainly co-authorship networks) to cluster publications. Levin and Heuser [5] introduced a set of matching functions based on the social network of authors and measured the strength of connections between the authors. Shin et al. [8] extracted the abstract and author's affiliation from the paper and considered the relation between authors to find similarities between publications. Wang et al. [9] proposed a unified semi-supervised framework to handle the synonym and homonym problem of author names.

In this paper we used an author grouping method (compare [3]) to cluster the publications of a set of random authors with the same name in the DBLP database. Considering the lack of rich bibliographic information in DBLP records, we applied co-authorship network analysis introduced by Levin and Heuser [5] to detect similarities between publications in order to investigate, how the amount of homonym names affects the disambiguation results. In the end, we employed a community detection algorithm (Louvain method) to reduce the effect of common names in our evaluation.

2   Disambiguation Approach

We use the author grouping method in order to assign all publications of each person to a certain group. For this purpose all publications belonging to the same ambiguous author name are categorized into one block. In a next step we compare any pair of publications in each block with each other to find a similarity between them. If we have n blocks and m_i publications in a block i, the number of comparisons for all blocks is:

$$\sum_{i=1}^{n} \frac{m_i(m_i - 1)}{2} \tag{1}$$

The result of each comparison is true or false. The *true* result means that two publications belong to one person and the same cluster. If one of them was compared with another one before and assigned to a cluster, the other one is added to that cluster too. If both of them were compared before and belong to different clusters, two clusters are rebuilt to one cluster. Otherwise a new cluster will be created and two publications are put in new cluster. In the next section we describe how to define the similarity indicator to build the clusters.

The bibliographic information that we can obtain from publications in DBLP is limited mainly to author names (the names of all co-author names are listed), title and publication venue. We chose the co-author names as our similarity indicator. Therefore we built a network of authors and documents. Figure 1 shows an example of the network. The continuous lines show the links between publications and authors in the network. As it was mentioned before each pair of documents within every block has to be compared.

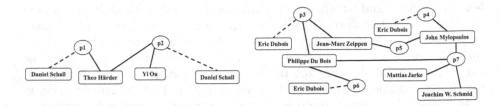

Fig. 1. An example of a co-authorship network

To compare the publications the relations in the network are analyzed. If there is a path between two publications, their distance is defined as the length of the shortest path between them, otherwise it would be infinite. The length of the shortest path is equal to the number of nodes between two nodes. For example the distance between publication p1 and p2 in Fig. 1 is 1; the distance between p3 and p4 is 3. The less distance between two publications means that more likely these publications were written by one person. So, the distance between two publications is assumed as the similarity measure. Different thresholds can be considered for the distance. For example, in Fig. 1, with the threshold = 1, p1 and p2 are two publications of one person with the name 'Daniel Schall', because they share the same co-author. Accepting the threshold = 3, p3 and p4 belong to same author with the name 'Eric Dubois'. In Sect. 3 we see the effect of selecting different thresholds on the evaluation results.

3   Evaluation

Gold Standard: In order to evaluate the output of the author disambiguation approach we need a gold standard of disambiguated author names. Many homonym author names in DBLP are disambiguated manually by the DBLP team and are identifiable with an id. For example, 'Wei Li' belongs to 59 different persons: 'Wei Li 0001', 'Wei Li 0002', etc. Thus, the set of publications for each person is recognizable. To build the gold standard [6] we selected these identified author names and compiled all their publications into one set. In our gold standard we provide a list of publications which have at least one disambiguated author name. Asian names, especially Chinese names are the most common names in DBLP and result in many homonym author names. These names are the most problematic names in author disambiguation and should be analyzed in particular. In total 1,578,316 unique author names exist in DBLP.

There are 5,408 authors who have an identification number (we mention them as disambiguated authors). These 5,408 authors and their publications form the gold standard. We got these numbers from DBLP, downloaded May/01 2015 from http://dblp.uni-trier.de/xml/. To measure the performance of our method 1,000 disambiguated author names have been randomly selected from the gold standard. In total we have 2,844 different authors and 32,273 publications in our random sample. In the next section we evaluate the performance of our method against the gold standard.

Evaluation Metrics: Bcubed metrics [1] are used to evaluate the quality of the algorithm. These clustering metric satisfy constraints on evaluation the clustering tasks [1] such as cluster homogeneity and cluster completeness. Therefore we applied them to evaluate our method. For this purpose Bcubed precision and recall are computed for each publication.

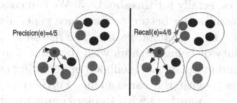

Fig. 2. Example BCubed precision and recall adapted from [1] (Color figure online)

Figure 2 shows an example how the precision and recall of one publication of an author are computed by BCubed metrics. In this Figure assume the circles in red, black and blue as the publications belonging to three different authors and our algorithm categorized them to three groups. The publication precision measures how many publications in its group belong to its author. The publication recall measures how many publications from its author appear in its group. Bcubed F as the combination of Bcubed precision and recall is computed as follows:

$$\frac{1}{\alpha(\frac{1}{P}) + (1 - \alpha)(\frac{1}{R})} \tag{2}$$

being P and R Bcubed precision and recall and being α and $(1 - \alpha)$ the relative weight of each metric (We assumed $\alpha = 0.5$). Bcubed precision, recall and F-measure were computed for every publication in any block. Then we consider their average as the Bcubed precision, recall and F of the block.

4   Results and Discussion

We clustered the publications with regard to the distance between them. For choosing the threshold we have checked the distances larger than 3, which results in a very low precision. Then we chose the threshold equal to 1 and 3. For the distance less than threshold (1 or 3), we assign two publications in the same cluster. The results of the evaluations for two thresholds are demonstrated in Table 1.

Table 1. Mean values of BCubed metrics for 1,000 blocks

	BCubed precision	BCubed recall	BCubed F
Threshold = 1	0.98	0.74	0.79
Threshold = 3	0.94	0.81	0.82

The results in Table 1 indicate that our co-author networks method performs well on the dataset and it can be utilized as author identification approach. No effort was made to define and compare against an external baseline. Comparing the results for two thresholds (1 and 3) we can conclude that using threshold = 3 provides us with the better balance between precision and recall and a higher F (slightly better BCubed recall of 0.81 and F of 0.82). We can shows that with the increasing number of publications in the blocks, the efficiency of our algorithm decreases, especially for threshold = 3. We can conclude that although using threshold = 3 results the better performance generally, it is less efficient than using threshold = 1 for common names. The reason is that common names enhance the probability of being authors with the same name in the same area of research activity and increase the likelihood of detecting the shared co-author for different researchers with the same name. Furthermore, it is more likely that these authors have co-authors with similar common names. This results in a higher probability of ambiguous co-authors and wrong connections between publications. Therefore, we should be more cautious to use the co-author of co-author as the similarity measure for these cases and will verify them more deeply. To remove the wrong connections that link two groups of publications from different authors community detection is a good solution. *Community detection* aims at grouping nodes in accordance with the relationships among them to form strongly linked subgraphs from the entire graph. Hence, we applied a community detection algorithm to optimize the results (threshold = 3) for the common names. We chose a subset of the author's names which have more than 200 publications (totally 28 names) in our DBLP dataset. To detect communities in the network we utilized the Louvain method in Pajek. This method maximizes the modularity of network. Single refinement is selected and the resolution parameter was set to 1. Because the less distance between publications increases the probability of being the same author, we gave the weight to connections. For the distances equal to 1 and 2 have weights with values 2 and 1 respectively. Table 2 shows that community detection improved the results for the most repeated names in our sample.

Table 2. BCubed metrics for author names with more than 200 publications, thr. = 3

	BCubed precision	BCubed recall	BCubed F
Before optimization	0.46	0.87	0.45
After optimization	0.79	0.61	0.58

5 Conclusions and Future Work

In this paper we implemented a method to identify authors with the same name based on co-authorship networks in DBLP. The results showed that although co-author networks have a substantial impact on author name disambiguation, but common names decrease the performance of our method and should be optimized in an extra step. For this reason, we implemented the community detection method which showed an improvement for highly frequent common names. Our approach can be applied to disambiguate author names in DBLP. In this way we create the network and link the publications automatically, then apply the community detection to find the suspicious connections and check them manually if they are a wrong connection. In this case, they will be removed from the network and increase the performance of algorithm. So, the speed of automatic disambiguating and the accuracy of manual checking can be combined. Our approach improves the disambiguation of common names, but this is not sufficient. To get better results we need to optimize the parameters such as resolution in the community detection method for different numbers of publications per name. We could also investigate the effect of changing the weights of links between publications depending on their distances. Because this method is based on co-author network, it is limited to multi-author papers. Therefore a multi-aspect indicator is required for single author papers. We can use the titles of publications to extract keywords and add this information to calculate similarity measures.

Acknowledgment. This work was funded by BMBF (Federal Ministry of Education and Research, Germany) under grant number 01PQ13001, the Competence Centre for Bibliometrics. We thank our colleagues at DBLP who helped with generating the testbed [6].

References

1. Amigó, E., Gonzalo, J., Artiles, J., Verdejo, F.: A comparison of extrinsic clustering evaluation metrics based on formal constraints. Inf. Retr. **12**(4), 461–486 (2009)
2. Caron, E., van Eck, N.J.: Large scale author name disambiguation using rule-based scoring and clustering (2014)
3. Ferreira, A.A., Gonçalves, M.A., Laender, A.H.F.: A brief survey of automatic methods for author name disambiguation. SIGMOD Rec. **41**(2), 15–26 (2012)
4. Gurney, T., Horlings, E., den Besselaar, P.V.: Author disambiguation using multi-aspect similarity indicators. Scientometrics **91**(2), 435–449 (2012)
5. Levin, F.H., Heuser, C.A.: Evaluating the use of social networks in author name disambiguation in digital libraries. JIDM **1**(2), 183–198 (2010)
6. Momeni, F., Mayr, P.: An Open Testbed for Author Name Disambiguation Evaluation (2016). http://dx.doi.org/10.7802/1234
7. Nguyen, H.T., Cao, T.H.: Named entity disambiguation: a hybrid statisticaland rule-based incremental approach. In: ASWC 2008
8. Shin, D., Kim, T., Jung, H., Choi, J.: Automatic method for author name disambiguation using social networks (2010)
9. Wang, P., Zhao, J., Huang, K., Xu, B.: A unified semi-supervised framework for author disambiguation in academic social network. In: DEXA 2014

Ten Months of Digital Reading: An Exploratory Log Study

Pavel Braslavski[1(✉)], Vivien Petras[2], Valery Likhosherstov[1], and Maria Gäde[2]

[1] Ural Federal University, Yekaterinburg, Russia
pbras@yandex.ru, v.lihosherstov@gmail.com
[2] Humboldt-Universität zu Berlin, Berlin, Germany
{vivien.petras,maria.gaede}@ibi.hu-berlin.de

Abstract. We address digital reading practices in Russia analyzing 10 months of logging data from a commercial ebook mobile app. We describe the data and focus on three aspects: reading schedule, reading speed, and book abandonment. The exploratory study proves a high potential of the data and proposed approach.

Keywords: Digital reading · Reading behavior · User modeling · Log analysis

1 Introduction

The volumes of digital reading have been steadily growing several years until recently. According to Pew Research, the share of Americans who read at least one ebook yearly grew from 17 % in 2011 to 28 % in 2014; there were 4 % of "ebook only" readers in 2014.[1] In Russia, where our data comes from, the figures for spring 2015 are quite similar: 25 % of adults have read fiction ebooks at least once in the last 12 months; 8 % read ebooks at least once a week.[2]

Distinct ebooks features can potentially influence everyday reading patterns and behaviors. For example, ebooks provide search, multimedia and hypertext functionalities that cannot be implemented in print. Another characteristic is content accessibility: a small mobile device can give the reader instant access to hundreds of thousands of book titles. Especially with streaming content delivery model (i.e. the user pays a flat rate and gets access to the whole collection) as studied in this paper, we assume that ebook readers follow a "try-and-drop" scenario more often.

This paper reports on an initial log study of ca. 3 million reading sessions of about 8,000 users during 10 months of 2015 from a mobile application in Russia. This work-in-progress demonstrates the unprecedented opportunities for low-level analysis of reading at scale that cannot be conducted based on surveys, controlled user studies, and book-level consumption data. In this paper, we describe the dataset and demonstrate its potential focusing on three aspects: (1) reading schedule, (2) reading speed, and (3) book abandonment. We are not

---

[1] http://pewrsr.ch/1LZOwBb.
[2] http://www.levada.ru/2015/05/19/rossiyane-o-chtenii/(in Russian).

© Springer International Publishing Switzerland 2016
N. Fuhr et al. (Eds.): TPDL 2016, LNCS 9819, pp. 392–397, 2016.
DOI: 10.1007/978-3-319-43997-6_32

aware of other large-scale reading log studies, except for solitary self-reported data from ebook services. Since reading is a central cultural practice, reading log analysis can be beneficial for many domains, e.g. schooling, second-language learning, creative writing, book publishing and recommendation, etc.

2 Related Work

Digital reading has been studied from HCI, educational and psychological perspectives. Previous research dealing with reading online or ebooks has focused on differences between screen and paper reading behavior as well as on contextual motivations, preferences and technological challenges in work-related or casual leisure situations [1,2,4,7]. Usage data of ebooks was studied focusing mainly on ebook selections or retrieval issues [3,5,12].

Due to the complexity of the reading process, studies investigating reading and in-book navigation patterns or reading strategies are rather underrepresented [8]. Some researchers report overlapping reading patterns that vary from linear, browsing to berry picking ebook sessions [9,13]. Comparing the contradictory results, study parameters and contextual issues seem to play an important role dealing with reading behavior [2].

The majority of research dealing with ebooks has focused on academic settings in English speaking countries [10] and used qualitative data such as interviews, diaries or observations focusing on individual differences and preferences reading online. While the usage logs of digital library ebooks have been investigated [6,11], non-academic genres are mainly represented by sales rates[3].

3 Bookmate Data

Bookmate[4] is a popular Russian digital reading service. Upon installing an application, users get instant access to the free collection. Standard paid subscription grants a user access to the entire Russian book collection, excluding new arrivals, bestsellers, and business books. Premium subscription provides unlimited access to the entire Bookmate collection. Bookmate logs used in the study correspond to almost 10 months – from January 1st to October 22nd, 2015. The data includes information about the users, books, and readings sessions.

Users. Title preferences of paying and non-paying subscribers are remarkably different – the latter seem to focus on classical novels, mostly by Russian authors. We speculate that these might be required reading material for high-school literature classes, indicating different reading behaviors than the general public. To reduce the variation in the sample, we focus our analysis on the behavior of the paying Bookmate users, who spent more than five hours in the app and read at least 10 different books during the study period. We refer to this group of 8,337 users as *CORE_USERS*. 6,897 (83%) of them indicated their gender;

---

[3] http://www.theguardian.com/media/2012/feb/05/ebook-sales-downmarket-genre.
[4] https://www.bookmate.com.

there is an almost equal number of female (3,445) and male (3,452) readers in the population. Out of 2,804 (34 %) *CORE_USERS* who indicated their year of birth, the majority were born in the 1980s (51 %) and 1990s (28 %).

Reading sessions. Approximately 172 million interactions were recorded for the *CORE_USERS*. Interactions contain user and book IDs, time stamps, and the character ranges that the user read or just browsed through in a certain book. Single interactions were aggregated into reading sessions comprising all subsequent interactions for one user with less then a 30 min pause between them.[5] This resulted in 3.1 million reading sessions. In addition, we isolated 'fast-forward' (faster than 300 words/min) and backward browsing sessions as navigational and did not consider them in statistics dealing with reading volumes. The median value of 305 sessions corresponds roughly to a daily usage pattern.

Books. The Bookmate collection contained 523,689 ebooks by the end of 2015. Roughly half of all books (243,264) are categorized according to 20 genre labels, such as *Love & Romance*, or *Politics & Society*. The book collection read by the *CORE_USERS* consists of 72,823 items; 10,316 of them are read by at least ten readers. It has to be noted that Bookmate's books do not always correspond one-to-one to printed editions: ebooks range from short verses (several hundred characters) to e-versions of multi-volume collections.

4   Digital Reading Behaviors

4.1   Reading Schedule

Reading logs allow us to uncover reading schedule at different scales: hours, days, and weeks. Figure 1 shows average weekly reading volumes over the entire period of observations. Although these volumes can be affected by instability of the user base and promotional campaigns, we can see a higher activity in the period of New Year holidays (January 1–11) and vacation season (July and August). There is also a noticeable increase in reading activity during spring holidays (first decade of May). This indicates more leisure-time reading pattern. Reading activity during the week shows a curious pattern: it increases from Monday (100 %) to Wednesday (102.9 %) and then drops, reaching the lowest point on Saturday (93.2 %). We can speculate that on Saturdays some Bookmate users prefer other leisure activities than reading. Figure 2 shows that most reading activities occur in the evening and night, which again corresponds to leisure reading pattern. Minor differences in fiction/non-fiction reading during the day are rather expected: non-fiction's relative figures surpass fiction's in the morning (10am–12pm), while fiction wins back afternoon and in the late night.

4.2   Reading Speed

Log-based analysis of reading speed can be seen as a large-scale reading proficiency test (when focusing on the users) and readability study (when looking

---

[5] We adopted 30-minute threshold widely used in search query log studies.

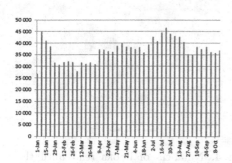

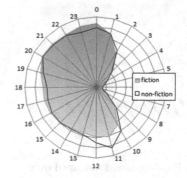

Fig. 1. Average user weekly reading volume.

Fig. 2. Relative volumes read by hours of the day for fiction vs. non-fiction.

at the books). Users' reading speeds have a bell-shaped distribution with mean around 150 words/min, which is the upper bound of the recommended reading speed for elementary school graduates. Figure 3 ranks genres by reading speed and thus reflects their 'difficulty', with recipes being most 'readable' and poetry – least easy to read. It is interesting to note that kids books that are expected to have an easy writing style appear in the 'difficult' subspectrum. This suggests that these books are either read by kids themselves or by parents aloud.

4.3   Book Completion and Abandonment

Reading logs provide us with an exceptional opportunity to determine book completion rates. It is also interesting to validate our hypothesis about higher abandonment rates for ebooks under streaming subscription models, although direct comparisons with printed books are hardly feasible.

We consider the book abandoned if a user does not recur to it within one month after the last reading (thus, a final reading must take place one month before the end of the period presented in the data). In addition, we require that the user started to read the book (i.e., there is a session corresponding to the first 10 % of the book) within our observation period. As a result of these limitations, we have 534,200 unique user–book pairs, and only 190,879 (35.7 %) have completion rate above 90 %. This score is lower than what was reported by the Kobo service that sells individual ebooks.[6] Another figure on book completion can be found in the Goodreads survey: 38.1 % of users reported that they always finish the book, when started.[7] Bookmate users show much lower persistence: only 111 (1.3 %) of them read at least 90 % of their books until 90 % of their length and beyond.

In addition, we calculated the average point of abandonment for a subset of popular books (8,274 books with 10+ readers that meet additional requirements stated above), distributed as shown in the Fig. 4. About half of the books is

---

[6] http://nyti.ms/1QFpcWz.
[7] http://bit.ly/1RAM32Y.

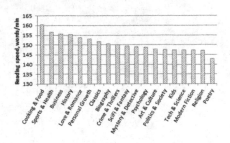

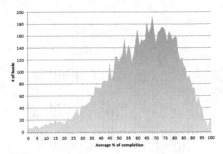

Fig. 3. Reading speed averaged over genres.

Fig. 4. Numbers of books read until the specified percent of their length.

abandoned upon reading 64 % or less of their length; only about 5 % of the books are read beyond 90 % of their length. For example, averaged completion rates for *Fifty Shades of Grey* by E. L. James, *Bridget Jones's Diary* by Helen Fielding, and *Martian* by Andy Weir are 69 %, 71 %, and 78 % of book length, respectively. It is interesting to note that the book length does not correlate with its completion rate.

5   Conclusion and Future Work

In the study, we described the Bookmate application log that corresponds to 10 months of reading of about 8,000 users and provided some initial statistics reflecting three aspects of reading behavior: (1) reading schedule, (2) reading speed, and (3) book abandonment. The main results of the current exploratory study are following:

- Bookmate users' reading corresponds to the leisure-type activity; reading is increasing during holidays and vacations; a moderate decrease on Saturdays can be associated with alternative leisure activities practiced on weekends; reading sessions are shifted towards evening and night hours;
- Ebook reading speeds correspond seemingly to those of printed books; log-based reading speed analysis enables large-scale readability and reading proficiency user studies;
- Flat-rate subscription model seems to promote "try-and-drop" pattern and lower book completion rates. Indirect comparisons suggest that these behaviors differ both from those of printed books readers and readers purchasing individual ebooks.

In our future work we will elaborate on the described aspects and address other characteristics of reading behavior, including title preferences and reading style (one-by-one vs. interleaved book reading), as well as investigate book genre and reader's gender facets. Moreover, we will attempt to map low-level navigational and reading interactions to book content, which opens new opportunities to reading analysis. The analyses can be beneficial for different domains and applications: digital libraries, creative writing and book publishing, as well as book recommendation.

Acknowledgments. We thank Bookmate and Samer Fatayri in person for preparing the dataset and granting access.Our study was performed using the computational cluster of the Ural Federal University.

References

1. Adler, A., et al.: A diary study of work-related reading: design implications for digital reading devices. In: CHI 1998, pp. 241–248 (1998)
2. Buchanan, G., McKay, D., Levitt, J.: Where my books go: choice and place indigital reading. In: JCDL 2015, pp. 17–26 (2015)
3. Hinze, A., et al.: Book selection behavior in the physical library: implications for ebook collections. In: JCDL 2012, pp. 305–314 (2012)
4. Hupfeld, A., Sellen, A., O'Hara, K., Rodden, T.: Leisure-based reading and the place of E-Books in everyday life. In: Kotzé, P., Marsden, G., Lindgaard, G., Wesson, J., Winckler, M. (eds.) INTERACT 2013, Part II. LNCS, vol. 8118, pp. 1–18. Springer, Heidelberg (2013)
5. Kim, J.Y., Feild, H., Cartright, M.: Understanding book search behavior on the web. In: CIKM 2012, pp. 744–753 (2012)
6. Littlewood, H., Hinze, A., Vanderschantz, N., Timpany, C., Cunningham, S.J.: A log analysis study of 10 years of ebook consumption in academic library collections. In: Tuamsuk, K., Jatowt, A., Rasmussen, E. (eds.) ICADL 2014. LNCS, vol. 8839, pp. 171–181. Springer, Heidelberg (2014)
7. Liu, Z.: Reading behavior in the digital environment: changes in reading behavior over the past ten years. J. Doc. **61**(6), 700–712 (2005)
8. Marshall, C.C.: Reading and writing the electronic book. Synth. Lect. Inf. concepts retrieval Serv. **1**(1), 1–185 (2009)
9. McKay, D.: A jump to the left (and then a step to the right): reading practices within academic ebooks. In: OzCHI 2011, pp. 202–210 (2011)
10. Staiger, J.: How e-books are used: a literature review of the e-book studies conducted from 2006 to 2011. Ref. User Serv. Q. **51**(4), 355–365 (2012)
11. Tucker, J.C.: Ebook collection analysis: subject and publisher trends. Collect. Build. **31**(2), 40–47 (2012)
12. Willis, C., Efron, M.: Finding information in books: characteristics of full-text searches in a collection of 10 million books. Proc. ASIST **50**(1), 1–10 (2013)
13. Zhang, T., Niu, X.: Final report for ER&L + EBSCO library fellowship research project. In: Libraries Reports, Paper 4 (2015)

A Case Study of Summarizing
and Normalizing the Properties of DBpedia
Building Instances

Michail Agathos$^{(\boxtimes)}$, Eleftherios Kalogeros, and Sarantos Kapidakis

Laboratory on Digital Libraries and Electronic Publishing,
Department of Archives and Library Science and Museology,
Ionian University, Corfu, Greece
{agathos,kalogero,sarantos}@ionio.gr

Abstract. The DBpedia ontology forms the structural backbone of DBpedia linked open dataset. Among its classes *dbo:Building* and *dbo:HistoricBuilding* entities, hold information for thousands of important buildings and monuments, thus making DBpedia an international digital repository of the architectural heritage. This knowledge for these architectural structures, in order to be fully exploited for academic research and other purposes, must be homogenized, as its richest source - Wikipedia infobox template system - is a heterogeneous and non-standardized environment. The work presented below summarizes the most widely used properties for buildings, categorizes and highlights structural and semantic heterogeneities allowing DBpedia's users a full exploitation of the available information.

Keywords: DBpedia · Wikipedia · Digital libraries · Semantic web · Historic buildings · Monuments · Cultural heritage · RDF

1 Introduction

Wikipedia has grown into one of the central knowledge sources of mankind and is maintained by thousands of contributors[1]. Up to now holds information for 58.720 various types of buildings, while a large number of these has been characterized as historic buildings (in fact 7.024). More specifically Wikipedia hosts information for many World Heritage buildings and monuments, thus forming a rich store of information about the past, some of it unique. As its known Wikipedia itself only offers very limited querying and search capabilities that are limited to keyword matching. For example, in the case of historic buildings, it is difficult to find all structures that meet a particular architectural style in a given period of time. One of the goals of the DBpedia project [1] is to provide these querying and search capabilities to a wide community by extracting structured data. Such information includes infobox templates, categorization information, images, geo-coordinates, links to external web pages, disambiguation pages, redirects between pages, and links across different language editions of Wikipedia, etc. The DBpedia extraction framework extracts this structured information

---

[1] http://dbpedia.org/about.

© Springer International Publishing Switzerland 2016
N. Fuhr et al. (Eds.): TPDL 2016, LNCS 9819, pp. 398–404, 2016.
DOI: 10.1007/978-3-319-43997-6_33

from Wikipedia and turns it into a rich knowledge base that can be used to answer expressive queries such as the one outlined above. The backbone of DBpedia is the DBpedia Ontology - a shallow, cross-domain ontology, which has been manually created based on the most widely used infoboxes within Wikipedia[2]. The classes *dbo: Building* and *dbo:HistoricBuilding* are 2 of the total 685 DBpedia's ontology classes. The DBpedia dataset is freely available as RDF dump files and can be queried through the DBpedia SPARQL Endpoint[3].

If you query DBpedia for all the properties of the instances of the class *dbo: Building* or its subclass *dbo:HistoricBuilding* a large amount of properties will be returned that comes from all the Wikipedia information representation structures and specifically from its richest source - the Wikipedia infobox template system, which is an heterogeneous and non-standardized environment. Although these infobox properties are not defined in the DBpedia Ontology, they are perfectly usable; for instance in SPARQL queries [2]. In this work in order for DBpedia users to be able to exploit all this available information we try to summarize all these infobox properties that refer to buildings, allowing users expressive queries relieved from heterogeneities such as these presented below. Through the summarization process structural and semantic heterogeneities concerning Wikipedia infobox properties for buildings are highlighted and isolated. Moreover the grouping of these properties based on their semantics could help designers and maintainers who work on Wikipedia infobox[4] consolidation in merging and eliminating name variations. Furthermore it could help the DBpedia community in the manual alignment between Wikipedia infoboxes and the DBpedia ontology which help to normalize name variations in properties and classes.

2   Summarizing DBpedia Building Entity Properties

Querying DBpedia for all the properties used for instances of the class *dbo:building* results in 2.946 properties, the majority of which (96 % - 2.827) are properties generated by the DBpedia extraction framework from Wikipedia infobox fields and can be distinguished because they have a distinct namespase (http://dbpedia.org/property/, *prefix:dbp*). A small percentage of the remaining properties derived from 8 other vocabularies presented in Table 1 below. A total of 410 of these properties are present (at least once) in almost 99.8 % of the DBpedia building instances. In order to classify these properties into distinct information categories we started exploring the function and use of each property separately, studying the semantics of its content value through many different types of building instances in order to verify these semantics. Having checked with this method the values of these properties against real world facts, in order to validate their accuracy, we classified these properties according to the above criteria (semantics and functionality) into 16 information categories. More specifically primary information for buildings such as names, dates associated with the creation of

---

[2] http://wiki.dbpedia.org/services-resources/ontology.

[3] http://dbpedia.org/sparql.

[4] WikiProject- https://en.wikipedia.org/wiki/Wikipedia:WikiProject_Infoboxes.

the building, properties describing the functional type or use of these resources were summarized in 11 main categories. Supplementary information, such as general info - specifications for buildings or persons associated with the management of a structure etc., was grouped in 5 other categories.

The summarization study for the properties of all the DBpedia instances of the class *dbo:Building* is available online[5]. For each property the rate of its usage ordered by their popularity is provided.

Table 1. Namespaces of DBpedia properties for *dbo:Building*

Namespace	Prefix	Num	%
http://dbpedia.org/property/	dbp:	2827	96
http://dbpedia.org/ontology/	dbo:	99	3.4
http://xmlns.com/foaf/0.1/	foaf:	9	0.3
http://www.w3.org/2000/01/rdf-schema#	rdfs:	4	0.1
http://www.w3.org/2003/01/geo/wgs84_pos#	geo:	3	0.1
http://www.w3.org/2002/07/owl#	owl:	2	0.1
http://purl.org/dc/terms/	dc:	1	0.0
http://www.w3.org/ns/prov#	prov:	1	0.0
Total		**2946**	**100**

For DBpedia instances of *dbo:HistoricBuilding,* a subset of the above 2.946 properties is used (in fact 818 properties). In order to explore differences in the use of these properties of DBpedia instances in the case of historic buildings, we followed the same methodology applied to the class *dbo:Building*. Finally we classified the most commonly properties (213) into 16 distinct information categories according to their function and their use. The grouping of these properties presents various differentiations compared to the grouping of the properties for *dbo:building* due to the different descriptive needs of historical buildings. A distinction between primary and supplementary information is provided as well. The summarization study for the properties of the entities of the *dbo:HistoricBuilding* is available online[6]. For each property the rate of its usage per building entity is provided. The following SPARQL query demonstrates how we retrieved all the interlinked DBpedia properties using DBpedia's SPARQL Endpoint.

```
SELECT DISTINCT ?property
WHERE
{?buildingrdf:typedbo:Building .
?building?property ?obj .}
```

---

[5] http://dlib.ionio.gr/standards/SUMDBbuildings/table1.pdf.
[6] http://dlib.ionio.gr/standards/SUMDBbuildings/table2.pdf.

3   Wikipedia Infobox Templates for Buildings

A key factor for the above analysis was the understanding of policies and practices of Wikipedia infobox template system and specifically their use in the case of buildings. Wikipedia's infobox is a quick ('at-a glance') and convenient summary of the buildings. Wikipedia's infobox is a quick ('at a glance') and convenient summary of the key facts about a subject, in a consistent format and layout. They quickly summarize important points in an easy-to-read format[7]. An infobox template mainly consists of particular properties representing a certain domain of interest and the combination or arrangement of these elements forming a certain infobox that appear in a Wikipedia article. Moreover the existence of definitions of the meaning of these attributes, as well as content component (declarations of what and how values should be assigned to these attributes), for some of these, leads to treat infoboxes as distinct metadata models.

Wikipedia provides the community of its users with 55 infobox templates about structures and buildings[8] which may be used to summarize information about a building or a structure. Wikipedia also uses different infobox templates for various building types that belong to the same functional type (e.g. monastery and church belongs to religious buildings). As a remark, Wikipedia does not provide an infobox template for historic buildings as a search will redirect you to the template: *infobox building*, which is intended mainly for modern buildings (skyscrapers, hotels etc.). Historic buildings are complex works, consisting of multiple parts and present specific descriptive needs [3] with a lot of these to require inclusion in an infobox. The absence of such an infobox template verified at the stage of the summarization process of the DBpedia's entity *dbo:HistoricBuilding*, as in many information categories, crucial attributes of core information are missing (e.g. from the Designation Category).

In reality there isn't a "one-size-fits-all" infobox template and it is hard to find a template that will cover all possible requirements for the various types of historic buildings. Even the infobox template system allows many infoboxes to include another infobox as a module (or child, or sub-template e.g. Designation list template in infobox building template), this method does not warranty the non-loss of information. Despite these practices many of the properties summarized in this study for historic buildings could be aligned, leading Wikipedia infobox template system into a more standardized environment for these structures. This could also be achieved eliminating the loss of core information, the existence of multiple templates for the various types of buildings, as well as heterogeneities such as the ones described in the next section.

---

[7] https://en.wikipedia.org/wiki/Help:Infobox.

[8] https://en.wikipedia.org/wiki/Category:Buildings_and_structures_infobox_templates.

4   Categorizing Heterogeneities in the Infobox Template System and Normalizing Raw Data

The DBpedia properties for buildings, studied for their function and semantics, in their majority come from Wikipedia infobox template system described above, which in turn presents various structural and semantic heterogeneities. Structural heterogeneities are caused mainly by the various ways that the infobox templates combine or arrange their properties. We categorized these structural heterogeneities as follows:

- *Property Naming Conflicts:* is one kind of structural heterogeneity. Distinct infobox templates for buildings assign lexically different or synonym names to properties that represent semantic equivalent real world concept. These naming conflicts include also spelling mistakes (e.g. dbp:architecturalStlye) and singular/plural irregularities (e.g. dbp:architecturalStyles/dbp:architecturalStyle).
- *Constraints Infobox Conflicts:* occur because distinct infobox templates provide different possibilities of defining constraints: The ability of one infobox template to be embedded into another, as a module, is not provided for all templates that are used for buildings.
- *Multilateral Properties Correspondences:* is another kind of structural heterogeneity as one property from an infobox template can correspond to multiple properties in another and vice versa.
- *Infobox Coverage conflicts:* are the structural heterogeneities that occur due to the fact that important and crucial properties reflected in one infobox template are left out in other. A typical example for this heterogeneity is the use of the property *dbp:materials* that although exists in the template: "infobox Church" is left out in the template: "infobox Monastery", despite the fact that both templates refer to the same building category (Religious Buildings).

Semantic Heterogeneities in the Wikipedia infobox-template system occur because of the different interpretations in the semantics of the infobox properties. We can categorize these semantic heterogeneities as follows:

- *Axiom Property Conflicts:* The fact that infobox properties come without any ontological axioms - in contrast to DBpedia properties (e.g. owl properties), leads to semantic confusion which results in properties that their semantics overlap, subsume, or aggregate other.
- *Misunderstood Content Conflicts:* occur for some properties that represent values irrelevant to the intended purpose of the property due to insufficient documentation of its use in the particular template page. These conflicts are reinforced by the fact that Wikipedia editors assign to some of the infobox properties abbreviated names, difficult to understand from Wikipedia Community (e.g. *dbp:nrhpType* to record the type of National Register Historic Place).
- *Scaling - Unit Conflicts:* occur because of conflicts in the intended meaning of the various properties or content values. In the various properties that we have studied, different scaling systems are used to measure content values. Due to the fact that many Wikipedia editors do not strictly follow the recommendations given on the

page that describes a template, attribute values are expressed using a wide range of different formats and units of measurement.

According to the above analysis in Table 2 presented below we have categorized various structural and semantic heterogeneities for the most widely used DBpedia properties of the instances of the class *dbo:Building*.

Table 2. Cases of structural and semantic heterogeneities in DBpedia properties of dbo: Building

Structural heterogeneities	
Property naming conflicts (semantic equivalent content)	**Names**: rdfs:label, foaf:name, dbp:name, dbp: buildingName
	Notes: dbo:abstact, rdfs:comment
	Style: dbp:architecture, dbp:style, dbo: architecturalStyle
	Date: dbo:yearOfConstruction, dbp:built, dbp: startDate
	Materials: dbp:materials, dbp:structuralSystem
	Type/Use: dbp:status, dbp:buildingType, dbp:functionalStatus
	Place: geo:lat- dbp:latitude, geo:long - dbp:longitude
	Designation: dbp:refnum, dbo: nrhpReferenceNumber
Multilateral properties correspondences	dbp:architectOrBuilder = dbp:architect + dbp: builder
Semantic heterogeneities	dbo:yearOfConstruction = dbp:startDate + dbp: yearCompleted
Scaling/unit conflicts	dbp:architecture, dbo:architecturalStyle, dbp:materials, dbp:structuralSystem, dbp:functionalStatus, dbp:buildingType, dbp: heritageDesignation
Misunderstood content conflicts	dbp:status, dbp:architecturalType

5   Conclusion

The summarization results presented in this study will act as a guide allowing DBpedia users for sophisticated queries about buildings. Also will act as a compass for Wikipedia designers - maintainers at the process stage of merging similar, redundant, and duplicate infoboxes and dealing with name variations of their properties. Moreover the analysis could be easily applied to other DBpedia instances to find and eliminate similar heterogeneities problems. Having isolated all the properties that present defective content values, our future work is to deal with Scaling/Unit Conflicts using semantic networks. Finally, proposing and assigning specific datatypes for these values, this will significantly increase the quality of infobox properties, thus transforming Wikipedia infobox template system into a more standardized environment.

References

1. Lehmann, J., et al.: DBpedia–a large-scale, multilingual knowledge base extracted from Wikipedia. Semant. Web **6**(2), 167–195 (2015)
2. Mihindukulasooriya, N., Rico, M., G-C, R., Gómez-Pérez, A.: An analysis of the quality issues of the properties available in the Spanish DBpedia. In: Puerta, J.M., et al. (eds.) CAEPIA 2015. LNCS, vol. 9422, pp. 198–209. Springer, Heidelberg (2015). doi:10.1007/978-3-319-24598-0_18
3. Agathos, M., Kapidakis, S.: Describing immovable monuments with metadata standards. Int. J. Metadata Semant. Ontol. **7**(3), 162–170 (2012)

What Happens When the Untrained Search for Training Information

Jorgina Paihama[✉] and Hussein Suleman

Department of Computer Science, University of Cape Town,
Private Bag X3, Rondebosch, Cape Town 7701, South Africa
{jpaihama,hussein}@cs.uct.ac.za

Abstract. Unemployed and information illiterate people often have the greatest need for information because it could change their lives. While a lot of information on jobs and training is available online, it is unclear if the target users are indeed able to find such information. This paper presents the findings of a study of the expectations of low skilled people with low information literacy when searching for information about training courses. The results indicate that users have access to technology and information is indeed available online but the users who need this information most are not able to find it using conventional search engines.

Keywords: Information retrieval · Information literacy · Low literacy · Low skills · Unemployment · Training

1 Introduction

Unemployment is a global problem, but one that particularly affects poor and developing countries. Specifically, this is a problem for people who are unskilled or low skilled and have low levels of literacy and information literacy. Gitau [2] showed that such people are unemployable due to their lack of skills for the jobs that are available. Although there are technological solutions to allow them to access information, a lot of people in that category are also either not technologically savvy or they use technology but are unaware of how to go about finding training information that may be useful to them.

Someone who is information literate is able to recognise the need for information, and given that need, he/she is able to adequately locate, analyse and use information [1]. The unemployed in poor countries are, however, often information illiterate.

In most modern societies, getting information on training opportunities using the Internet is a straight-forward process. However, for the information illiterate, there are the following potential problems: lack of knowledge of what information to look for; lack of knowledge about where to look for such information; difficulties in understanding and analysing results; and lack of knowledge to modify the query in order to improve on the results.

© Springer International Publishing Switzerland 2016
N. Fuhr et al. (Eds.): TPDL 2016, LNCS 9819, pp. 405–410, 2016.
DOI: 10.1007/978-3-319-43997-6_34

Similar problems could be faced by literate users performing complex and exploratory searches, but it is unlikely that literate users would face these issues if given a training information search task[1]. In an attempt to better understand this and other issues that information seekers in the target group face, we conducted interviews and studied their expectations with regards to the search engine and the information obtained. The details and outcomes of this study are presented in this paper.

2   Literature Review

The process of searching for information that fully satisfies ones needs may seem simplistic, but in fact it can be a laborious and time consuming process for novices as well as experts [3–5]. For years researchers have studied how different groups of people search for information, their search strategies &patterns, as well as their interactions with the results set [5,6]. That in an attempt to provide fitting solutions to groups that may somehow be limited in using standard IR systems. Cultural factors [7], language barriers, young age [4,8,9], physical ability [10], along with illiteracy, limited subject knowledge and lack of systems expertise, are some of the constraints affecting users ability to find relevant results. Multiple avenues for possible solutions have been explored, including text free, visual &audio based solutions for illiterate and semi-literate users [10,11], collaborative search [12], simplification or translation of language used [10], and creating bundles of aggregated information [13] to name a few. While not all showed statistically significant improvements, most efforts towards personalised solutions for disadvantages information seekers lead to similar findings; given how distinctive users are, there is no one-size-fits-all solution to providing access to information, and therefore it is important to tend to individual needs.

The details of our data gathering study are presented in Sect. 3.

3   Methodology

The goal of this study was to understand the search experience of users in the target population. There were 15 participants: 12 in Cape Town – South Africa and 3 in Luanda – Angola. The participants in Cape Town were approached through the Fisantekraal Centre for Development (www.fisantekraal.org.za); and the participants in Luanda were randomly selected in a township. The criteria was that participants had to be semi skilled and preferably in search of training opportunities. They were interviewed and their use of search systems was observed during a simple information seeking task.

This was a three part process. First, the users were given an explanation about the purpose of the exercise, and asked to sign a consent form and fill out a questionnaire. This initial questionnaire covered information about demographics, devices use and ownership, employment status, level of education, familiarity with search engines, difficulties they experience in their searches for training

---

[1] A study analysing the search behaviour of different users has since been conducted.

courses and possibilities for assistance in the process. Secondly, the participants were asked to use Google to search for information about a training course of their interest. Finally, based on the results obtained from the second task, participants were given another questionnaire to complete. This time the questions were related to the results they obtained in their search. We asked them to rate the results obtained as relevant, slightly relevant, or irrelevant, and justify their choices. We asked whether the result set was satisfactory for their information needs, what information was deemed important but was not available, suggestions for possible improvements to the system and problems they may have faced during the process.

Twelve of the participants were females, and three were males, with ages ranging from 16 to 42 years old. Prior to this exercise, all 15 participants had Internet access and had already used it to search for information.

4   Results and Analysis

4.1   Platforms and Preparation

In terms of hardware, 53 % of participants use only their cellphones, 40 % use both cellphone and computer, and 7 % use only a desktop computer. In terms of search services, 60 % of participants use Google, and 7 % use Bing. The remaining 33 % (5 participants) referred to the browser rather than the search service itself, indicating some confusion on how Internet services work.

Participants were asked about their highest form of formal education; some chose multiple responses as the levels overlap to some degree. 60 % of participants completed high school, 33 % completed a professional training course, 20 % completed primary school, and 20 % completed a course at a Further Education and Training (FET) college[2]. Those who chose the other option also specified courses from professional training centres.

Most of the participants (93 %) were formally unemployed, 87 % were searching for employment, and 67 % were searching for training opportunities.

4.2   Finding Jobs and Training

Participants were asked where they searched for training course information. The Internet is the most popular option, chosen by 33 % of users, followed by Newspapers and Door to Door search with 13 % each. The options of looking everywhere and not yet looking were chosen by 7 % each. 12 of our 15 participants were engaged in a training course during this exercise. In conversation with the researcher, all 12 said that the found out about that specific course via the newspapers. However, only 2 participants stated that they use newspapers to search for information about training courses. Thus it is clear that relevant information is in the public domain, but there is no clear link between where users search for information and where that information can be found. It is also

---

[2] FET colleges in South Africa are equivalent to Community Colleges in the USA.

notable that, while a third of users search for opportunities online, none had enrolled in the current training because of online information.

Participants were asked about the constraints they have in mind when searching for training courses. 6 participants noted that the cost of training is a key constraint. All other constraints (location, search know-how, technology literacy, etc.) were mentioned by only 1–2 participants.

Table 1 displays what the participants indicated as means of improving the current search engines. Not all issues raised relate to the quality of the search engine itself. Some issues were related to the source of the information. However, from the perspective of the information seeker, there is no difference between the two.

Table 1. Suggestions by users on how to improve current search engines

Improvements to current search engines
User friendly, simple, not unnecessarily complicated
Allow for online application
Provide basic information upfront, correct &updated information
Provide data relevant to user location,
Location and direction information, direct user to real people or institution
Aggregate information, no link hopping

Figure 1 displays the common factors that users indicated to be crucial in results that they rated as being relevant. Complete information, prerequisites and details of the application process were considered to be the most informative details. The users performed a single query[3] and were then asked to analyse, rate results (as relevant, slightly relevant and irrelevant) and justify their ratings for the different results. Users found this task to be onerous and most users did not complete the task - when the number of results was reduced from 10 to 4, more users were able to complete the task. This suggests that the assumptions search interface designers make about the ideal form and quantity of information to present to users does not hold true for all users, especially novice users, even those with a clearly defined need for relevant information.

Figure 2 shows pieces of information that the participants indicated were missing from most of the results in the set, but which they consider essential for training course information. It is clear from this data that users had very specific expectations that from their perspective were not met by the technology used. Again, given the users lack of knowledge on how search engines work, they could not make a clear distinction between the expectations for the search engines services and the expectations for information providers, i.e. the ability to find the information v.s. the quality of the information.

A deeper understanding of IR tools is not required for online information seekers. The problems identified here stem from the fact that the target users

---

[3] With the first group the whole exercise took over 1.5 h to complete, and although the users were not told not to refine or change their queries, none of them did so.

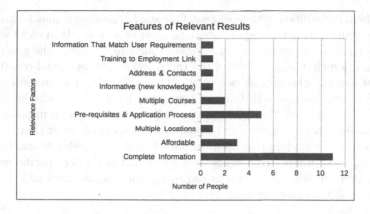

Fig. 1. Factors that make participants rate a result as relevant

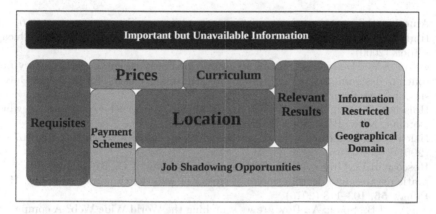

Fig. 2. Missing, but important information

are both information illiterate and search novices. Hoewver that should not deter them from finding information online. It is therefore up to researchers to provide a seamlessly integrated system that fulfills all their needs, by amongst other things steering information providers into adhering to standardised formats to safeguard data quality.

5   Conclusions and Future Work

We have conducted an information seeking exercise to find out what are the expectations of people with low levels of literacy and information literacy when searching for information about training courses. The aim was to assess if the current search technology meets the needs of such users.

60 % of participants (9 people) stated that they were not fully satisfied with the result set obtained during the exercise and that, in their opinion there is a

need for help in facilitating this process. Our study has shown that technology is indeed available but not yet appropriate to meet crucial needs in some societies.

The goal of future work is to investigate solutions to bridge the gap between the information that is available on the Internet and the potential consumers of such information, who appear not to be able to find the information just yet. Potential solutions include training of users or some form of social engineering, such as the recommendation by Lazonder [12] that users search in pairs instead of individually. Given the specific responses from users on what they value and their difficulties in finding this information, it may be possible to combine various techniques in information processing to extract and offer specific results to users. Further studies will be conducted to support an informed and structured approach to solving this problem.

References

1. American Library Association: Information Literacy Competency Standards for Higher Education. The Association of College and Research Libraries, Chicago Illinois (2000)
2. Gitau, S.W.: Designing Ummeli a case for mediated design, a participatory approach to designing interactive systems for semi-literate users. Ph.D. thesis, University of Cape Town (2012)
3. Holscher, C., Strube, G.: Web search behavior of internet experts and newbies. Comput. Netw. Book **33**(1–6), 337–346 (2000)
4. Hsieh-Yee, I.: Research on web search behaviour. Libr. Inf. Sci. Res. **23**, 167–185 (2001)
5. Roscoea, R.D., Grebitusb, C., O'Brianc, J., Johnsond, A.C., Kulae, I.: Online information search and decision making: effects of web search stance. Comput. Hum. Behav. **56**, 103–118 (2016)
6. Jansen, J.B., Spink, A.: How are we searching the World Wide Web? A comparison of nine search engine transaction logs. Inf. Process. Manag. **42**, 248–263 (2006)
7. Kralisch, A., Berendt, B.: Cultural determinants of search behaviour on websites. In: Proceedings of the IWIpPS 2004 Conference on Culture, Trust, and Design Innovation, Vancouver, Canada, 8–10 July 2004
8. Bilal, D., Kirby, J.: Differences and similarities in information seeking: children and adults as Web users. Inf. Process. Manag. **38**(5), 649–670 (2002)
9. Bilal, D., Bachir, I.: Childrens interaction with cross-cultural and multilingual digital libraries. II. information seeking, success, and effective experience. Inf. Process. Manag. **43**(1), 65–80 (2007)
10. Deo, S., Cunningham, S.J., Nichols, D.M., Witten, I.A.: Digital Library Access for Illiterate Users. Technical report (2002). ISRN UIUCLIS-/4+ISRL
11. Medhi, I., Sagar, A., Toyama, K.: Text-free user interfaces for illiterate and semi-literate users. In: Proceedings of the International Conference in Information and Communications Technologies and Development (2006)
12. Lazonder, W.: Do two heads search better than one? Effects of student collaboration on web search behaviour and search outcomes. In: Proceedings of the British Journal of Education Technology, Vol. 36, No 3, pp. 465–475 (2005)
13. Bota, H., Zhou, K., Jose J. M.: Exploring composite retrieval from users perspective. In: Proceedings of the 37th European Conference on Information Retrieval, pp. 13–24. Springer International Publishing, Switzerland (2015)

InterPlanetary Wayback: Peer-To-Peer Permanence of Web Archives

Mat Kelly[✉], Sawood Alam, Michael L. Nelson, and Michele C. Weigle

Department of Computer Science, Old Dominion University, Norfolk, VA 23529, USA
{mkelly,salam,mln,mweigle}@cs.odu.edu

Abstract. We have integrated Web ARChive (WARC) files with the peer-to-peer content addressable InterPlanetary File System (IPFS) to allow the payload content of web archives to be easily propagated. We also provide an archival replay system extended from pywb to fetch the WARC content from IPFS and re-assemble the originally archived HTTP responses for replay. From a 1.0 GB sample Archive-It collection of WARCs containing 21,994 mementos, we show that extracting and indexing the HTTP response content of WARCs containing IPFS lookup hashes takes 66.6 min inclusive of dissemination into IPFS.

Keywords: Web archives · Memento · Peer-to-peer · IPFS

1 Motivation

The recently created InterPlanetary File System (IPFS) [2] facilitates data persistence through peer-to-peer content-based address assignment and access. While web archives like Internet Archive (IA) provide a system and means of preserving the live web, the persistence of the archived web data over time is dependent on the resilience of the organization and the availability of the data [5]. In this paper we introduce a scheme and software prototype[1], InterPlanetary Wayback (ipwb), that partitions, indexes, and deploys the payloads of archival data records into the IPFS peer-to-peer "permanent web" for sharing and offsite massively redundant preservation and replay.

2 Background and Related Work

The Web ARChive (WARC) format is an ISO standard [4] to store live web archive content in a concatenated record-based file. IA's web crawler, Heritrix [7], generates WARC files to be read and the content re-experienced in an archival replay system. OpenWayback[2] (written in Java) and pywb[3] (written in Python) are two such replay systems. We leverage and extend on pywb in this work.

---

[1] https://github.com/oduwsdl/ipwb.
[2] https://github.com/iipc/openwayback.
[3] https://github.com/ikreymer/pywb.

© Springer International Publishing Switzerland 2016
N. Fuhr et al. (Eds.): TPDL 2016, LNCS 9819, pp. 411–416, 2016.
DOI: 10.1007/978-3-319-43997-6_35

To access the representations stored by an archival crawler, a replay system must refer to an index that maps the records in a WARC file to the original URI (or, URI-R in Memento [9] terminology). The CDX format is one such index along with the extended CDXJ format [1], which allows for arbitrary JSON objects within each index record. InterPlanetary Wayback takes advantage of pywb's native support for CDXJ and uses the arbitrary JSON data to store metadata about WARC records within IPFS (i.e., the content digest needed for lookup in IPFS).

IPFS is a peer-to-peer distributed file system that uses a file's content as the address for lookup [2]. By extracting the HTTP message body [3] (henceforth "payload") from the records within a WARC file, IPFS allows our prototype to generate a signature uniquely representative of this content. This payload can then be pushed into the IPFS system, shared to the network, and then retrieved at a later date when the URI-M is queried. This unique signature of WARC content provides de-duplication of identical content in web archives where other efforts to prevent redundant information within a web archive at crawl time has mostly been addressed [8]. While these efforts focus on intra-archive de-deduplication, ipwb additionally provides inter-archive de-deduplication of archival content.

3 Methodology

We utilize the JSON object of a CDXJ record to store digests of the HTTP response payloads, HTTP response headers, original status code when the URI-R was crawled, and the MIME-type of the payload content. A CDXJ record also contains a Sort-friendly URL Reordering Transform (SURT) URI-R and a datetime in the fields before the JSON. The digests are encoded into a field called locator as a Uniform Resource Name (URN) [6] (Fig. 1).

In designing ipwb, it was critical to consider the HTTP headers returned at crawl time separately from the payload. The HTTP response headers will change with every capture, as the datetime returned from a server includes the current

```
1  SURT_URI DATETIME {
2         "id": "WARC-Record-ID",
3         "url": "ORIGINAL_URI",
4         "status": "3-DIGIT_HTTP_STATUS",
5         "mime": "Content-Type",
6         "locator": "urn:ipfs/HEADER_DIGEST/PAYLOAD_DIGEST"
7  }
```

Fig. 1. A CDXJ index allows a memento to be resolved to a WARC record in a playback system. In the ipwb prototype we extract the relevant values from the HTTP response headers at time of index and include the IPFS hashes as the means for a replay system to obtain the HTTP headers and payload corresponding to the URI-M requested.

time. Compare this to the payload, which often contains the same content on each access. Were the HTTP headers and payload combined then added to IPFS, every IPFS hash would be unique, nullifying the potential for de-duplication of identical content. Furthermore, ipwb only retains HTTP response records. The rationale for this design decision is that the state of the art of web archive replay systems do not consider the WARC request record upon replay. While including request records may be useful in the future (for instance, to take into account the user agent originally used to view the live website), WARC content is currently able to be replayed without preserving the request records.

4   Implementation

Our prototype implements an indexer (shown as the red circles in Fig. 2) that extracts HTTP headers and payload from a WARC record and stores them in the IPFS storage as separate objects. The two object references returned from the IPFS along with a URI-R, datetime, HTTP status code, content-type, etc. are then used to construct an index record, which is stored in CDXJ format (Fig. 1), per Sect. 3.

InterPlanetary Wayback also implements an archival replay system (shown as the blue circles in Fig. 2) using components from pywb. When a client requests a TimeMap, all corresponding records are fetched from the index and a TimeMap is constructed. When a client requests a memento, a single corresponding record

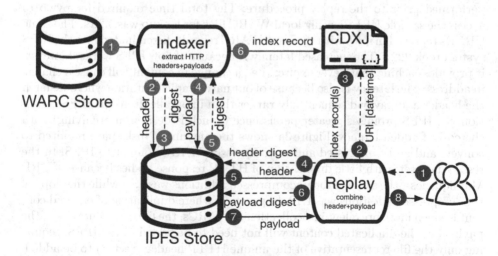

Fig. 2. Pushing WARC records to IPFS (red circles) requires the WARC response headers and payloads to be extracted (red 1), pushed to IPFS to obtain digest hashes (red 2–5), and hashes to be included in an index (red 6). The replay process (blue circles) has a user querying a replay system as usual (blue 1) that obtains a digest for the URI-datetime key from the index (blue 2 and 3), which is used as the basis for retrieving the content associated with the digests from IPFS (blue 4–7). The replay system can then process these payloads as if they were in local WARC files and return the content to the user (blue 8). (Color figure online)

is fetched from the index. This record will contain IPFS reference hashes for the HTTP headers and payloads corresponding to the URI requested. Using these two hashes, HTTP headers and payloads are fetched from the IPFS store and a response is constructed.

5   Evaluation

We tested our ipwb prototype on a data set from an Archive-It collection[4] about the 2011 Japan Earthquake consisting of 10 WARC files, each about 100 MB when compressed, totaling 1.0 GB on disk.

We indexed the WARCs using pywb's cdx-indexer and ipwb's prototype indexer to generate a standard CDXJ file and one containing the IPFS-hashes, respectively, as described in Sect. 3. The experiment was run on a late 2013 MacBook Pro running OS X version 10.11.3 with a 2.4 GHz Intel i5 processor, 8 GB of RAM, and a 250 GB SSD disk. Generating ipwb's CDXJ file for 21,981 mementos in the data set took 66.6 min including the time required to push the data into the IPFS network, producing the IPFS hashes to be included in the CDXJ. The average indexing rate inclusive of the data dissemination to IPFS was 9.48 files per second. Because IPFS is in the early stages of development, performance when adding files to the IPFS network[5] contributes a large part of the latency in the replay procedure.

To evaluate the replay time, we fetched 600 sample URI-Ms from each of pywb and ipwb independently, both using the same WARC basis for CDXJ generation, performed prior to the replay procedure. The total time required for pywb to access the sample URI-Rs using local WARC files for lookup was 5.26 s. The same URI-Rs replayed in ipwb with the same WARC records disseminated into the IPFS system took 222 s. The increased latency is because of how IPFS works, however, it provides caching that never expires, i.e., a change in content will cause a change in address. The latency is also because of our naive implementation where we fetch the header and payload sequentially rather than in parallel from the IPFS. Additionally, IPFS promises greater persistence (which is desired in archiving) with the cost of added latency. Figure 3a shows the amount of disk space required to convert and add compressed and uncompressed WARC content to IPFS. In the tested data set, with little duplication of HTTP response bodies because of URI-M uniqueness, the slope of the uncompressed additions was 1.10 while the slope of the compressed additions was 1.12. In practice, where duplication of payload content is much more prevalent in a collection of WARCs, the file representative of the payload of the duplicated content will not need to be added to the IPFS, requiring only the file representative of the unique HTTP header (Sect. 3) to be added. This would result in a significantly smaller slope were the experiment extended to a larger collection. Figure 3b shows that as more files (extracted from the WARCs) are added to the IPFS system, the time required to do so correlates linearly with the number of files (not necessarily the size of the files for small files) with a slope of 1.74 (on average, 570 files per minute).

---

[4] https://archive-it.org/collections/2438.
[5] https://github.com/ipfs/go-ipfs/issues/1216.

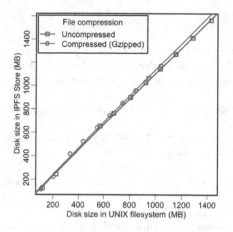
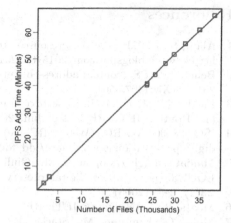

(a) Disk space required using both compressed and uncompressed WARC headers and payloads as compared to the file system and the IPFS Store.

(b) As the number of files added to the IPFS network increases, the time required to do so linearly increases (on average, 570 files per minute).

Fig. 3. IPFS Storage space and Time cost analysis

6 Future Work and Conclusions

Because of the novelty of IPFS, particularly relative to web archiving, there are numerous applications to expand this work. Collection builders can share their collections by just exchanging the index while keeping the data in the IPFS network and others can optionally replicate the data in their storage for redundancy. Further considerations of access control can also be addressed to encrypt and restrict content based on privacy and security mechanisms. Another model of IPFS-based archiving system can be built entirely using IPFS and IPNS technologies without the need of external indexes.

In this work we developed a prototype to partition, disseminate, and replay WARC file records in the InterPlanetary File System (IPFS). Through experimentation on a 1.0 GB data set containing 21,994 URI-Ms, we found that extracting and indexing records from WARC files took 66.6 min inclusive of dissemination into the IPFS system. The average indexing rate inclusive of the data dissemination to IPFS was 570 files per minute on average.

Acknowledgements. We would like to thank Ilya Kreymer for his feedback during the development of the ipwb prototype and guidance in interfacing with the pywb replay system. This work was supported in part by NSF award 1624067 via the Archives Unleashed Hackathon[6], where we developed the prototype.

---

[6] http://archivesunleashed.ca.

References

1. Alam, S.: CDXJ: an object resource stream serialization format, September 2015. http://ws-dl.blogspot.com/2015/09/2015-09-10-cdxj-object-resource-stream.html
2. Benet, J.: IPFS - content addressed, version, P2P file system. Technical report, July 2014. arXiv:1407.3561
3. Fielding, R., Reschke, J.: Hypertext Transfer Protocol (HTTP/1.1): Message Syntax and Routing. IETF RFC 7230, June 2014
4. ISO 28500. WARC (Web ARChive) file format, August 2009. http://www.digitalpreservation.gov/formats/fdd/fdd000236.shtml
5. Maniatis, P., Roussopoulos, M., Giuli, T.J., Rosenthal, D.S.H., Baker, M.: The LOCKSS peer-to-peer digital preservation system. ACM Trans. Comput. Syst. **23**(1), 2–50 (2005)
6. Moats, R.: URN Syntax. IETF RFC 2141, May 1997
7. Mohr, G., Kimpton, M., Stack, M., Ranitovic, I.: Introduction to Heritrix, an archival quality web crawler. In: Proceedings of the 4th International Web Archiving Workshop (IWAW 2004), September 2004
8. Sigurðsson, K.: Managing duplicates across sequential crawls. In: Proceedings of the 6th International Web Archiving Workshop (IWAW 2006), September 2006
9. Van de Sompel, H., Nelson, M., Sanderson, R.: HTTP Framework for Time-Based Access to Resource States - Memento. IETF RFC 7089, December 2013

Exploring Metadata Providers Reliability and Update Behavior

Sarantos Kapidakis[✉]

Laboratory on Digital Libraries and Electronic Publishing,
Department of Archive, Library and Museum Sciences, Ionian University,
72, Ioannou Theotoki Str., 49100 Corfu, Greece
sarantos@ionio.gr

Abstract. Metadata harvesting is used very often, to incorporate the resources of small providers to big collections. But how solid is this procedure? Are the metadata providers reliable? How often are the published metadata updated? Are the updates mostly for maintenance (corrections) or for improving the metadata? Such questions can be used to better predict the quality of the harvesting. The huge amount of harvested information and the many sources and metadata specialists involved makes prompt for answers by examining the actual metadata, rather than asking about opinions and practices. We examine such questions by processing appropriately collected information directly from the metadata providers. We harvested records from 2138 sources in 17 rounds over a 3-year period, and study them to explore the behaviour of the providers. We found that some providers are often not available. The number of metadata providers failing to respond is constantly increasing by the time. Additionally, the record length is slightly decreasing, indicating that the records are updated mostly for maintenance/corrections.

Keywords: OAI · Metadata · Harvesting · Reliability · Services · Record enrichment

1 Introduction

In this work we examine the harvesting of metadata and how it evolves over time. Metadata harvesting is used very often, to incorporate the resources of small or big providers to large collections. The metadata *harvesters,* like National Science Digital Library (NSDL) and Europeana, accumulate metadata from many *collections* (or *sources*), belonging to *metadata providers* mostly memory institutions, by automatically contacting their *servers* and storing the retrieved metadata locally. Their goal is to enable searching on the huge quantity of heterogeneous content, using only their locally store content. Metadata harvesting is very common nowadays and is based on the Open Archives Initiative Protocol for Metadata Harvesting (OAI-PMH). As examples, we mention the Directory of Open Access Repositories (OpenDOAR) that provides an

© Springer International Publishing Switzerland 2016
N. Fuhr et al. (Eds.): TPDL 2016, LNCS 9819, pp. 417–425, 2016.
DOI: 10.1007/978-3-319-43997-6_36

authoritative directory of academic open access repositories[1], and the OAIster[2] database of OCLC with millions of digital resources from thousands of providers.

In [6] Lagoze et al. discuss the NSDL development and explains why OAI-PMH based systems are not relatively easy to automate and administer with low people cost, as one would expect from the simplicity of the technology. It is interesting to investigate the deciciencies of the procedure.

National or large established institutions consistently try to offer their metadata and data reliably and current and to keep the quality of their services as high as possible, but local and smaller institutions often do not have the necessary resources for consistent quality services – sometimes not even for creating metadata, or for digitizing their objects. In small institutions, the reliability and quality issues are more prominent, and decisions often should also take the quality of the services under consideration.

The evaluation and quality of metadata is examined as one dimension of the digital library evaluation frameworks and systems in the related literature, like [2, 7, 10]. Fuhr et al. in [2] propose a quality framework for digital libraries that deal with quality parameters. The service reliability falls under their System quality component, and the metadata update under their Content quality component.

In [9] Ward describes how the Dublin Core is used by 100 Data Providers registered with the Open Archives Initiative and shows that is not used to its fullest extent. In [4] Kapidakis presents quality metrics and quality measurement tool, and applied them to compare the quality in Europeana and other collections, that are using the OAI-PMH protocol to aggregate metadata. In [5] Kapidakis further studies the responsiveness of the same OAI PMH servers, and the evolution of the metadata quality over 3 harvesting rounds between 2011 and 2013.

From the different aspects of the quality of digital library services, the quality of the metadata is the one that has been mostly studied. Some approaches are applied on OAI-PMH aggregated metadata: Yen Bui and Jung-Ran Park in [1] provide quality assessments for the National Science Digital Library metadata repository, studying the uneven distribution of the one million records and the number of occurrences of each Dublin Core element in these. Another approach to metadata quality evaluation is applied to the open language archives community (OLAC) in [3] by Hughes that is using many OLAC controlled vocabularies. Ochoa and Duval in [8] perform automatic evaluation of metadata quality in digital repositories for the ARIADNE project, using humans to review the quality metric for the metadata that was based on textual information content metric values.

In this paper we want to examine how solid is the metadata harvesting procedure. Are the metadata providers really reliable, responding when they are accessed, so that we have current information? How often are the published metadata updated? Are the updates mostly for maintaining (correcting) or for improving the metadata? The rest of the paper is organized as follows: In Sect. 2 we explore the reliability of the servers over many harvesting rounds. In Sect. 3 we study the frequency and the nature of the updates of the harvested metatata, and we conclude on Sect. 4.

---

[1] http://www.opendoar.org.

[2] http://www.oclc.org/oaister.en.html.

2 Reliability of the Servers Over Time

The reliability of the servers is important for ensuring current information. If the metadata harvesting service is not responding, the corresponding metadata records will not be updated at that time. This is not a very serious deficiency, as the updates will eventually be performed on the next successful metadata exchange, and will be normally used afterwards. Nevertheless, the unreliability - downtime of the metadata harvesting server usually indicate a proportional unreliability or downtime of the resource providing service, which always resides on the local sites, where both the local and the harvested metadata link to. When the resources are not available, the corresponding user requests are not satisfied, affecting the quality of the service.

In order to see if the metadata provider services are reliable and work over the years, we organized 17 *harvesting rounds*, over three years, from 2014 to 2016 (27 months). Each round took a few days to complete and was held apart of the previous round between one and two months. We tried to harvest the 2138 OAI sources listed in the official OAI Registered Data Providers[3] on January of 2014. We expected the providers listed there to be the most used and from the most seriously involved ones, and seriously considering their content and services.

In the first round, in January 2014, 1221 servers responded on all our rounds, only 1338 of the 2138 servers (63 %) responded at least one time. All servers were harvested on all rounds, but the remaining 800 servers never responded. Any OAI valid response was considered a satisfying server response, even in the rare cases that the communication with the server failed later on, because subsequent communication could eventually force the server to provide all its records, in most cases. It seems that the initial list, although official, was not up-to-date. Most failures were attributed to the OAI server not been found, to the OAI server not responding, or to protocol errors/incompatibilities. We performed the harvesting rounds by developing an application based on the pyoai[4] python library version 2.4.5. In order to have comparable results, we did not change the harvested servers in any of our rounds. On Fig. 1 we can

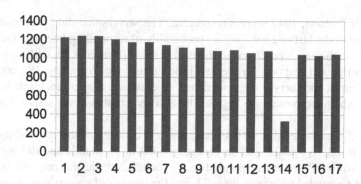

Fig. 1. The number of responses (y-axis) on each of the 17 harvesting rounds (x-axis).

---

[3] https://www.openarchives.org/Register/BrowseSites.

[4] https://pypi.python.org/pypi/pyoai.

see the number of servers responding on each round. We observe that temporary issues may affect the harvesting:

In one round (December 2015) there were much fewer responsive servers (330) than all other rounds. There must have been network connection problems closer to our harvesting client, which may not affect the harvesting from other clients. Therefore, we ignore this round in the rest of our study. The numbers of responses on all other rounds seem natural: In the first round (January 2014) 1221 servers responded, and on the last one (March 2016) 1047 servers. The most servers (1238) responded on the second round (February 2014) and the fewest (1033) on the one prior to the last (February 2016).

In Fig. 1 we observe that the number of servers responding on each round (excluding the ignored round) decreases almost linearly (and decreased by 174 between the 16 rounds), so we assume that many servers stopped working or responding permanently. In order to verify that, we depict in Fig. 2, for each round (except the last), the number of servers that responded for the last time during that round, and never responded afterwards.

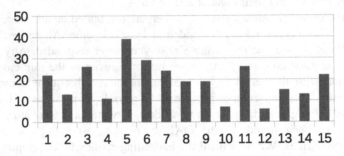

Fig. 2. The number of servers (y-axis) that responded for the last time in each round (but the last) (x-axis)

In the 16th round, 1047 servers responded. The servers that did not respond on the rounds close to the last one have higher probability of responding later on, and may not be dead. Nevertheless, on each round there were some (from 6 to 39, with an average of 20) servers that did respond for the last time during our 16 rounds. We counted the number of times each server responded and we clustered the servers by their count of responses, which we present on Fig. 3. The 800 servers that never responded are also included.

In Fig. 3 we can see that 721 servers, the vast majority of the responding servers, responded in all 16 rounds, 223 servers responded in 15 rounds and did not respond only once while 69 servers responded in 14 rounds and did not respond twice. The servers that responded only once were 27, and the number of servers responding from 1 to 13 times are similarly distributed, with 25 servers on average and a minimum of 14 servers (those responding 10 times) and a maximum of 39 servers (those responding 5 times).

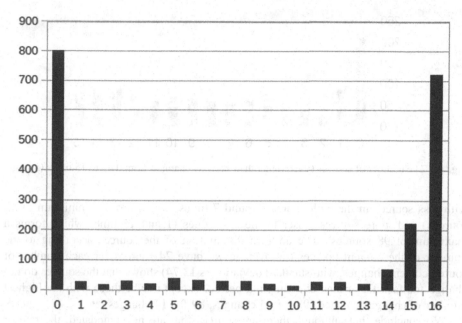

Fig. 3. The number of servers that responded exactly from 0 up to 16 rounds.

It is not expected that some servers fail accidentally many times, and this fact usually indicates a more permanent reason. In Fig. 3 we observe the accumulation of two quantities, the servers that responded few times and ceased working afterwards (and these should normally be about equal for each round and form a horizontal line on Fig. 3) and those that only failed occasionally a few times – and these are accumulated on the last few columns, which are sharply increasing, possibly forming an exponential line.

3 Frequency and Nature of the Updates

We want to get estimation on how often the published metadata are updated, which should affect the harvesting round, and also can indicate how important it is to reliably harvest successfully. For that we tried to harvest 1000 metadata records on each of the previously mentioned rounds. We examined the records of each collection and we recorded how many times they change during successive harvesting rounds, counting even the smallest change in any of their harvested records. This represents the frequency of update of the sample records, and is an indication of the update frequency of the whole collection. We depict that on Fig. 4, where the x-axis represents the number of rounds with value changes of the examined records and the y-axis represents the number of sources with that property.

We see that the largest group is the one with the records that never change. We counted 228 such sources that their content is therefore always the same. The remaining sources, which are the majority of them, occasionally update their content – some of them quite frequently. The records of the rest of the sources changed from 1 to 15 times,

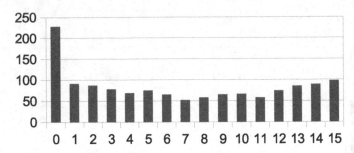

Fig. 4. The number of sources (y-axis) that their records changed from 0 up to 15 times (x-axis).

with less sources in the middle area (around 7 times, which forms a minimum of 52 sources) and more sources closer to the two edges (1 and 15 times which forms a maximum of 98 sources). The uniform distribution of the sources according to the number of the content updates (on average we have 74 sources for each number of source content changes, with standard deviation is 13.74) shows that the sources do not change too fast, faster than our sampling period, except maybe those with the highest number of changes counted (98 for 15 changes, 89 for 14 changes, etc.).

We conclude that, although there are sources that are never updated, the rate or quantity of the updates is ranging almost uniformly to any value and there are not any other obvious update patterns.

The harvested metadata records can always be mapped to Dublin Core. Thus, in addition to examining the whole record, we can examine individual Dublin Core elements. In Fig. 5 we can see in separate columns the number of sources that their Dublin Core elements title, creator, subject, description, type and rights changed, sorted by the number of times they change.

The columns in Figs. 4 and 5 are related, but not derived one from the other, as they count the records or the individual elements when they change. Nevertheless, they have a similar shape. In Fig. 5 we can see that most metadata elements do not change many times.

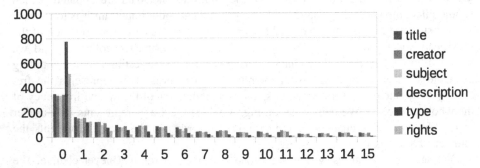

Fig. 5. The number of sources (y-axis) that some Dublin Core elements changed from 0 up to 15 times (x-axis) (Color figure online)

The resulting curve is decreasing, which means that the number of sources is decreasing by the number of times they change during the harvesting rounds. The sources that do not change at all are even more. The situation is similar for all examined Dublin Core elements, but more intense for the rights and type elements than with the other elements, the ones that are mostly used: title, creator, subject, description. This can be explained as the rights and type elements contain more standard information that does not get enriched and more rarely need improvement or correction than the more descriptive elements In general, the changes on the contents of the elements vary by the elements. Furthermore, the changes to the individual descriptive elements seem to have similar pattern and therefore to take place at the same round, so most changes are done record-wise rather than element-wise.

Finally, we examine if the size of the record, or the number or the content of its elements increases in an obvious (statistically significant) way, then we can assume that the metadata updates are actually enrichments, adding new information to the corresponding records. On the other hand, small, insignificant, changes in the size are most probably just corrections, that do not add any additional information. Therefore we examine the size of the records on their first and last harvesting round, to see if the updates are mostly for maintenance (corrections) or for improving the metadata.

Figure 6 shows the difference of the size of the record content, measured in words, between the last and first harvesting round for each source. In most cases, the changes/additions are zero or very small, and many times they have a negative sign: more words are removed than added. Very few sources have higher differences, and in most cases they are negative! The average increase of words per source is -14.6, but the standard deviation is 101.7, that indicates that we cannot derive a specific pattern.

We conclude that most changes are not record enrichments but rather maintenance changes, that happen to contain more removal of words. In most other cases the quality improvement was small, if any, indicating that it is hard to make extensive or consistent improvements on large metadata collections.

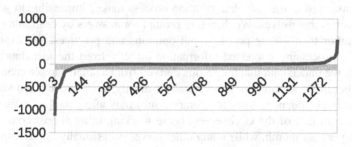

Fig. 6. The average increase of words (y-axis) contained in the records, for each source (x-axis)

This conclusion is also supported from Table 1, where we can see the minimum, maximum and average increase of words per collection. We can also see the standard deviation, which is always much higher than the corresponding average, and the number of sources involved in each calculation. We present in different rows the data for the whole record and those for the individual Dublin Core elements. The average

Table 1. Statistics on the increase of words in the source records and elements.

	Min	Max	Mean	SDev.	Number
Record	−1128.3	542.1	−14.6	101.7	1330
Title	−21.1	15.4	0.0	1.5	1327
Creator	−46.1	21.0	−0.4	3.4	1296
Subject	−31.9	44.4	0.1	3.4	1198
Type	−6.2	4.2	0.0	0.6	1264
Date	−6.5	5.2	0.0	0.4	1303
Language	−0.8	2.9	0.0	0.1	1080
Identifier	−27.4	32.0	0.1	2.1	1327
Description	−718.9	540.1	0.1	41.6	1289
Contributor	−39.6	24.5	0.1	2.7	760
Format	−7.9	6.4	0.0	0.5	1044
Publisher	−8.0	8.0	0.1	0.9	1209
Relation	−178.6	167.5	0.0	8.8	840
Coverage	−11.0	6.8	0.0	1.2	249
Source	−268.9	23.3	0.2	10.1	763
Rights	−1118.0	205.0	−23.7	108.8	856

increase is almost always minimal, close to 0 (and mostly negative), except in the element "rights" and in the whole records, where there is a clean decrease in size – but still with a much higher standard deviation.

4   Conclusions and Future Work

In this work we tried to find answers that will help predicting the quality of the harvesting. The huge amount of harvested information and the many sources and metadata specialists involved in the metadata creation process makes impossible to get answers from the sources themselves. We therefore prompt for answers by examining the actual metadata, rather than asking people about opinions and practices. We examine such questions by processing harvested information directly from the metadata providers, even though we understand that the results derived from a small sample cannot be very accurate. Therefore, we presented our numbers mostly in charts, and not verbatim, to show the derived tendency. Also, temporary issues may affect the harvesting.

A significant part of the OAI servers cease working in an almost constant rate of about 14 servers per month, while many other serves occasionally fail to respond. Over the years, the sources are updated, and the updates are mostly record-wise rather than element-wise, because changes to the individual descriptive elements seem to have similar pattern and therefore to take place at the same round. The rights and type elements change less often than the other, more descriptive, elements. Most changes seem to be rather maintenance changes and not record enrichments, as they do not increase the words in the content.

In the future, we can also examine the non responding servers in more detail, and derive and cluster the failure reasons.

References

1. Bui, Y., Park, J.: An assessment of metadata quality: a case study of the national science digital library metadata repository. In: Moukdad, H. (ed.) CAIS/ACSI 2006 Information Science Revisited: Approaches to Innovation. Proceedings of 2005 Annual Conference of the Canadian Association for Information Science Held with the Congress of the Social Sciences and Humanities of Canada at York University, Toronto, Ontario (2005)
2. Fuhr, N., Tsakonas, G., Aalberg, T., Agosti, M., Hansen, P., Kapidakis, S., Klas, P., Kovács, L., Landoni, M., Micsik, A., Papatheodorou, C., Peters, C., Sølvberg, I.: Evaluation of digital libraries. Int. J. Digit. Libr. **8**(1), 21–38 (2007). Springer
3. Hughes, B.: Metadata quality evaluation: experience from the open language archives community. In: Chen, Z., Chen, H., Miao, Q., Fu, Y., Fox, E., Lim, E.-p. (eds.) ICADL 2004. LNCS, vol. 3334, pp. 320–329. Springer, Heidelberg (2004). doi:10.1007/b104284
4. Kapidakis, S.: Comparing metadata quality in the Europeana context. In: Proceedings of 5th ACM International Conference on PErvasive Technologies Related to Assistive Environments (PETRA 2012), Heraklion, Greece, 6–8 June 2012. ACM International Conference Proceeding Series, vol. 661 (2012)
5. Kapidakis, S.: Rating quality in metadata harvesting. In: Proceedings of the 8th ACM International Conference on PErvasive Technologies Related to Assistive Environments (PETRA 2015), Corfu, Greece, 1–3 July 2015. ACM International Conference Proceeding Series (2015). ISBN 978-1-4503-3452-5
6. Lagoze, C., Krafft, D., Cornwell, T., Dushay, N., Eckstrom, D., Saylor, J.: Metadata aggregation and "automated digital libraries": a retrospective on the NSDL experience. In: Proceedings of 6th ACM/IEEE-CS Joint Conference on Digital Libraries (JCDL 2006), pp. 230–239 (2006)
7. Moreira, B.L., Goncalves, M.A., Laender, A.H.F., Fox, E.A.: Automatic evaluation of digital libraries with 5SQual. J. Inform. **3**(2), 102–123 (2009)
8. Ochoa, X., Duval, E.: Automatic evaluation of metadata quality in digital repositories. Int. J. Digit. Libr. **10**(2/3), 67–91 (2009)
9. Ward., J.: A quantitative analysis of unqualified dublin core metadata element set usage within data providers registered with the open archives initiative. In: Proceedings of 3rd ACM/IEEE-CS Joint Conference on Digital Libraries (JCDL 2003), pp. 315–317 (2003). ISBN 0-7695-1939-3
10. Zhang, Y.: Developing a holistic model for digital library evaluation. J. Am. Soc. Inf. Sci. Technol. **61**(1), 88–110 (2010)

Posters and Demos

TIB|AV-Portal: Integrating Automatically Generated Video Annotations into the Web of Data

Jörg Waitelonis[1]([✉]), Margret Plank[2], and Harald Sack[3]

[1] Yovisto GmbH, August-Bebel-Str. 25-53, Potsdam 14482, Germany
joerg@yovisto.com
[2] German National Library of Science and Technology,
Welfengarten 1 B, Hannover 30167, Germany
margret.plank@tib.eu
[3] Hasso-Plattner-Intitute, Prof.-Dr.-Helmert-Str. 2-3, Potsdam 14482, Germany
harald.sack@hpi.de

Abstract. The German National Library of Science and Technology (TIB) aims to promote the use and distribution of its collections. In this context TIB publishes metadata of scientific videos from the TIB|AV Portal as linked open data. Unlike other library metadata the TIB|AV-Portal deploys automated metadata extraction and named entity linking to provide time-based semantic metadata. By publishing this metadata, TIB is offering a new service involving the provision of quarterly updated data in RDF format which can be reused by third parties. In this paper the strategy and the challenges regarding the linked open data service are introduced.

1 Introduction

In order to improve the accessibility, citability and the sustainable use of scientific videos, the German National Library of Science and Technology (TIB) in cooperation with the Hasso Plattner Institute (HPI) has developed the TIB|AV-Portal[1][4]. The portal provides access to high grade scientific videos from the fields of technology/engineering, architecture, chemistry, information technology, mathematics and physics in English and German. In addition to reliable authoritative metadata (Dublin Core[2]) time-based metadata is generated by automated media analysis. Based on text-, speech- and image recognition text-based terms are extracted and mapped to subject specific GND[3] subject headings by the means of named entity linking (NEL). The cross-lingual retrieval uses interlanguage links based on an ontology mapping (DBpedia[4], Library of Congress

---

[1] http://av.tib.eu/.
[2] http://dublincore.org/.
[3] http://www.dnb.de/EN/gnd.
[4] http://dbpedia.org/.

© Springer International Publishing Switzerland 2016
N. Fuhr et al. (Eds.): TPDL 2016, LNCS 9819, pp. 429–433, 2016.
DOI: 10.1007/978-3-319-43997-6_37

Subject Headings[5], e. a.). These technologies improve the search for and the re-use of scientific videos by e.g. enabling pinpoint access to individual video segments. Individual film segments can be cited by means of DOI and Media Fragment Identifier (MFID).

For enabling efficient metadata re-use and interlinking, TIB decided to revise its underlying metadata model, make the data mashine-readable and publish it in RDF format[6] according to Linked Data principles. The fundamental difference to other Linked Data based publications, e.g. [1–3,6,7] is that the videos of the TIB|AV-Portal are subject to automated metadata extraction and semantic analysis. This results in time-based metadata which contains extracted content descriptions including provenance and – if available – confidence information.

2   RDF-Export for the TIB|AV-Portal

For the benefit of simplicity, clarity and avoidance of redundancies, linked data standards recommend the usage of a single suitable subset of all potentially available vocabularies. The following general requirements had to be considered:

1. How many metadata items can be covered by which vocabulary?
2. How well does a vocabulary item's semantics fit the intended meaning of the metadata item?
3. A high degree of generality should be achieved, but which items should be picked for a better accuracy?
4. How well do datatypes match? Are there any conversions necessary or is there any loss in accuracy?
5. How popular is the vocabulary (in a certain community only or in general)?
6. When only using individual parts of a vocabulary in combination with other vocabularies, how can we ensure that no logical contradictions occur?

Table 1 summarizes vocabularies used for the mapping of standard as well as temporal metadata of the TIB|AV-Portal. `dcterms` and `dctypes` are generic and widely used vocabularies that were chosen to represent generic items as e. g. the subject or the language of a resource. All major search engines support the `schema.org` vocabulary for structured data. It defines a standard set of type and property names covering a wide range of topics. The TIB|AV-Portal website embeds RDFa annotations with `schema.org` to improve the accessibility for search engines. Since all TIB|AV-Portal resources are referenced by DOIs, the library specific `bibframe` vocabulary is used to model DOI as well as specific title properties, e. g. subtitle to reference subheadings. The `foaf` vocabulary is mainly used to describe persons as well as organizations. As a result of the TIB|AV-Portal's automated multimedia analyses different kinds of spatio-temporal metadata is generated. In order to annotate a video segment the Open Annotation (OA) Data Model[7] was chosen. The primary aim of the

---

[5] http://id.loc.gov/authorities/subjects.

[6] https://www.w3.org/RDF/.

[7] http://www.openannotation.org/spec/core/.

Table 1. Overview of namespaces and prefixes used for the RDF export of standard metadata (top), and spatio-temporal metadata (bottom)

Prefix	Namespace	Name
bibframe	http://bibframe.org/vocab/	Bibframe Vocabulary
dbp	http://dbpedia.org/resource/	DBpedia Resources
dcterms	http://purl.org/dc/terms/	DCMI Metadata Terms
dctypes	http://purl.org/dc/dcmitype/	DCMI Type Vocabulary
foaf	http://xmlns.com/foaf/0.1	Friend of a Friend Vocabulary
gnd	http://d-nb.info/gnd/	Integrated Authority File (GND)
schema	http://schema.org/	Schema.org Vocabulary
tib	http://av.tib.eu/resource/	TIB|AV-Portal Resources
itsrdf	http://www.w3.org/2005/11/its/rdf#	Internationalization Tag Set (ITS)
nif	http://persistence.uni-leipzig.org/ nlp2rdf/ontologies/nif-core#	NLP Interchange Format
oa	http://w3.org/ns/oa#	Open Annotation Data Model

OA Data Model is to provide a standard description mechanism for sharing annotations between systems. In connection with W3C's Media Fragment Identifier spatio-temporal segments of a video can easily be cited. Another challenge was the inclusion of semantic text annotations, which was solved by means of NLP interchange format (NIF2) vocabulary for interoperability between Natural Language Processing (NLP) tools, language resources, and annotations. Figure 1 illustrates a simplified example as the combination of OA Model and NIF2. The `tib:annotation/yxyx` is the dedicated RDF resource of the annotation referring to the body as well as the target. The annotation target refers to the URI of the video including media fragment identifier(s). The annotation body

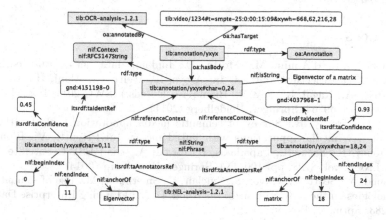

Fig. 1. Example of temporal annotation of a video frame with NIF2 and OA.

refers to a resource `tib:annotation/yxyx#char=0,24` of type `nif:Context`. The URI is extended by a fragment identifier specifying the character range of text. This resource is the reference context for the semantic annotations typed by `nif:Phrase`. Each annotation holds all the additional information created by the generating algorithms (e. g. NEL), such as the text ranges, confidence values, the knowledge-base entity URI, as well as the version of the algorithm itself.NIF2 enables instantly to connect the dataset to other NEL annotators, which allows to asses the quality of the semantic analysis objectively compared to other systems [5].

3   Further Challenges and Conclusion

Automated extraction methods always produce potential errors. In order to maximize data quality, manual data cleansing is conducted regularly before publishing. Inaccuracy and potential faults of the automated metadata extraction are subject of current and ongoing research. Multimedia analysis technology as well as semantic analysis are constantly refined to obtain higher quality results. However, the possibility of a manual override is currently evaluated. Manual correction of faulty low level results, as e.g. in the speech-to-text analysis imply a subsequent reevaluation of the thereby changed context. Since the context has changed, the semantic analysis might reconsider previously taken decision for the NEL. Furthermore, a manually confirmed or changed result should be considered for future context analysis and entity disambiguation. The general goal of these efforts is to allow for high quality automatically generated metadata, which in turn enable more precise and complete search results.

In this paper, we have presented the strategy for and implementation of the linked open data service of the TIB|AV-Portal. The authoritative as well as time-based, automatically generated metadata of the TIB|AV-Portal are published in standard RDF format[8]. The underlying data model was revised according to linked data requirements to enable easy accessibility, readability and re-usage by third parties.

References

1. Fernandez, M., d'Aquin, M., Motta, E.: Linking data across universities: an integrated video lectures dataset. In: Aroyo, L., Welty, C., Alani, H., Taylor, J., Bernstein, A., Kagal, L., Noy, N., Blomqvist, E. (eds.) ISWC 2011, Part II. LNCS, vol. 7032, pp. 49–64. Springer, Heidelberg (2011)
2. Nixon, L., Troncy, R.: Survey of semantic media annotation tools for the web: towards new media applications with linked media. In: Presutti, V., Blomqvist, E., Troncy, R., Sack, H., Papadakis, I., Tordai, A. (eds.) ESWC Satellite Events 2014. LNCS, vol. 8798, pp. 100–114. Springer, Heidelberg (2014)
3. Raimond, Y., Scott, T., Oliver, S., Sinclair, P., Smethurst, M.: Use of semantic web technologies on the BBC web sites. In: Wood, D. (ed.) Linking Enterprise Data, pp. 263–283. Springer, Boston (2010)

---

[8] http://av.tib.eu/opendata/.

4. Sack, H., Plank, M.: AV-Portal - the German National Library of Science and Technology's semantic video portal. ERCIM News **96**, 2014 (2014)
5. Usbeck, R., et al.: GERBIL - general entity annotation benchmark framework. In: Proceedings of 24th WWW Conference (2015)
6. Waitelonis, J., Sack, H.: Augmenting video search with linked open data. In: Proceedings of International Conference on Semantic Systems (2009)
7. Yu, H.Q., Pedrinaci, C., Dietze, S., Domingue, J.: Using linked data to annotate, search educational video resources for supporting distance learning. IEEE Trans. Learn. Technol. **5**(2), 130–142 (2012)

Supporting Web Surfers in Finding Related Material in Digital Library Repositories

Jörg Schlötterer$^{(\boxtimes)}$, Christin Seifert, and Michael Granitzer

University of Passau, Innstraße 32, Passau 94032, Germany
{joerg.schloetterer,christin.seifert,michael.granitzer}@uni-passau.de
http://www.uni-passau.de

Abstract. Web surfers often face the need for additional information beyond the page they are currently reading. While such related material is available in digital library repositories, finding it within these repositories can be a challenging task. In order to ease the burden for the user, we present an approach to construct queries automatically from a textual paragraph. Named entities from the paragraph and a query scheme, which includes the topic of the paragraph form the two pillars of this approach, which is applicable to any search system, that supports keyword queries. Evaluation results point towards users not being able to find optimal queries and needing support in doing so.

Keywords: Just-in-Time Retrieval · Zero effort queries · User study

1 Introduction

Reading a paragraph in a web page often triggers the need for additional information. Then a user has to visit a digital library or general search engine and express this information need as a query in order to retrieve related resources. Proactive retrieval simplifies this process by presenting related material according to the current context without explicit user interaction. This approach was first made popular by Rhodes as Just-in-Time Retrieval [7] and has recently been continued under the topic of zero effort queries [1]. Zero effort queries require minimal, ideally no effort from the user in expressing her information need and obtaining relevant results. However, most of the existing systems either treat the retrieval system as an integral part of the application or focus on domain-specific sets of information needs [9]. We present a proactive retrieval approach that is agnostic of the underlying retrieval system and can be applied to any search system whose contents are searchable via keyword queries. The de-coupling from the retrieval engine is achieved by focusing on the query-side of retrieval: our aim is to construct queries that yield results relevant to the current paragraph.

2 Problem Definition

The aim of our work is to find relevant results for a paragraph of text. More formally, we define the paragraph as P, the query as Q and the result set as R. The

N. Fuhr et al. (Eds.): TPDL 2016, LNCS 9819, pp. 434–437, 2016.
DOI: 10.1007/978-3-319-43997-6_38

mapping from a paragraph to a query is defined by $h: P \rightarrow Q$ and the retrieval of results by $g: Q \rightarrow R$. We then aim to optimize the function $f = g \circ h: P \rightarrow R$ towards relevant results. Less formal, the whole process is defined as $P \xrightarrow{h} Q \xrightarrow{g} R$. In just-in-time retrieval systems, that treat the search engine as integral part, f is not a composition of g and h, but results are retrieved directly according to the paragraph ($f: P \rightarrow R$). In this paper, we treat the search engine as black box and focus on the query side of retrieval. That is, we have no influence on g, but rather seek to optimize $f = g \circ h$ by optimizing h.

3   Approach

Query Representation. P is represented by its sequence of words and Q by a set of keywords, which provide a compact representation of the paragraph. Q can be represented in two principled ways: either as a keyword query or as a Boolean query. We use the boolean representation, as it provides richer expressiveness and most digital libraries, which expose their contents via a search API, support boolean queries. In order to avoid over- or underspecified queries, we propose to formulate a boolean query of the following structure:

$$(\text{``main topic''}) \; AND \; (\text{``keyword 1''} \; OR \; \text{``keyword 2''} \; OR \; ...)$$

where the main topic is defined as the overall topic of the paragraph and the right part of the conjunction are additional keywords. This way, we can be sure, that a keyword triggers only results which are connected to the overall topic of the paragraph. Even though, from the perspective of the search engine, all of the query terms are keywords, we will refer to the left part of the conjunction as *main topic* and to the right part as *keywords* in the further course.

Extraction of Keywords. Keyword extraction algorithms, that represent the keywords in terms of a subset of terms from the original text are available in the literature [5,8]. However, query log analysis research revealed, that over 71 % of (user generated) search queries contain named entities [2]. In addition, named entities have been shown to be beneficial to query segmentation [3], a technique that is used to optimize queries. Also, named entity extraction can be seen as some kind of keyword extraction task, as the original text is represented by a smaller set of terms. Therefore, we base our query generation on named entities, which are obtained via DBpedia Spotlight[1].

Extraction of the Main Topic. To extract the main topic, we utilize Doc2Vec [4]. Based on Word2Vec [6], Doc2Vec produces a word embedding vector, given a sentence or document. Hence, we use the entire input paragraph and compute a vector representation given a Doc2Vec model created on a Wikipedia corpus. Each entity in DBpedia spotlight also corresponds to a Wikipedia page, from

---

[1] http://spotlight.dbpedia.org/.

Table 1. Comparison of optimal, automatic and best user queries.

	Own index			Original index		
	Optimal	Automatic	USER*	Optimal	Automatic	USER*
Precision	0.55	0.34	0.37	0.23	0.29	0.33
Recall	0.49	0.33	0.37	0.22	0.25	0.33
F1-score	0.49	0.31	0.35	0.21	0.24	0.31

which we obtain the Doc2Vec representation of the extracted entities. We compare the Doc2Vec representations of the paragraph and extracted entities by computing the cosine similarity. The named entity with the highest similarity to the input paragraph represents the main topic. The remaining named entities are used as keywords in the right part of the boolean conjunctive query.

4   Evaluation

We evaluated the approach by means of a user study with university students. Due to space constraints, we only report the principle setup of the study[2]. For a given piece of text, i.e. a paragraph, an automatic query was generated by the approach described in the previous section and participants had to rate the results. Then, participants were asked to adapt the query, in order to retrieve more relevant results. The evaluable study data consisted of 251 paragraphs and 558 associated queries, performed by 69 users.

The collection of a repository is subject to change (new items may be added or the ranking may change), which would render future comparability of the results infeasible. To counter for this fact, we set up an Elasticsearch[3] index with the results retrieved during the study. On this index, we collected the optimal queries, that can be posed, based on the extracted main topic, keywords and query scheme as defined in Sect. 3. In principle, we collected those queries with a brute-force approach, testing all possible combinations of keywords and choosing the best performing in terms of F1-score as optimal query.

Results. We evaluated the query quality of our approach for automatic queries with all extracted keywords and the structure described in Sect. 3 (automatic), the optimal queries and the best queries, users were able to formulate (USER*) on the original data and our own index with the results depicted in Table 1. The reported precision-, recall- and F1-scores are macro averaged over all queries.

For the evaluation of user queries, we took the best query, a user was able to formulate for a paragraph (USER*). This means, that if the initial automatic query scored better in terms of F1-score than all subsequent modifications by the

---

user, we take the initial query. As can be seen from the table, the performance of automatic and user queries (USER*) is quite low, with the user generated queries performing slightly better (0.35) than the automatic queries (0.31). If we consider, that automatic queries are restricted to the extracted main topic and keywords, while users can provide arbitrary values for these two, the automatic queries still perform quite well. As one would expect, the performance on our own index is higher, as the amount of potential false positives is reduced. Also, the performance of optimal queries (0.21) on the original index is lower than on our custom index (0.49) (and even lower than the performance of automatic queries and USER*), since the optimal queries trigger results, which are not contained in the original result sets. Hence, as we do not have relevance feedback for those results, they are treated as irrelevant, even though they may be relevant in fact.

5   Summary

In this paper, we presented a search-engine agnostic approach for the automatic generation of boolean queries from a paragraph, based on named entities. We evaluated the performance against optimal achievable queries and the best queries, users were able to formulate. Results indicate, that users are able to formulate better queries than the automatic approach, but still below the optimal achievable performance and hence need support in query formulation.

Acknowledgments. The presented work was developed within the EEXCESS project funded by the European Union Seventh Framework Programme FP7/2007-2013 under grant agreement number 600601.

References

1. Allan, J., Croft, B., Moffat, A., Sanderson, M.: Frontiers, challenges, and opportunities for information retrieval: report from SWIRL 2012. SIGIR Forum **46**(1), 2–32 (2012)
2. Guo, J., Xu, G., Cheng, X., Li, H.: Named entity recognition in query. In: SIGIR 2009, pp. 267–274. ACM (2009)
3. Hagen, M., Potthast, M., Beyer, A., Stein, B.: Towards optimum query segmentation: in doubt without. In: CIKM 2012, pp. 1015–1024. ACM (2012)
4. Le, Q.V., Mikolov, T.: Distributed representations of sentences and documents. In: ICML 2014, pp. 1188–1196 (2014)
5. Mihalcea, R., Tarau, P.: Textrank: bringing order into texts. In: EMNLP 2004 (2004)
6. Mikolov, T., Sutskever, I., Chen, K., Corrado, G.S., Dean, J.: Distributed representations of words and phrases and their compositionality. In: NIPS 2013, pp. 3111–3119. Curran Associates, Inc. (2013)
7. Rhodes, B.J.: Just-in-Time information retrieval. Ph.D. thesis, Massachusetts Institute of Technology (2000)
8. Rose, S., Engel, D., Cramer, N., Cowley, W.: Automatic Keyword Extraction from Individual Documents, pp. 1–20. Wiley, New York (2010)
9. Shokouhi, M., Guo, Q.: From queries to cards: re-ranking proactive card recommendations based on reactive search history. In: SIGIR 2015, pp. 695–704 (2015)

Do Ambiguous Words Improve Probing for Federated Search?

Günter Urak[✉], Hermann Ziak, and Roman Kern

Know-Center GmbH, Inffeldgasse 13, 8010 Graz, Austria
{gurak,hziak,rkern}@know-center.at

Abstract. The core approach to distributed knowledge bases is federated search. Two of the main challenges for federated search are the source representation and source selection. Different solutions to these problems were proposed in the literature. Within this work we present our novel approach for query-based sampling by relying on knowledge bases. We show the basic correctness of our approach and we came to the insight that the ambiguity of the probing terms has just a minor impact on the representation of the collection. Finally, we show that our method can be used to distinguish between niche and encyclopedic knowledge bases.

Keywords: Federated search · Source selection · Query based sampling

1 Introduction

Today's Web is full of specialized knowledge sources that are scattered throughout the WWW but could be of potential usefulness to answer a user's information need. In order to satisfy this information need one typically relies on a single search interface under which all of these remote knowledge sources are unified. Hence the problem known as "aggregated search" or "federated search" arises.

At the core of this problem are the challenges of "source representation" and "source selection". The first deals with the question how one obtains a proper representation of a document collection. The latter asks how one chooses the relevant subset of sources among those representations. In our setting knowledge sources did not automatically disclose information about their content, which is called an "uncooperative setting". Therefore the main focus of this work is on the collection representation problem in an uncooperative setting. In addition our work is also relevant to challenge one, as it provides information for an informed decision based on the thematic proximity.

The main problem of source selection for aggregated search boils down to the question whether a certain source is able to deliver reasonable answers for certain queries [4].

2 Related Work

In the setting of uncooperative environments the collection's key statistics have to be approximated by sampling documents from the collection, hence the name

© Springer International Publishing Switzerland 2016
N. Fuhr et al. (Eds.): TPDL 2016, LNCS 9819, pp. 438–441, 2016.
DOI: 10.1007/978-3-319-43997-6_39

"query-based sampling" (QBS) [1]. A typical problem that arises in QBS is the question of where and how to draw the samples from [2,4].

Methods include sampling from a static and separate dictionary, as well as sampling from a dictionary that was constructed during querying. In addition a stopping criterion can be defined in the number of queries or number of returned documents. This particular number heavily depends on the collection in question but can be estimated through a method called "adaptive sampling" [3].

Over the past decade QBS has been analyzed and refined in a wide variety of settings [3] One particular method that hasn't been explored yet is the question whether "ambiguous" queries help to establish good coverage of knowledge sources.

3   Approach

The goal of our approach is to bias the samples towards ambiguity. For this we use two libraries: WordNet[1] and WordNet Domains[2]. WordNet is an English language dictionary and is organized in so called synsets. Each word is attributed by one or multiple synsets, that reflect a certain semantic meaning. WordNet Domains augments words by a set of topical domains.

The domains themselves are organized in a hierarchical manner, creating tree of domains of 164 nodes. Each domain represents a category, e.g. Art, Plastic Art, Ceramic Arts. In case none of the available classes fits the label factotum is chosen. For a given input term set our mapping function conducts a look-up within WordNet Domains. Each word is initially given the same weight. The weights are propagated to the respective parent node using a decay function and are then aggregated over all words of the input. The result is a weighted set of domains, which is finally pruned. Up to 5 different domains are included, if their weight exceeds at least 0.5 of the weight of the top ranked domain.

Instead of using a plain English dictionary we used only nouns that belonged either to the class "factotum" or to multiple classes and contrasted this to plain random queries. In addition we contrasted this by discarding drawn words between consecutive runs. In order to compare different source representations we made use of two datasets: (i) FedWeb Greatest Hits[3] collection and (ii) ZBW[4], the world's largest research infrastructure for economic literature and conforms to our definition of a niche source.

4   Evaluation

We asked three important questions: The first is whether our implementation works in general. For this we constructed 3 blocks of 20 queries, where each block belonged to a specific domain (math, religion and health). These queries

---

[1] http://wordnet.princeton.edu/.
[2] http://wndomains.fbk.eu/.
[3] https://fedwebgh.intec.ugent.be/.
[4] http://www.zbw.eu/.

were chosen in a way that even a non-expert could easily categorize the queries in their respective classes. The results can be seen in Table 1.

Table 1. The table shows the detected domains in relation to their corresponding unnormalized weights. All of the three original domains are recovered except that health and medicine are switched due to their similarity.

Rank	1	2	3	4	...	16
Class	Religion	Math	Med.	Geo	...	Health
Weight	18	13	9	3	...	1

The second question is whether our algorithm yields stable results. By stable we mean whether the list of detected domains stabilized over time or starts to fluctuate, when fed with increasingly more queries. In order to measure the change between consecutive runs we decided to use "rank biased overlap" as it penalizes changes in the head of the list stronger than changes in the tail.

The third and last question is how useful the implementation is when it comes to the detection of sources where the true underlying domains are not known beforehand. We drew words (100 to 2k, steps in hundreds) from our dictionary according to the strategies introduced in the section before Sect. 3. In Addition we tested whether our approach deviates from a pure random baseline. We divided the sources into niches and encyclopedias and averaged across those classes. The results can be seen in Fig. 1.

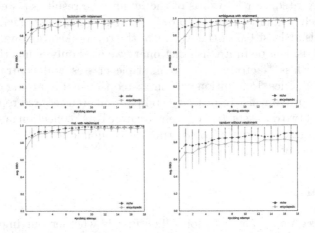

Fig. 1. Comparison of the 4 different query generation methods with each other. Each diagram shows the mean performance on niche and encyclopedic sources with corresponding standard deviation. All 4 methods show a similar mean performance, except the last method exhibits extreme deviations from the mean.

5   Conclusions and Future Work

From our results we can conclude that categorization seems to get relatively stable between 300 and 700 queries depending on the source in question. Encyclopedic sources need more queries to reach the same level as niche sources. Upon examination of the results we saw that the quality of the mapping does not benefit from "ambiguous" nor "factotum" queries. From manual inspection of the assigned classes we can state that the algorithm seems to work at least on niche sources, while on encyclopedic sources the differences are less pronounced. Finally we can state that results for encyclopedic sources appear to change the most from a pure random selection without retainment of keywords. In the future we will focus on evaluating the precision and accuracy of the categories mapping, especially on queries. Given our findings we further plan to work on a reliable way to distinguish between niche and encyclopedic sources relying just on random probing. This distinction between niche and encyclopedic sources might be useful for a number of scenarios.

Acknowledgments. The presented work was developed within the EEXCESS project funded by the European Union Seventh Framework Programme FP7/2007-2013 under grant agreement number 600601. The Know-Center is funded within the Austrian COMET Program - Competence Centers for Excellent Technologies - under the auspices of the Austrian Federal Ministry of Transport, Innovation and Technology, the Austrian Federal Ministry of Economy, Family and Youth and by the State of Styria. COMET is managed by the Austrian Research Promotion Agency FFG.

References

1. Callan, J., Connell, M.: Query-based sampling of text databases. ACM Trans. Inf. Syst. (TOIS) **19**(2), 97–130 (2001)
2. Markov, I.: Uncertainty in distributed information retrieval. Ph.D. thesis, Università della Svizzera Italiana (2014)
3. Shokouhi, M., Scholer, F., Zobel, J.: Sample sizes for query probing in uncooperative distributed information retrieval. In: Zhou, X., Li, J., Shen, H.T., Kitsuregawa, M., Zhang, Y. (eds.) APWeb 2006. LNCS, vol. 3841, pp. 63–75. Springer, Heidelberg (2006)
4. Shokouhi, M., Si, L.: Federated search. Found. Trends Inf. Retr. **5**(1), 1–102 (2011)

Automatic Recognition and Disambiguation of Library of Congress Subject Headings

Rosa Tsegaye Aga[1], Christian Wartena[1(✉)], and Michael Franke-Maier[2]

[1] Hochschule Hannover - University of Applied Sciences and Arts,
Hannover, Germany
christian.wartena@hs-hannover.de
[2] Freie Universität Berlin, Universitätsbibliothek, Berlin, Germany

Abstract. In this article we investigate the possibilities to extract Library of Congress Subject Headings from texts. The large number of ambiguous terms turns out to be a problem. Disambiguation of subject headings seems to have potentials to improve the extraction results.

1 Introduction

Library of Congress (LoC) Subject Headings (LCSH) constitute a huge collection of terms that can be used for indexing. LCSH should not be confused with the much smaller and more structured Library of Congress Classification system. Yi and Chan [10] give a structural analysis of LCSH. We investigate possibilities to assign LCSH terms automatically by extracting them from the abstracts of records manually annotated with LCSH. A specific problem in LCSH is the ambiguity of many labels. LoC subjects may have a number of variants or alternative labels. Two subjects can have the same word as a variant. If this variant occurs in a text, it is not clear which subject should be used for indexing. E.g. the term *plants* is a label of lcsh:sh85102839[1] (green-growing things) as well as of lcsh:sh85046823 (manufacturing facility). Usually, different LCSHs for a term refer to different aspects of the same concept. E.g. lcsh:sh85000800 (adaptation) denotes the adaptation of plants and animals to ecosystems, while lcsh:sh85000892 refers to behavioral adaptation in psychology.

Given the large amount of LCSH concepts, we cannot use classification, but should rather use keyword extraction. Pouliquen et al. [8] distinguish between conceptual thesauri and natural language thesauri. In the first case, they argue, keyword *extraction* is not suited, since most concepts never will be found in literal form in texts. Here we should use keyword *assignment*. LCSH clearly has characteristics of a conceptual thesaurus. For keyword assignment Pouliquen et al. build signatures for each concept, where a signature is basically a vector of word weights. As for classification, the amount of training material needed for each concept is the bottle neck of the approach. Wartena et al. [9] use distributional similarity between potential keywords and abstracts to rank candidates. Below we use a similar approach to disambiguate ambiguous keywords.

---

[1] We use the namespace lcsh for http://id.loc.gov/authorities/subjects/

© Springer International Publishing Switzerland 2016
N. Fuhr et al. (Eds.): TPDL 2016, LNCS 9819, pp. 442–446, 2016.
DOI: 10.1007/978-3-319-43997-6_40

Medelyan et al. [5] report on a system that can work with LCSH. However, they did not evaluate the system for LCSH. Paynter [7] automatically assigns LCSH to a text by collecting LCSH terms that are assigned to similar texts.

2 Data

The LCSH collection (http://id.loc.gov/download/) consists of 414 355 subject headings. 162 569 of these concepts are pre-combined concepts, like e.g. lcsh:sh94004310 with the label *Voyages and travel – Mythology*. The labels with dashes will never occur in running texts. Thus we removed these labels but kept eventual variants without dashes.

There are 3605 headings that are meant to be used as subdivisions of main headings. We removed these headings as well as all Children's Subject Headings and all inverted labels were removed. 497 427 labels now are left. Since many headings are in plural we lemmatized all headings and added singular forms if that did not lead to ambiguity. Thus we added 75 270 labels.

Many (ambiguous and non-ambiguous) headings have some additional terms that specify the intended sense. These terms are put in brackets as a part of the label, like e.g. *Taxis (Biology)* and *Taxis (Vehicles)*. We removed these disambiguating terms, of course increasing the number of ambiguous labels. Now there are 15 661 ambiguous labels. Most ambiguous labels (12 010 to be precise) just have two senses, over 1 000 have three meanings. The most extreme cases are the words *cooking*, *suites* and *concertos* that each has over 400 senses. Finally, there are highly frequent words with a very infrequent meaning: e.g. the word *in* is a label for the subject lcsh:sh85069880 (Confucian Philosophy). Thus, we removed labels found in a common stop word list.

For keyword assignment we use data from the catalog of 200 scientific libraries of the German Länder Bavaria, Brandenburg and Berlin, called B3Kat. The catalog can be downloaded or queried over a SPARQL Endpoint (https://lod. b3kat.de/doc/sparql-endpoint/). We retrieved about 16 000 records of works in English, with a title, a subtitle and an abstract of at least 200 characters. After filtering out records with French and German abstracts, 15 629 records remain.

Next we retrieved for all records all available LoC subject headings and all keywords from the authority file of the German National Library, the *Gemeinsame Normdatei* (GND). Using the mapping from from GND to LCSH that is part of the GND, we could add more subject headings. The number of occurrences of original and mapped subject headings and the number of pre-combined headings and subdivision headings is give in Table 1.

Table 1. Occurrences of subject headings in our data set.

	Records	Subject headings	Unique subj. head	Pre-combined	Subdivisions
Original	12543	73528	11546	45857	104
Mapped	9982	48790	4259	30289	108
Combined	14267	89440	13428	47576	212

3   Experimental Setup

In order to disambiguate ambiguous headings, we compare the words in the title, subtitle and abstract of a record with the "glosses" of each potential subject. As a gloss for the subject we take all preferred and alternative labels of the subject and all disambiguating terms used in the labels. The subject with the largest overlap between its gloss and the abstract is selected. In many cases the text and the words related to the correct subject heading will have no overlap. If the words are not the same but at least semantically related, we can use distributional similarity to find the most likely subject heading. Co-occurrence vectors present the meaning of a word quite well (see e.g. [3]). Thus, for each word we build a vector of co-occurrence frequencies with 17 400 mid frequency words from the UkWaC Corpus [1]. We compute the average of all co-occurrence vectors of all words from the gloss. For this average we give the words found in the scope notes a smaller weight than the labels themselves. We also compute the average co-occurrence vector of all words in the text. Now we can compute the cosine between the vectors and select the subject heading with the largest cosine.

To extract the labels we lemmatize the text using the Natural Language Toolkit (http://nltk.org/) and collect all occurrences of labels in the words and in the lemmata. We did 4 runs. In the first run we considered only labels without disambiguation information. In the second run we included all these labels, but without doing any kind of disambiguation. The third run is identical to the second one, but we added all pre-combined subject headings for which each of the components was found. The last run again is the same as the second one, but now we used the disambiguation to select the most probable concept for each ambiguous term. For words for which we did not collect enough co-occurrence values in the UkWaC Corpus, we choose the subject without disambiguating terms.

4   Results and Conclusion

Table 2 gives the average results for the extraction of subject headings for all 14 267 records. We evaluated against the original LCSH annotations (12 543 records), against LoC headings obtained by mapping from GND-terms (9 982 records) and against the merged annotations (14 267 records).

It becomes clear that extracting LoC subjects from text is a very challenging task: we have over 400 000 subjects and even our small test set of 14 267 records actually uses almost 90 000 different subjects. These subjects range from abstract classes like *science* to the name of specific buildings on the fairground in Hanover (*Halle 13*) or to complex concepts like *Behavior therapy for children*). It should be noted that keyword extraction often yields a low recall, even if much smaller vocabularies are used. Gazendam et al. [2] reach an average recall@5 of 0, 23 using a thesaurus of 3800 concepts and Medelyan et al. [6] report a recall@5 of 0, 197 using the Agrovoc-thesaurus (16 000 terms). Kim et al. [4] report on a shared task for keyword extraction from scientific articles in which 27 systems were evaluated. In this task keywords were freely assigned by authors and readers. The highest recall@5 was about 0, 13; the highest recall@15 was 0, 27. Paynter [7]

Table 2. Average results, evaluated on original, mapped and merged annotations

	# predictions	Original		Mapped		Merged		
		Recall	Precis.	Recall	Precis.	Recall	Precis.	F_1
Unamb. headings	12,6	0,160	0,0535	0,195	0,0585	0,219	0,0728	0,109
All headings	37,0	0,180	0,0225	0,210	0,0236	0,242	0,0302	0,0537
+Pre-combined	43,3	0,182	0,0210	0,213	0,0217	0,245	0,0279	0,0501
Disambiguated	18,1	0,166	0,0374	0,192	0,0388	0,224	0,0501	0,0819

does not use keyword extraction but assigns LCSH found in similar documents. He reaches a precision of $0,19$ and recall of $0,21$. Also the overlap between the original LCSH and the mapped LCSH is very low: The average Jaccard coefficient for 8258 records with both original and mapped LCSH annotations, is 0.18. If we treat the original annotations as our gold standard and the mapped headings as predictions, we have a recall of $0,15$ and a precision of $0,27$.

If we use the disambiguation, we see that the best F-score is reached when we do not consider ambiguous labels at all. If we take ambiguous labels into account, the best F-score is obtained when the labels are disambiguated.

References

1. Baroni, M., Bernardini, S., Ferraresi, A., Zanchetta, E.: The wacky wide web: a collection of very large linguistically processed web-crawled corpora. Lang. Resour. Eval. **43**(3), 209–226 (2009)
2. Gazendam, L., Wartena, C., Brussee, R.: Thesaurus based term ranking for keyword extraction. In: Tjoa, A.M., Wagner, R.R. (eds.) Database and Expert Systems Applications, DEXA, International Workshops, Bilbao, Spain, August 30 - September 3, 2010, pp. 49–53. IEEE Computer Society (2010)
3. Kiela, D., Clark, S.: A systematic study of semantic vector space model parameters. In: 2nd Workshop on Continuous Vector Space Models and their Compositionality (CVSC) (2014)
4. Kim, S.N., Medelyan, O., Kan, M., Baldwin, T.: Automatic keyphrase extraction from scientific articles. Lang. Resour. Eval. **47**(3), 723–742 (2013)
5. Medelyan, O., Perrone, V., Witten, I.H.: Subject metadata support powered by maui. In: Proceedings of the 10th Annual Joint Conference on Digital Libraries, pp. 407–408. ACM (2010)
6. Medelyan, O., Witten, I.H.: Thesaurus-based index term extraction for agricultural documents. In: Proceedings of the 6th Agricultural Ontology Service Workshop (2005)
7. Paynter, G.W.: Developing practical automatic metadata assignment and evaluation tools for internet resources. In: Proceedings of the 5th ACM/IEEE-CS Joint Conference on Digital Libraries, pp. 291–300. ACM (2005)
8. Pouliquen, B., Steinberger, R., Ignat, C.: Automatic annotation of multilingual text collections with a conceptual thesaurus. In: Workshop in Ontologies and Information Extraction (EUROLAN 2003) (2003)

9. Wartena, C., Brussee, R., Slakhorst, W.: Keyword extraction using word co-occurrence. In: Tjoa, A.M., Wagner, R.R. (eds.) Database and Expert Systems Applications, DEXA, International Workshops, Bilbao, Spain, August 30 - September 3, 2010, pp. 54–58. IEEE Computer Society (2010)
10. Yi, K., Chan, L.M.: Revisiting the syntactical and structural analysis of library of congress subject headings for the digital environment. J. Am. Soc. Inf. Sci. Technol. **61**(4), 677–687 (2010)

The Problem of Categorizing Conferences
in Computer Science

Suhendry Effendy$^{(\boxtimes)}$ and Roland H.C. Yap

National University of Singapore, Singapore, Singapore
{effendy,ryap}@comp.nus.edu.sg

Abstract. Research in computer science (CS) is mainly published in conferences. It makes sense to study conferences to understand CS research. We present and discuss the problem of categorizing CS conferences as well as the challenges in doing so.

1 Introduction

Most research in scholarly analysis focus on authors, e.g. in analysis of scientific collaboration networks – on collaboration trends among researchers, finding similar authors, etc. However, little research has been done on publication venues. This is particularly important in computer science (CS) where conferences are the prime venue for research publication [4,5].

Many interesting questions which can be answered by studying conferences, e.g., how to automatically predict a conference rating [2], what conferences are similar to a particular conference, research trends in conferences, how to categorize conferences, etc. In this paper, we focus on the last one – how to categorize conferences. Conference categorization can have many uses, we sketch some from various perspectives. For researchers, it is useful for studying (possibly a new) area and also for finding publication venues. For organizations, it may be useful when evaluating research/individuals. For search engines and (digital) libraries, it is useful as part of search and for presenting/classifying results. In this paper, we explain the problem of conference categorization which has been little studied and discuss the challenge of conference categorization.

2 Conference Categorization

We use the term *"categorization"* deliberately to have a neutral name for assigning some kind of label, area, context to a conference. As the examples in the later sections show, while categorization is useful, it can have multiple (but related) definitions. What we call categorization is closely linked to the topic(s) and areas of interest for a conference. However, what constitutes a categorization can be complex. For example, categorization based on the conference's name or its CFP (call for paper) may not be accurate. We observe that while it appears that conference forms a community, the opposite direction in this relationship has more

This work has been supported by Microsoft Research Asia.

N. Fuhr et al. (Eds.): TPDL 2016, LNCS 9819, pp. 447–450, 2016.
DOI: 10.1007/978-3-319-43997-6_41

impact, i.e. a conference is formed by its community. Conferences are where researchers exchange ideas and present on going works to their peers. Thus, it is also important to understand the community before categorizing the conference, i.e. studying papers published in the conference or the researchers publishing in that conference.

3   Existing Categorization

We discuss below some existing data sources which can be considered as giving categorizations:

Google Scholar (GS) – https://scholar.google.com. Google Scholar is an academic search engine and bibliographic database from Google. It indexes scientific publications across various disciplines from the web and also other data sources. For categories, it only lists "top 20" venues based on their h5-index for each category, as such the categories are sparse and few.

Microsoft Academic Search (MAS) – http://academic.research.microsoft.com. Formerly known as 'Microsoft Libra', Microsoft Academic Search is an academic search engine and bibliographic database developed by Microsoft Research. MAS groups conferences into categories, with 24 categories for computer science in MAS. In addition to categorization, MAS also gives a conference ranking based on citation count and field rating. Recently, Microsoft has released the Microsoft Academic Graph (MAG) which contains scholarly data on papers and venues. Papers in MAG are given topics but without conference categories – note that categorizing as a union of the topics may lead to too many topics.

AMiner [6] – http://aminer.org. AMiner is an online service which provide an analysis based on the academic social network. In addition, it provides a list of conferences and journals in computer science with their topic categorization. We observed that only some conferences and journals are categorized and their categorization is similar to CCF (perhaps as AMiner is also from China).

China Computer Federation Ranking List (CCF) – http://www.ccf.org. cn/sites/ccf/paiming.jsp. China Computer Federation publishes a list of peer-reviewed journals and conferences (currently 309) in CS with 10 categories. This list is widely used in China. Rather than using a single topic, a category in CCF may group several topics into a single label, e.g., SOFTWARE ENGINEERING, SYSTEM SOFTWARE, AND PROGRAMMING LANGUAGE as one category.

4   Limitations of Existing Categorization

It is not clear how the various forms of categorizations in Sect. 3 are obtained. A conference ranking list such as CCF is likely to be manually curated. GS categorization may be manual since its rather small. We suspect that the others may have significant manual input. Nevertheless, as we discuss below, categorization is not a simple problem and multiple approaches are used but the (most useful) categorization may not be so clear.

Potentially Miscategorized Conferences. We discuss how the categorizations from Sect. 3 may be better categorized. Some representative examples are given below:

- CP (Principles and Practice of Constraint Programming) is categorized as SOFT-WARE ENGINEERING and PROGRAMMING LANGUAGES in MAS and CCF. Although there is a programming language element to CP, AI may be a better label within their categories.
- HICCS (Hawaii Intl. Conf. on System Sciences) is categorized as DATABASES in MAS. This result may seem surprising since HICCS consists of many tracks which can change over time and can be quite diverse. While giving a single categorization can always be limiting, an analysis of papers in HICSS would suggest that it is more in the information systems area. In some studies, HICSS regarded as a high ranking conference among conferences in management information system [3].

Incomplete Categorization. The limitation of a category label is precisely that the label itself may miss much information. Thus, there may be a better label or perhaps the scope of a label may need to be broadened if more completeness is desired. Some representative examples of this problem follow:

- Most of the above categorizations label PODC (ACM Symp. on Principles of Distributed Computing) as a computer system related conference: MAS gives DISTRIBUTED & PARALLEL COMPUTING, CCF gives COMPUTER SYSTEM AND HIGH PERFORMANCE COMPUTING, and HIGH PERFORMANCE COMPUTING in AMiner. While such a categorization is not wrong, a closer look at typical papers in PODC would suggest that it is closer to theory conferences than to the areas covered by other computer system related conferences.
- NDSS (Network and Distributed System Security Symp.) is categorized as SECU-RITY & PRIVACY in MAS. However, as the name suggests and this can be seen in the papers, NDSS also covers networking.
- SPAA (ACM Symp. on Parallel Algorithms and Architectures) is categorized as THEORETICAL COMPUTER SCIENCE in GS, and DISTRIBUTED & PARALLEL COM-PUTING in MAS. Both categorizations are possible. However, what is better is their union, thus SPAA could be considered as a conference on the theory/algorithmic aspects of parallel computing.
- CRYPTO (International Cryptology Conf.) is categorized as SECURITY & PRI-VACY in MAS, NET AND INFORMATION SECURITY in AMiner, and NETWORK AND INFORMATION SECURITY in CCF. These strict categorizations miss the point that CRYPTO is also concerned with theoretical CS.
- The GS categorization for CS papers is rather broad. It does not have software engineering or programming language, rather, they are merged in the SOFTWARE SYSTEM category which also includes TACAS (Intl. Conf. on Tools and Algorithms for the Construction and Analysis of Systems). Perhaps due to short list, OSDI (Operating Systems Design and Implementation) a top systems conference is not listed in under COMPUTING SYSTEMS although NSDI is.

5   Discussion

We have seen that the categorization problem has many challenges. It depends on what the intended usage is but that might not be easily formalized. A label chosen for a particular categorization may be too specific, it is unclear what "choosing the best" means which may be highly usage dependent. It might also be too broad. It may not be clear how many kinds of categorizations are needed.

The other important question is how to create the categorization. One could apply a manual/semi-manual approach but that may not be able to necessarily address the challenges. A manual approach is also likely to have its biases. It is also not scalable given the large and ever increasing number of conferences in computer science. Another approach may be to semi-automatically categorize conferences with the help of semantic mining tools, e.g., run the semantic mining tools against (all) papers in a conference and use this to decide what category best describes the conference. However, such semantic mining may have its own challenges. Processing the data in the papers, e.g. PDFs, has its own problems and may lead to some amount of noise in the extracted information.

A recent approach suggests that looking at the social network induced by the conferences [1] can give a reasonable relatedness metric between conferences. We believe that this can be promising direction for investigation since it can give a community approach to categorizing conferences without the need to rely too much on semantics.

Ideally the categorization should be automated. Some other challenges for an automated categorization are: How strict is the categorization – there is a tension between being too specific and too general. The categorization process may need to flexible to be adjust to different uses/goals. Thus, there could be some kind of tuning/parameters to the categorization algorithm. Scholarly data may not be so easy to obtain in a comprehensive fashion and the data itself may be noisy, thus, the techniques should be robust. Finally, new resources like MAG are likely to be useful as they contain more information than what is in bibliographic databases such as DBLP.

References

1. Effendy, S., Jahja, I., Yap, R.H.C.: Relatedness measures between conferences in computer science - a preliminary study based on DBLP. In: WWW Workshop on Big Scholarly Data (2014)
2. Effendy, S., Yap, R.H.C.: Investigations on rating computer sciences conferences: an experiment with the Microsoft Academic Graph dataset. In: WWW Workshop on Big Scholarly Data (2016)
3. Hardgrave, B.C., Walstrom, K.A.: Forums for MIS scholars. Commun. ACM **40**(11), 119–124 (1997)
4. Konstan, J.A., Davidson, J.W.: Should conferences meet journals and where? A proposal for 'PACM'. Commun. ACM **58**(9), 5 (2015)
5. Rosenblum, D.S.: The pros and cons of the 'PACM' proposal: counterpoint. Commun. ACM **58**(9), 44–45 (2015)
6. Tang, J., Zhang, J., Yao, L., Li, J., Zhang, L., Su, Z.: ArnetMiner: extraction and mining of academic social networks. In: SIGKDD (2008)

Germania Sacra Online – The Research Portal of Clerics and Religious Institutions Before 1810

Bärbel Kröger[✉] and Christian Popp

Germania Sacra, Göttingen Academy of Sciences and Humanities,
Geiststraße 10, Göttingen 37073, Germany
{bkroege, cpopp}@gwdg.de

Abstract. The research project Germania Sacra provides a comprehensive prosopographical database, that makes structured and comparable data of the Church of the Holy Roman Empire available for further research. The database contains approximately 38,000 records of premodern persons, new data is continuously added. This digital index of persons is supplemented by the "Database of Monasteries, Convents and Collegiate Churches of the Old Empire". The access through ecclesiastical institutions offers a broad variety of visualization possibilities for the prosopographical data. In order to make as much information as possible accessible for scholarly use the next steps that will be undertaken are cross-institutional collaboration and integration of scientific data resources of other research projects.

Keywords: Prosopography · Linked data · Authority files · Cross-database query

Within Europe, the Germania Sacra is a unique research project that deals with the history of dioceses, monasteries, convents and collegiate churches in the Holy Roman Empire of the German Nation. The clerical institutions included date from late antiquity until the Reformation or, respectively, the secularization at the beginning of the 19th century. Geographically, research includes the present Federal Republic of Germany and the border regions of neighboring countries. The aim of the project is to present the source material of the respective archives in a way that makes structured and comparable data of the Church of the Holy Roman Empire available for further research. A key component of this is the prosopography; specifically an inventory of clerical members of the Roman Empire, not just by name, but with all the information that can be collected from local archives and from Vatican sources. Contained within the reference books published by Germania Sacra are biographical lists of the bishops of a diocese, the canons or canonesses of collegiate churches, the monks and nuns of numerous monasteries etc. According to current research which considers approximately a thousand years of church history about 10,000 monasteries and convents existed. If we assume that an average institution over time had approximately 500 clerical staff members, this means we are discussing millions of individuals.

© Springer International Publishing Switzerland 2016
N. Fuhr et al. (Eds.): TPDL 2016, LNCS 9819, pp. 451–453, 2016.
DOI: 10.1007/978-3-319-43997-6_42

New technologies lead to new possibilities; the networking opportunities in the digital world breathe new life into the vision of a comprehensive prosopographical overview of the Church of the Roman Empire. Consequently, Germania Sacra has created a broad portfolio of digital resources in the field of the Church of the Holy Roman Empire in recent years. The key facility of the online portal is a scholarly "Digital Index of Persons" with a focus on clerics. Another primary feature is constituted by the digitized reference books concerning the history of ecclesiastical institutions compiled in the context of the long-term project which began in 1917. A "Database of Monasteries, Convents and Collegiate Churches of the Old Empire" was recently released. The biographical database is continuously updated. Currently, it contains approximately 42,000 records (June 2016). Similarly, the monastic database is a developing resource with new data continuously added (approximately 2,300 records, June 2016). In order to enable a deeper examination of the prosopographical information found in Germania Sacra volumes, we have compiled a digital index of persons who appear in the published reference books. This has led to a scholarly database which provides comprehensive information about the clerics of the Church of the Holy Roman Empire and other individuals associated with clerical institutions. The "Digital Index of Persons" provides targeted access to the biographical and prosopographical information in the reference books. Queries for individuals can be made via a variety of search options: name, surname and its variants, personal data, including office data, institutional affiliations, and geographic impact areas. The identification and merging of data sets of identical persons has been done manually so far and is mainly undertaken for high ranking church and secular officials (such as bishops and abbots). The extensive amount of data contained in our prosopographical database enables reliable statistical and empirical studies that were – based solely on the monographic publications – formerly not possible to this extent. Plurality of benefices and offices or family and social networks of church officials can be reconstructed and visualized much more easily by using the data pool. By combining various search criteria, correlations can be uncovered which go beyond the information provided in the printed reference books.

Alternative forms of access and a broad variety of visualization possibilities for the prosopographical data are made possible by access through the ecclesiastical institutions. For this purpose, we recently released the "Database of Monasteries, Convents and Collegiate Churches of the Old Empire". Through the monastic database, it is easy to get an overview of the abbots of a particular monastery or the abbesses of a particular group of monasteries from the same order in a particular region. All query results can be displayed on interactive maps that visualize the monastic landscape of the medieval and early modern periods. Temporal aspects as well as regional criteria such as medieval dioceses or religious orders can be modified by the user. The entries of religious houses in this database are linked to those clerics associated with the respective religious houses that appear in Germania Sacra's prosopographical database. For all institutions and individuals, the datasets provide interactive links which lead directly to the corresponding pages in digitized Germania Sacra reference books.

In order to realize the vision of a comprehensive prosopographical database, the preparation and presentation of the Germania Sacra project data achieves only the first step; additionally, cross-institutional collaboration and integration of scientific data of other research projects are essential. An important building block for reliable networking is the systematic enrichment with authoritative data. For names of persons in German speaking countries, the authority file GND (Gemeinsame Normdatei) of the German National Library is particularly relevant. Automatic generation of links to external data sources by using GND has become an established standard for a large number of scholarly biographical and prosopographical databases. For the Germania Sacra Portal, we utilize this authority file to enrich our data and connect it to external web sources via automated links. Besides from other biographical databases especially library catalogs, inventory overviews of archives, editions of sources, bibliographies and portrait collections are relevant sources.

In addition to authoritative data for personal names, we enrich our records with authoritative data for corporate bodies such as monasteries, orders and dioceses as well as geographic entities (GeoNames). In order to make structured access to our data available, entries in Wikipedia and DBpedia, the biggest player in the linked data cloud, are referenced. The records of the monastic database are already available in Linked Data format. Data enrichment and semantic web technologies offer a high potential for condensing the information network of the relations of persons and clerical institutions for medieval and early modern times. Automated linking of individuals mentioned in scholarly databases of medieval and early modern times is, nevertheless, a largely unsolved problem in the field of digital humanities.[1] One reason is that most of these individuals are not recorded in authority files. For many individuals, we simply don't have enough data to provide reliable authentication (date of birth and death, offices and official data). The identification process is often complicated by name variations, translation and transcription errors, Latinized forms and the late-onset use of surnames.

There are a number of historical databases that are of interest for the prosopography of the Church of the Holy Roman Empire. These provide biographical information about clerics of medieval and early modern times, and information about educational careers or benefice systems. Integrating that data by means of automatic mapping has proved difficult and often the identification of individuals remains hypothetic. A pioneering contribution to completing this complex task is provided by an international cross-institutional project. Germania Sacra, together with the German Historical Institute in Rome (DHI) and the Repertorium Academicum Germanicum (RAG) is developing a cross-database query. We are developing technical solutions – for example the integration of algorithms or the use of thesauri – to enhance the search for phonetic and orthographic name variants as well as Latinized names. We are going to provide an interactive system, which gives qualified identifying suggestions to the user.

---

[1] van Hooland, S., de Wilde, M., Verborgh, R., Steiner T., and van de Walle, R. (2015): Exploring entity recognition and disambiguation for cultural heritage collections. In: Digital Scholarship Humanities (Vol. 30 Iss. 2), S. 262–279.

Open Digital Forms

Hiep Le, Thomas Rebele[(✉)], and Fabian Suchanek

Télécom ParisTech, Paris, France
{hle,rebele}@enst.fr

Abstract. The maintenance of digital libraries often passes through physical paper forms. Such forms are tedious to handle for both senders and receivers. Several commercial solutions exist for the digitization of forms. However, most of them are proprietary, expensive, centralized, or require software installation. With this demo, we propose a free, secure, and lightweight framework for digital forms. It is based on HTML documents with embedded JavaScript, it uses exclusively open standards, and it does not require a centralized architecture. Our forms can be digitally signed with the OpenPGP standard, and they contain machine-readable RDFa. Thus, they allow for the semantic analysis, sharing, re-use, or merger of documents across users or institutions.

1 Introduction

The world's digital libraries contain millions of documents, visual artwork, and audio material. The maintenance of these libraries often passes through physical paper forms. Out of the 50 first English digital library projects on Wikipedia[1], more than half offer a service that requires sending a paper form by letter. Such paper forms are used for membership requests, for licensing, for revocation requests, etc. In one case, 3 different forms have to be filled out and sent physically to the library[2]. The dependence on paper forms is particularly surprising for projects that aim to digitize paper documents, but the problem is of course in no way restricted to digital libraries.

Paper forms require substantial manual work for the sender and the receiver. Therefore, several commercial solutions exist for the digitization of forms. However, most of these are proprietary, pricy, centralized, or require software installation (see Sect. 2). Thus, the challenge is to develop a solution so that

- no additional software is required to display and fill forms.
- only open standards are used, so everyone can use the format and software.
- the user remains in control of their data.

With this demo, we propose a free, secure, and lightweight framework for digital forms that fulfills these desiderata:

---

[1] https://en.wikipedia.org/wiki/List_of_digital_library_projects.
[2] http://www.ahds.ac.uk/depositing/how-to-deposit.htm.

© Springer International Publishing Switzerland 2016
N. Fuhr et al. (Eds.): TPDL 2016, LNCS 9819, pp. 454–458, 2016.
DOI: 10.1007/978-3-319-43997-6_43

- Our solution is based on HTML documents, and requires only a browser.
- Our solution uses exclusively open standards, which makes the forms easy to parse, verify, produce, or modify.
- Our solution does not require a centralized architecture. The user remains in complete control of their data.
- Our forms can be digitally signed with the OpenPGP standard, which can be verified by open software.
- Our forms are machine-readable in RDFa, thus allowing the semantic analysis, sharing, re-use, or merger of forms.

2   Related Work

Commercial Solutions. Adobe Acrobat, LibreOffice and others support the creation of Portable Document Format (PDF) forms. The full range of capabilities (such as electronic signature, digital signature, or importing from other forms) requires the installation of one or several proprietary software packages. Microsoft Word documents can also be used as forms, and can be digitally signed. The certificates are mostly not free[3]. In addition, such forms require the installation of large software packages. The same goes for IBM Lotus.

Online Services. Several cloud-based services (such as Google Forms) offer online polls or online collaboration on documents. However, they typically do not allow digital signing. Other services provide form creation and filling capabilities[4]. The user creates forms and sends a link to the person, who fills out the form online. However, the form creator needs to pay per form. Furthermore, the user data has to pass through a central architecture.

Open Standards. Various form description formats exist, such as XForms, XUL, Vexi, ZF, XAML, and others[5]. However, all of them either require programming experience or special software to create and fill the forms.

XML-based Solutions. Kuo et al. generate HTML forms from a XML schema, but do not allow signing the documents [3]. Boyer proposed a combination of ODF and XForms, which allows saving and sending the forms per email, filling them offline, and signing them with XML Signatures [1,2]. However, users would need additional software to use or edit the combination of ODF and XForms.

3   Open Digital Forms

Infrastructure. We decided for a decentralized solution, where forms are not stored in a central repository, but kept and sent by the users themselves. This has two advantages: First, the users remain in total control of their data, and they remain independent of a service provider. Second, our format can be used

---

[3] https://help.libreoffice.org/Common/Applying_Digital_Signatures.

[4] http://www.jotform.com/, http://www.moreapp.com/.

[5] https://en.wikipedia.org/wiki/List_of_user_interface_markup_languages.

in a grass-root fashion. It can easily complement other types of forms that are mandated by the host institution.

Format. Our forms are in HTML, because this format allows basic formatting while at the same time being human-readable. HTML can also be displayed on almost any digital device with no additional software. A form consists of a title and several sections (DIVs). Each section describes one particular entity of interest, such as a client, a digital library, or a licensed content. Each section contains a list of attributes of that entity, such as the birth date and address of a client, the name of a library, or the type and the price of the licensed content. An HTML INPUT field for each attribute stores its value.

Semantic Annotation. Our forms are semantically annotated, so that machines can read and analyze them. We use RDFa for this purpose, the W3C standard for semantical annotation of Web documents. This makes our forms compatible with the Semantic Web and Linked Data standards. By default, we use the vocabulary from schema.org, which is designed, supported, and maintained by Google, Bing, Yandex, and Yahoo[6]. Each section of our document defines a new entity (by help of typeof), and each attribute in that section creates a new property of that entity (by help of property).

Signing. Our documents can be electronically signed by an image of a scanned signature. For this purpose, the form contains a button which allows the user to select the image file from their hard drive. The image is then embedded as a base-64 encoded string in a data:image URL directly into the HTML document. This way, the form remains self-contained, and the user has to send only a single file if they want to share the signed document.

Digital Signing. The forms can also be digitally signed. We opted for PGP for this purpose, because unlike S/MIME it does not require a central certification authority. PGP allows signing documents with *ASCII armor*, i.e. a plain text header and footer containing the signature. GnuPGP accepts a signed document enclosed by additional text. We sign only the content of the HTML tag, so that we retain a valid HTML document. CSS styling hides the ASCII armor.

4   Demo

Creating a Form. Our tool for creating forms is an open source JavaScript program[7]. If Alice wants to create a form, she can just visit our Web page[8]. She can drag-and-drop section elements from the toolbar into the form (Fig. 1 left). For each section, Alice can choose the section type (book, institution, etc.) from a drop-down menu from the schema.org vocabulary. She can also name each section (which creates an instance of the type in the underlying RDFa code). Alice can then drag controls such as text fields and check boxes into the sections. She can assign a caption and an attribute name to each control. We

---

[6] http://schema.org.
[7] https://github.com/lenguyenhaohiep/formless3.
[8] https://rawgit.com/lenguyenhaohiep/formless3/master/.

Fig. 1. Our tool (left) and the resulting form (right)

offer predefined attributes such as `birthDate`, `email`, and other properties from schema.org. Alice can also create her own property, which will then live in the local namespace of the document. In the end, Alice can download the form as a self-contained HTML document, which she can store for later modification, send by email or, if need be, print.

Filling a Form. If Bob receives the form, he can display it on any HTML-enabled device (Fig. 1 right). He can fill out the form by typing into the `INPUT` elements of the form. The input elements use standard HTML 5 validation to check the format of dates, emails, etc. Filling and saving the form is done by embedded JavaScript, and does not need an Internet connection.

Importing. Since our forms are semantically annotated, Bob can also import data from other forms he already filled. This functionality is provided by a JavaScript program that is loaded from our Web page on demand if Bob clicks on the *Import* button. The program proposes a match between the existing and the incoming sections, and Bob can accept, reject, or modify this proposed match. Since the semantic annotations are taken from a standard vocabulary, such imports work also across forms from different institutions.

Signing the Form. Finally, Bob can sign the document by attaching the image of a scanned physical signature to the document. This functionality is embedded in the document, and thus available offline. Bob can also add a digital signature with his private PGP key. This functionality is provided by JavaScript code that is loaded on demand. The private key never leaves the computer, and is used just to generate the signature. After this step, no modifications of the form are possible, and Bob can send around the signed document. Alice can check the validity of the signature in our tool, by providing the form and the public key of Bob. She can also verify the signature with any standard tool such as OpenPGP, Gpg4win, or GPGsuite. Since the document is self-contained and nearly human-readable, users can also always inspect the workings of the form.

5 Conclusion

Our demo proposes a free and secure format for digital forms, which is based completely on open standards. It provides a mechanism of information integration even across institutions due to the use of Semantic Web standards, and therefore supports the infrastructure of digital libraries.

Acknowledgments. This research was partially supported by Labex DigiCosme (project ANR-11-LABEX-0045-DIGICOSME) operated by ANR as part of the program "Investissement d'Avenir" Idex Paris-Saclay (ANR-11-IDEX-0003-02).

References

1. Boyer, J.M.: Interactive office documents. In: Document Engineering (2008)
2. Boyer, J.M., Wiecha, C.F., Akolkar, R.P.: Interactive web documents. Comput. Sci.-R&D **27**(2), 127–145 (2012)
3. Kuo, Y.-S., Shih, N.C., Tseng, L., Hu, H.-C.: Generating form-based user interfaces for XML vocabularies. In: Document Engineering (2005)

A Text Mining Framework for Accelerating the Semantic Curation of Literature

Riza Batista-Navarro[1]([✉]), Jennifer Hammock[2], William Ulate[3],
and Sophia Ananiadou[1]

[1] School of Computer Science, University of Manchester, Manchester M13 9PL, UK
riza.batista@manchester.ac.uk
[2] Smithsonian Institute, Washington, D.C., USA
[3] Missouri Botanical Garden, Missouri, USA

Abstract. The Biodiversity Heritage Library is the world's largest digital library of biodiversity literature. Currently containing almost 40 million pages, the library can be explored with a search interface employing keyword-matching, which unfortunately fails to address issues brought about by ambiguity. Helping alleviate these issues are tools that automatically attach semantic metadata to documents, e.g., biodiversity concept recognisers. However, gold standard, semantically annotated textual corpora are critical for the development of these advanced tools. In the biodiversity domain, such corpora are almost non-existent especially since the construction of semantically annotated resources is typically a time-consuming and laborious process. Aiming to accelerate the development of a corpus of biodiversity documents, we propose a text mining framework that hastens curation through an iterative feedback-loop process of (1) manual annotation, and (2) training and application of statistical concept recognition models. Even after only a few iterations, our curators were observed to have spent less time and effort on annotation.

1 Introduction

The Biodiversity Heritage Library (BHL) holds the largest collection of legacy literature on biodiversity, contributed by a number of botanical and natural history libraries striving to make their content available to the public through digitisation efforts (e.g., scanning, optical character recognition, bibliographic indexing). As a result, more than 100,000 titles are now publicly accessible through the digital library, equivalent to around 40 million pages of taxonomic literature.

To allow its users to find information of interest, BHL provides a search system that facilitates keyword-based matching of user-supplied queries against bibliographic metadata, e.g., book titles. It is not rare, however, for keywords to have ambiguous senses. Methods which are more sophisticated than keyword-based matching are thus necessary for ensuring that the results returned by BHL are relevant to a user's query. For example, to help distinguish between a word's several senses, one could incorporate a tool that will automatically mark up mentions of concepts within text and assign them into predefined categories: an

N. Fuhr et al. (Eds.): TPDL 2016, LNCS 9819, pp. 459–462, 2016.
DOI: 10.1007/978-3-319-43997-6_44

information extraction task known as named entity recognition (NER). Amongst the many NERs which have emerged, machine learning-based ones have demonstrated superior performance owing to their stronger ability to generalise on unseen cases, based on samples presented to them during model training. The drawback, however, is their reliance on the availability of textual corpora with gold standard semantic annotations. Whilst such corpora proliferate in a few domains, e.g., newswire, biomedicine, some subject areas such as biodiversity are relatively low-resourced. To the best of our knowledge, there has only been one corpus of biodiversity documents which was semantically enriched through manual annotation of taxonomic names[1]. This scarcity in resources can be attributed to the costly time and human effort required by corpus construction, especially in cases where the target domains are specialised.

In order to support the development of advanced NER techniques for the purpose of automatically categorising concepts in BHL documents, we pursued the construction of a gold standard annotated corpus of biodiversity literature. We however innovate the annotation process to reduce the time and effort required from curators. Specifically, we propose a text mining-based framework driven by an iterative feedback-loop mechanism that (1) gathers manual annotations from curators, (2) employs the conditional random fields (CRF) algorithm [1] to train NER models on manual annotations gathered so far, and (3) applies the trained models on unannotated documents which will then be manually validated by curators. The framework is implemented as processing workflows in Argo[2], a Web-based text mining workbench [2]. Following the Unstructured Information Management Architecture (UIMA) standard, Argo integrates various natural language processing (NLP) tools into workflows. Furthermore, it supports the execution of semi-automatic workflows which allow curators to manually provide or revise annotations through a user-interactive annotation editor.

2   Methods

As a preliminary step, we asked our partners from the Missouri Botanical Garden, one of the botanical libraries behind BHL, to select 10 titles in the BHL collection to form the basis of the corpus. Based on this process, a total of 145 pages were collected, comprising our new corpus which was annotated by two curators according to the approach outlined below.

2.1   Preparation of the Annotation Schema

With the guidance of our partner biodiversity specialist, we identified the semantic categories which define concept types that should be annotated by the curators. The following seven categories were selected: (1) taxonomic entity, (2) geographic location, (3) habitat, (4) temporal expression, (5) person, (6) anatomical

---

[1] https://github.com/VBRANT/vibrantcorpus.

[2] http://argo.nactem.ac.uk.

entity, and (7) quality. A detailed set of annotation guidelines was prepared to elucidate cases that should be included and how they should be demarcated[3]. Since Argo is a generic text mining workbench and does not cater to any specific application or domain, we configured its annotation editor by means of a new UIMA type system formally representing our semantic categories of interest.

2.2   Selection of Controlled Vocabularies

The NER task was cast as a sequence labelling problem, making use of sentence tokens as the fundamental units of analysis. Features representing tokens are therefore key in the training and subsequent application of conditional random field models for NER. Our previous work on developing CRF-based NERs showed that a boost in performance can be obtained through the incorporation of semantic features, e.g., matches between tokens in text and entries in controlled vocabularies [3]. We generated our semantic features based on the Catalogue of Life and the following ontologies: the Ontology of Biological Attributes, and the Gazetteer, Environment, Uber Anatomy, Phenotype and Trait, and Flora Phenotype Ontologies. The names and synonyms contained in these resources were compiled using the NERsuite package[4], our chosen implementation of CRFs. The manner in which our framework utilised the compiled dictionaries is described in the next section.

2.3   Machine Learning-Based Training and Tagging Workflows

The framework is initiated by curators upon their creation of new annotations for a small number of documents, e.g., two pages, using Argo's annotation editor. These seed annotations are saved in XML Metadata Interchange (XMI) files, UIMA's native serialisation format. Our feedback-loop mechanism is realised as the iterative execution of two workflows: (1) a completely automatic workflow for training CRF models, and (2) a semi-automatic one for applying the newly trained models on unseen documents and presenting their results to human curators for revisions. The first workflow (Fig. 1a), begins with an XMI Reader which loads documents from a directory containing manually provided annotations (i.e., training documents). After the Reader is a series of NLP components for pre-processing, i.e., sentence splitting, tokenisation, lemmatisation and part-of-speech tagging. These components are succeeded by instances of the NERsuite Dictionary Feature Generator, which load pre-compiled dictionaries (described in the previous section) to generate semantic features based on token matches. Next in the workflow are instances of the NERsuite Trainer component, each of which generates character and word n-grams according to the outputs of pre-processing tools. Together with the previously captured dictionary matches, these features are learned by the NER models during training.

---

[3] http://wiki.miningbiodiversity.org/doku.php?id=guidelines.
[4] http://nersuite.nlplab.org.

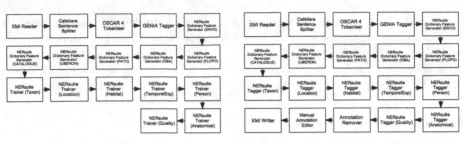

(a) Training of NER models

(b) Application of NER models and manual revision

Fig. 1. Training of NER models

The second workflow (Fig. 1b) similarly begins with an XMI Reader, albeit one for loading unannotated documents. Similar components for pre-processing and semantic feature generation follow. However, these are instead succeeded by NERsuite Tagger components which apply the previously trained models to annotate semantic concepts in the documents of the current set. These automatically annotated documents are then presented by the editor to the curators for revision. The revised annotations are saved as XMI files in the curators' directory of training documents, thus incrementally increasing the amount of gold standard annotations used for model retraining in succeeding iterations.

3   Discussion and Conclusion

As the framework progresses with the iterative execution of the workflows described above, the amount of annotated data increases, allowing the models to become better at learning the NER task. In turn, the quality of annotations generated by the models on unseen documents also improves. There is therefore a progressive decrease in the amount of time and effort required from the curators in revising automatically generated annotations. Our two curators empirically confirmed this, observing that whilst in the first two to four iterations they had to revise a huge proportion of the automatically generated annotations, by the seventh or eighth iteration they were supplying substantially less corrections.

References

1. Lafferty, J.D., McCallum, A., Pereira, F.C.N.: Conditional random fields: probabilistic models for segmenting and labeling sequence data. In: Proceedings of the Eighteenth International Conference on Machine Learning, ICML 2001, pp. 282–289. Morgan Kaufmann Publishers Inc., San Francisco (2001)
2. Rak, R., Rowley, A., Black, W., Ananiadou, S.: Argo: an integrative, interactive, text mining-based workbench supporting curation. Database: J. Biol. Databases Curation **2012**, bas010 (2012)
3. Batista-Navarro, R., Rak, R., Ananiadou, S.: Optimising chemical named entity recognition with pre-processing analytics, knowledge-rich features and heuristics. J. Cheminformatics **7**(Suppl. 1), S6 (2015)

Erratum to: CERN Analysis Preservation: A Novel Digital Library Service to Enable Reusable and Reproducible Research

Xiaoli Chen[1,5], Sünje Dallmeier-Tiessen[1(✉)], Anxhela Dani[1,2],
Robin Dasler[1], Javier Delgado Fernández[1,4], Pamfilos Fokianos[1],
Patricia Herterich[1,3], and Tibor Šimko[1]

[1] CERN, Geneva, Switzerland
{xiaoli.chen,sunje.dallmeier-tiessen,anxhela.dani,
robin.dasler,javier.delgado.fernandez,
pamfilos.fokianos,patricia.herterich,
tibor.simko}@cern.ch
[2] Alexander Technological Educational Institute of Thessaloniki,
Thessaloniki, Greece
[3] Humboldt-Universität zu Berlin, Berlin, Germany
[4] Universidad de Oviedo, Oviedo, Spain
[5] University of Sheffield, Sheffield, UK

Erratum to:
Chapter 27: N. Fuhr et al. (Eds.)
Research and Advanced Technology for Digital Libraries
DOI: 10.1007/978-3-319-43997-6_27

In an older version of the paper starting on page 347 of this volume, Fig. 2 was incorrect. This has been corrected.

The updated original online version for this chapter can be found at 10.1007/978-3-319-43997-6_27

Tutorials

Introduction to Fedora 4

David Wilcox(⊠)

DuraSpace, 9450 SW Gemini Drive #79059, Beaverton, OR 97008, USA
dwilcox@duraspace.org

Abstract. Fedora is a flexible, extensible repository platform for the preservation, management and dissemination of digital content. Fedora 4, the new, revitalized version of Fedora, includes vast improvements in scalability, linked data capabilities, research data support, modularity, ease of use and more. Both new and existing Fedora users will be interested in learning about and experiencing these new features and functionality first-hand.

This tutorial will provide an introduction to and overview of Fedora 4, with a focus on the latest features. After an initial review of the most significant new features, participants will learn about best practices and community standards for modeling data in Fedora 4. With a strong focus on interoperability, the Fedora community has been collaborating on shared data models that can be used across a variety of platforms. Fedora 4 also implements the W3C Linked Data Platform 1.0 specification, so a section of the tutorial will be dedicated to a discussion about LDP and the implications for Fedora 4 and linked data. Linked data is at the core of Fedora 4, so participants will be introduced to several current use cases for new linked data capabilities in Fedora 4.

Attendees will be given pre-configured virtual machines that include Fedora 4 bundled with the Solr search application and a triplestore that they can install on their laptops and continue using after the workshop. These virtual machines will be used to participate in a hands-on session that will give attendees a chance to experience Fedora 4 by following step-by-step instructions. This section will demonstrate how to create and manage content in Fedora 4 in accordance with linked data best practices. Finally, participants will learn how to search and run SPARQL queries against content in Fedora using the included Solr index and triplestore.

General Terms

Connecting Digital Libraries, Practice of Digital Libraries, Digital Libraries in Science.

Keywords: Fedora · Repository · Linked data · Open source

© Springer International Publishing Switzerland 2016
N. Fuhr et al. (Eds.): TPDL 2016, LNCS 9819, pp. 465–467, 2016.
DOI: 10.1007/978-3-319-43997-6

1 Outline

The tutorial will include three modules, each of which can be delivered in 1 h.

1.1 Introduction and Feature Tour

This module will feature an introduction to Fedora generally, and Fedora 4 in particular, followed by an overview of the core and non-core Fedora 4 features. It will also include a primer on data modeling in Fedora 4, which will set the audience up for the next section.

1.2 Linked Data and LDP

The Fedora community is deeply invested in linked data best practices; this is exemplified by our alignment with the W3C Linked Data Platform recommendation in Fedora 4. This section will feature an introduction to linked data and LDP, with a particular focus on the way Fedora implements linked data. Attendees will have an opportunity to create and manage content according to linked data best practices using the Fedora 4 virtual machine.

1.3 Fedora 4 Integrations

Fedora 4 is fundamentally a middleware application – it is meant to be used in conjunction with other applications. This section will provide an overview of the most common integrations, such as Solr and triplestores. Attendees will learn how to use these tools to index and query content in Fedora.

2 Duration

Half-day (3 h)

3 Audience

This tutorial is intended to be an introduction to Fedora 4 - no prior experience with the platform is required. Repository managers and librarians will get the most out of this tutorial, though developers new to Fedora would likely also be interested.

4 Outcomes

Tutorial attendees will:

- Learn about the latest and greatest Fedora 4 features and functionality
- Discover new opportunities enabled by LDP and linked data

- Learn how to create and manage content in Fedora
- Understand how to index and query content in Fedora

5 Presenters

David is the Product Manager for the Fedora project at DuraSpace. He sets the vision for Fedora and serves as strategic liaison to the steering committee, leadership group, members, service providers, and other stakeholders. David works together with the Fedora Technical Lead to oversee key project processes, and performs international outreach to institutions, government organizations, funding agencies, and others.

Building Digital Library Collections
with Greenstone 3 Tutorial

David Bainbridge(✉)

Department of Computer Science, University of Waikato,
Hamilton, New Zealand
davidb@waikato.ac.nz

Abstract. This tutorial is designed for those who want an introduction
to building a digital library using an open source software program. The
course will focus on the Greenstone digital library software. In particular,
participants will work with the Greenstone Librarian Interface, a flexible graphical user interface designed for developing and managing digital library collections. Attendees do not require programming expertise,
however they should be familiar with HTML and the Web, and be aware
of representation standards such as Unicode, Dublin Core and XML. The
Greenstone software has a pedigree of approaching two decades, with over
1 million downloads from SourceForge. The premier version of the software has, for many years, been Greenstone 2. This tutorial will introduce
users to Greenstone 3—a complete redesign and reimplementation of the
original software to take better advantage of newer standards and web
technologies that have been developed since the original implementation
of Greenstone. Written in Java, the software is more modular in design to
increase the flexibility and extensibility of Greenstone. Emphasis in the
tutorial is placed on where Greenstone 3 goes beyond what Greenstone 2
can do. Through the hands-on practical exercises participants will, for
example, build collections where geo-tagged metadata embedded in photos is automatically extracted and used to provide a map-based view in
the digital library of the collection.

Keywords: Digital libraries · Open source software · Tutorial

Overview

The librarian interface to Greenstone provides a graphical interface to the underlying digital library infrastructure, and allows users to gather sets of documents,
import or assign metadata, build them into a Greenstone collection, and serve
it from their web site. It supports eight basic activities:

- opening an existing collection or defining a new one;
- copying documents into it, with metadata attached (if any);
- mirroring documents from the Web or else protocols such as OAI and SRW/U
  if required;

© Springer International Publishing Switzerland 2016
N. Fuhr et al. (Eds.): TPDL 2016, LNCS 9819, pp. 468–470, 2016.
DOI: 10.1007/978-3-319-43997-6

- enriching the documents by adding further metadata to individual documents or groups;
- designing the collection by determining its appearance and the access facilities it will support;
- building it using Greenstone;
- previewing the newly created collection from the Greenstone home page; and
- exporting the collection to other formats such as MARCXML, FedoraMETS, and DSpace.

Collections built with Greenstone automatically include effective full-text searching and metadata-based browsing. They are easily maintainable and can be rebuilt entirely automatically. Searching is full-text, and different indexes can be constructed (including metadata indexes). Browsing utilizes hierarchical structures that are created automatically from metadata associated with the source documents. Collections can include text (including PDFs), pictures, audio and video. Geo-tagged content can be seamless integrated with an integrated Map-based view. The interface can be extensively customized. Documents can be in any language: the interface has been translated into over thirty languages. The Greenstone software runs on all versions of Windows, and Unix based systems such as Mac, Linux, and Solaris. There's even an experimental version available that runs self-contained on Android devices.

Objectives

Participants will learn to:

- Install the software
- Set up a digital library system
- Build and export their own collections
- Learn about capabilities and features of the software

About the Greenstone Digital Library Software

Greenstone is a suite of software for building and distributing digital library collections. It is not a digital library but a tool for building digital libraries. It provides a way of organizing information and publishing it on the Internet in the form of a fully-searchable, metadata-driven digital library. It has been developed and distributed in cooperation with UNESCO and the Human Info NGO in Belgium. It is open-source, multilingual software, issued under the terms of the GNU General Public License. Its developers received the 2004 IFIP Namur award for "contributions to the awareness of social implications of information technology, and the need for an holistic approach in the use of information technology that takes account of social implications."

Attendees will learn enough to install the software, set up a digital library system, build their own collections, customize them, and export them to other formats. Those with programming skills should be able to extend and tailor the system extensively. Moreover, all attendees will be equipped with extensive course material that is freely redistributable. Intended audience

The tutorial is designed for those who want to build their own digital library but do not want to write their own software. It is intended for librarians, archivists, and other information workers who are interested in building their own digital collections.

The Greenstone Librarian Interface is designed for end users. No programming ability is required. Attendees should be familiar with HTML and the Web, and be aware of representation standards such as Unicode, Dublin Core and XML.

Topics

Overview

- What does Greenstone do?
- Greenstone facts; standards
- Reader's Interface: examples of collections

Librarian interface

- Build a collection in 30 sec
- Add metadata, classifiers, search indexes
- Working with HTML, Word, PDF, Photos, and Scanned Documents

Going further

- Extracting embedded metadata
- Utilizing GPS metadata to provide a map-enhanced end-user interface
- Under the hood: the collection configuration file

Customizing and controlling the look

- Different interface languages
- Selecting a different look using jQuery-UI themes
- Greenbug: a browser embedded interactive graphical editor for Greenstone

Concluding discussion

Presenter

David Bainbridge is a professor of Computer Science at the University of Waikato in New Zealand where he leads the New Zealand Digital Library research project. His research interests include multimedia content analysis, and human computer interaction in addition to digital libraries. He has published widely on these areas, including the book How to build a digital library (2010), with colleagues Witten and Nichols, now into its second edition. He graduated with a Bachelors of Engineering in Computer Science from Edinburgh University, UK as the class medalist, and undertook his PhD in Computer Science at Canterbury University, New Zealand as a Commonwealth Scholar.

Text Mining Workflows for Indexing Archives with Automatically Extracted Semantic Metadata

Riza Batista-Navarro[1]([✉]), Axel J. Soto[1], William Ulate[2],
and Sophia Ananiadou[1]

[1] School of Computer Science, University of Manchester, Manchester, UK
{riza.batista,axel.soto,sophia.ananiadou}@manchester.ac.uk
[2] Missouri Botanical Garden, St. Louis, MO, USA
william.ulate@mobot.org

Abstract. With the vast amounts of textual data that many digital libraries hold, finding information relevant to users has become a challenge. The unstructured and ambiguous nature of natural language in which documents are written, poses a barrier to the accessibility and discovery of information. This can be alleviated by indexing documents with semantic metadata, e.g., by tagging them with terms that could indicate their "aboutness". As manually indexing these documents is impracticable, automatic tools capable of generating semantic metadata and building search indexes have become attractive solutions. In this tutorial, we demonstrate how digital library developers and managers can use the Argo text mining platform to develop their own customised, modular workflows for automatic semantic metadata generation and search index construction. In this way, we are providing digital library practitioners with the necessary technical know-how on building semantic search indexes without any programming effort, owing to Argo's graphical interface for workflow construction and execution. We believe that this in turn will allow various digital libraries to build search systems that will enable their users to find and discover information of interest more efficiently and accurately.

1 Introduction

The need to preserve cultural and scientific heritage has prompted various museums, libraries and archives around the world to engage in extensive digitisation efforts. This in turn has made huge amounts of content accessible via digital libraries, e.g., Europeana, the Digital Public Library of America (DPLA) and the Biodiversity Heritage Library (BHL). A significant portion of such digitised content consists of textual collections containing e.g., archival records, books, monographs and articles which are all written in natural language. Data contained by such collections is thus largely unstructured, posing barriers to the accurate retrieval of relevant information. Basic keyword-based search systems

© Springer International Publishing Switzerland 2016
N. Fuhr et al. (Eds.): TPDL 2016, LNCS 9819, pp. 471–473, 2016.
DOI: 10.1007/978-3-319-43997-6

usually fail to take into account the ambiguity inherent in natural language and are likely to return spurious results for polysemous queries. One way to alleviate this issue is by indexing documents with semantic metadata—a task that can be feasibly carried out through automated means.

2   Approach

To enable digital library developers and managers to develop their own solutions for indexing textual collections with semantic metadata, we present a tutorial organised around the activities outlined below. It is worth noting that in this context, semantic metadata refers to terms contained in text whose semantic categories have been automatically recognised using text mining methods.

Introduction to Semantic Metadata Generation over Textual Collections. An overview of the challenges to information discovery from textual archives will be given. This will be followed by a brief discussion of potential solutions, e.g., semantic metadata generation using natural language processing and text mining methods. We present an approach based on pipelining of diverse tools, e.g., part-of-speech taggers, syntactic parsers, named entity recognisers.

Argo Walk-Through. The pipelining of various text mining and indexing tools can be facilitated by the workflow construction platform Argo[1], a Web-based application that conforms with an interoperability standard known as the Unstructured Information Management Architecture (UIMA). The audience is given an introduction to its various features such as its library of interoperable tools, workflow editor and interface for annotation visualisation. The construction of simple toy workflows is demonstrated to boost familiarisation with the system.

Constructing Semantic Metadata Generation Workflows. The tutorial provides an overview of the types of tools which typically constitute semantic metadata generation workflows. Using Argo, custom workflows are designed and subsequently executed on sample textual corpora. The text-mined semantic metadata is then incorporated into a search index (e.g., Solr, Elasticsearch). Before the population of the search index, however, text-mined results can be visualised in order to provide a better appreciation of the semantic metadata being generated.

Querying Automatically Generated Search Indexes. An exploration of the generated search index is carried out to provide an understanding of the underlying schema. The different types of queries that can be accommodated by the index are presented in order to demonstrate various search use cases.

---

[1] http://argo.nactem.ac.uk.

Applications. Potential applications of semantic search indexes are demonstrated. One particular use case of interest is the enhancement of the search functionalities of the Biodiversity Heritage Library (BHL). This provides the audience with an appreciation of the various advanced features that can be potentially enabled in other digital libraries through the incorporation of semantic search indexes.

Author Index